Fifth Business

synthesis

A series in the history of chemistry, broadly construed,
edited by Carin Berkowitz, Angela N. H. Creager, John E. Lesch,
Lawrence M. Principe, Alan Rocke, and E. C. Spary, in partnership
with the Science History Institute

Fifth Business

A LIFE OF THE CHEMIST
AND EDUCATIONIST
HENRY EDWARD ARMSTRONG

William H. Brock

The University of Chicago Press   CHICAGO AND LONDON

The University of Chicago Press, Chicago 60637
The University of Chicago Press, Ltd., London
© 2025 by The University of Chicago
All rights reserved. No part of this book may be used or reproduced in any manner whatsoever without written permission, except in the case of brief quotations in critical articles and reviews. For more information, contact the University of Chicago Press, 1427 E. 60th St., Chicago, IL 60637.
Published 2025
Printed in the United States of America

34 33 32 31 30 29 28 27 26 25   1 2 3 4 5

ISBN-13: 978-0-226-83958-5 (cloth)
ISBN-13: 978-0-226-83959-2 (e-book)
DOI: https://doi.org/10.7208/chicago/9780226839592.001.0001

Library of Congress Cataloging-in-Publication Data

Names: Brock, W. H. (William Hodson), author.
Title: Fifth business : a life of the chemist and educationist Henry Edward Armstrong / William H. Brock.
Other titles: Synthesis (University of Chicago. Press)
Description: Chicago : The University of Chicago Press, 2025. | Series: Synthesis | Includes bibliographical references and index.
Identifiers: LCCN 2024045637 | ISBN 9780226839585 (cloth) | ISBN 9780226839592 (ebook)
Subjects: LCSH: Armstrong, Henry Edward, 1848–1937. | Chemists—England—Biography. | Chemistry—Study and teaching.
Classification: LCC QD22.A67 B76 2025 | DDC 540.92 [B]—dc23/eng/20241211
LC record available at https://lccn.loc.gov/2024045637

♾ This paper meets the requirements of ANSI/NISO Z39.48-1992 (Permanence of Paper).

Contents

Notes on Abbreviations, References, Bibliographies,
and Illustrations * vii

Preface * ix

Part I

ONE
Becoming a Chemist * 3

TWO
Cobbling a Career in London * 19

THREE
Finsbury College * 30

FOUR
The Central Chemist * 50

FIVE
Chemical Research at the Central Technical College * 67

SIX
Running the Chemical Society * 94

SEVEN
The Admission of Women into the Chemical Society * 112

EIGHT
The Heuristic Method * 128

NINE
Ionomania * 153

Part II

TEN
Semi-Retirement * 191

ELEVEN
The Great War * 201

TWELVE
Heurism Denigrated * 224

THIRTEEN
The 1920s * 237

FOURTEEN
Campaigns Old and New * 256

FIFTEEN
The Lewis Carroll of Chemistry * 271

SIXTEEN
The Final Years * 296

CONCLUSIONS * 307

Acknowledgments * 313
Archives Consulted * 315
Cited Works by Henry Edward Armstrong * 317
General Bibliography * 331
Index * 349

Notes on Abbreviations, References, Bibliographies, and Illustrations

In-text citations to Henry Edward Armstrong's works, compiled in the Armstrong bibliography at the end of the main text, are given in the form "(HEA [year], [page])"; the year is followed where necessary by a lower-case letter of the alphabet (the letters "l" and "o" are not used, as they may be potentially confused with numbers). The Armstrong bibliography is supplemented by a general bibliography comprising all other cited works from the primary and secondary literatures, except for items mentioned in passing. Copyright-free print sources for images are specified in the relevant accompanying texts; photograph portraits, for which every effort has been made to identify possible copyright holders, have been kindly supplied by Imperial College Archives.

ACS	American Chemical Society
AP1	Jeanne Pingree, *List of the Correspondence and Papers of Henry Edward Armstrong, FRS, Preserved in the Imperial College Archives* (London: Imperial College Archives, 1967). The abbreviation is followed by the item number.
AP2	Jeanne Pingree, *List of Correspondence and Other Papers of Henry Edward and Edward Frankland Armstrong [in the] Imperial College Archives*, second series (London: Imperial College Archives, 1974). The abbreviation is followed by the item number.
APSSM	Association of Public School Science Masters
BAAS	British Association for the Advancement of Science
BASF	Badische Anilin- und Soda-Fabrik
BJHS	*British Journal for the History of Science*
Brit. Ass. Reports	Annual Reports of the British Association for the Advancement of Science. The date that follows, by

convention, refers to the meeting year and not the actual year of publication.

DSA	Department of Science and Art
DSIR	Department of Scientific and Industrial Research
FRS	Fellow of the Royal Society
HEA	Henry Edward Armstrong
IACS	International Association of Chemical Societies
IRC	International Research Council
IUPAC	International Union of Pure and Applied Chemistry
JCS	*Journal of the Chemical Society.* This listing incorporates the title *Transactions of the Chemical Society.*
J. Inst. Brewing	*Journal of the Institute of Brewing*
J. Roy. Soc. Arts	*Journal of the Royal Society of Arts*
J. Soc. Chem. Industry	*Journal of the Society of Chemical Industry.* From 1923 onwards, this journal included *Chemistry and Industry.*
LAT	Lawes Agricultural Trust
LI	London Institution
OxfordDNB	*Oxford Dictionary of National Biography* (Oxford: Oxford University Press, 2004), https://www.oxforddnb.com/
PRI	*Proceedings of the Royal Institution*
PRS	*Proceedings of the Royal Society*
PCS	*Proceedings of the Chemical Society*
RCC	Royal College of Chemistry
RCS	Royal College of Science
RI	Royal Institution of Great Britain
RRL	Rothamsted Research Library
RS	Royal Society of London
RSC	Royal Society of Chemistry
SCI	Society of Chemical Industry
UCL	University College London

Preface

"Somewhat arrogant and extreme perhaps in his denunciations"—the opinion penned by George Eliot on Savonarola in the proem to *Romola*—will probably be the criticism passed on me by my friends. The situation is saved by the remark: "But a *Frate predicatore* who wanted to move the people, how could he be moderate?"
—HEA 1932–1933, part 2, 449

Write heresy, pure heresy. Rouse tempers, goad, lacerate and raise whirlwinds.
—Attributed to Kenneth Tynan, 1960s

The organic chemist Henry Edward Armstrong was a regular contributor to the pages of the *Journal of the Society of Chemical Industry* in the early part of the twentieth century. In May 1932, his readers were faced by a typical tirade:

> Whilst the academically minded unadventurous Fellows of the Chemical Society are comatose—lost in dreams of tadpole formulae, leaning protons, electron sinks, and other frippery for which the world cares not a jot—without any proper educational spirit and scarce an original idea, let alone any sense of proportion or of practical value: whilst too the Royal Society, equally high-brow'd and withdrawn from the world, is developing the new exciting game of atomic skittles (let us hope it may not prove to be "all beer" also): the Royal Society of Arts, the one practical body left to us, mindful of great men like Prince Albert and Lord Playfair who ever had the public welfare in mind, true to its traditions, seizes on an all-chemical topic, that of "vitamins"—the most important topic that could possibly be considered—for its spring-cleaning. (HEA 1932b)

The paragraph is vintage Armstrong. Penned at the age of eighty-four, it snarls at many of his twentieth-century bogies: ions and electronic theory in chemistry; nuclear bombardment in physics; scientific jargon; the loading of education toward theory instead of laboratory and workshop practice and investigation; and the fact that scientific societies had ceased to concern themselves with really pressing problems like the supply of food and coal. Three years earlier he had dismissed a paper by the eminent Christopher Ingold as "jargonthropus." As for Nobelist Robert Robinson's and Ingold's innovative looping arrows that indicated electron shifts, still used today by chemists, they were dismissed with the brilliant witticism that "bent arrows never hit their mark" (quoted in Eyre 1958, x–xi).

Historians and popular-science writers love to quote Henry Armstrong: the man who got it wrong about the noble gases by insisting that William Ramsay had mistaken argon for an allotrope of nitrogen; the chemist who (while not entirely wrong) set his heart against the three musketeers of the new physical chemistry—Svante Arrhenius, Wilhelm Ostwald, and Henry van 't Hoff—and their theory of ionic dissociation; and the severe critic of learned societies, governments, and education. As a journalist might say, he was good copy. He is only too easily dismissed like his German *Doktorvater*, Hermann Kolbe, as an old fogey, gadfly, and stick-in-the-mud. But just as Kolbe has been brilliantly rehabilitated and contextualized by Alan Rocke in his engrossing biography of the German chemist (Rocke 1993), the time has come to do something similar for Henry Armstrong.

The comparison between Kolbe and Armstrong is a little unfair, since Armstrong was never cold-shouldered by his fellow chemists or felt to be an embarrassment in the way that his German teacher was. Instead, the scientific community generally welcomed Armstrong's sharp and witty comments at scientific meetings and was delighted to read his highly literate, often even scintillating letters and essays in the pages of the *Times*, *Chemical News*, *Chemical Age*, *Nature*, and the *Journal of the Society of Chemical Industry*. His last paper, in 1937, the year he died at the age of eighty-nine, was on the value of weeds—a reminder of the strong interest he took in agriculture and horticulture from his time as a trustee of the Lawes Agricultural Trust at Rothamsted.

A bewhiskered Falstaffian figure, this genial English chemist, born near London in 1848, was a student of August Kekulé's contemporary rivals Kolbe and Edward Frankland. Armstrong ranked the Lancastrian Frankland, with whom he studied at the Royal College of Chemistry from 1865 to 1867, and after whom he christened his firstborn son, Edward Frankland Armstrong, as equal to Louis Pasteur and Joseph Lister as "a saviour of the world's health." He regarded Kolbe, with whom he studied between 1867 and 1870, as "the real parent of the modern system of structure and formulae." As a student he also attended the lectures of Thomas Henry Huxley and John Tyndall at the Royal School of Mines. In his youth he was, therefore, closely associated with both of the institutions that would be joined in 1890 to form London's Royal College of Science (RCS). In 1907, the RCS began a merger with the City and Guilds of London Central Technical College in South Kensington, where Armstrong had been teaching for many years, to form a new institution, the Imperial College of Science and Technology (now named Imperial College London).

Today, Armstrong (who saw himself as born an Englishman, but "made in Germany") is more remembered for his campaigns to improve the teaching of chemistry, and science generally, by the deployment of the heuristic, or self-discovery, method of instruction, and less for his theory of residual affinity and reverse electrolysis—or for his hostility toward physical chemistry. However, right up until his retirement from the Central Technical College, he was a significant and prolific organic chemist as well as a major figure in the academic and social life of the Chemical Society. Indeed, he did much to raise the profile of the Chemical Society, of which he was a fellow for nearly seventy years, serving as its president from 1893 to 1896. His successor as president, Vernon Harcourt (1896), could say confidently that "probably no occupant of the [Presidential Chair] had ever done so much work for the Chemical Society" as Armstrong had done. Yet, in retrospect, the historian can hardly admire the only-too-successful campaign Armstrong waged against the admission of women to the Society.

In the following study of this remarkable chemist, educationist, and scientific critic, I have used a roughly chronological approach in reconstructing his life and achievements as an active chemist in part I (1848–1911) and as an acidic (but benevolent) critic in his long retirement in part II (1911–1937). Such an approach faces a familiar problem: many events, activities, and publications occur concurrently and not one at a time, and many also extend over decades. This problem is particularly acute for any account of Armstrong's life because he combined activities both as a research chemist and as an educationist while simultaneously acting as a critic of contemporary science teaching, nutrition, and environmental studies. The reader will notice the overt chronological orientation of the first four chapters of part I and the last four chapters of part II. The middle chapters of this book are more thematic than chronological, in order to accommodate Armstrong's impressive propensity to move effectively in several circles at once.

Because Armstrong is such a reliable and entertaining bridge to the past for historians, I have been generous in quoting from his writings. As his son-in-law wrote: "He was a vigorous man physically and mentally, a keen critic, a great worker, a good friend, and a good companion, a veritable Dr. Johnson of modern times, with many of the good qualities, and some of the faults of the great lexicographer" (Miall 1937, 669).

When I was a young lecturer at the University of Leicester in the 1960s, Brian Simon (1915–2002), a professor of education, persuaded me to contribute to the teaching of a new diploma (later an MA) in the history and philosophy of education, aimed at practicing schoolteachers. Together with Jack Meadows (1934–2016) I developed a module on the development of science teaching in British schools. It was in this way that I first came across Henry Armstrong and the heuristic method. A talk given to the new History of Education Society in 1969 on how heurism contributed to the changing curriculum led to an invitation in 1973 from Professor A. V. Judges to edit a selection of Armstrong's writings on heurism for Cambridge University Press's useful series of Texts and Studies in the History of Education. By then I was completely hooked on Armstrong's extraordinary career as a chemist, educator, critic, and polemicist. Over the subsequent years I published many articles on him, but it was only in retirement from university teaching that I began to contemplate writing a new biography.

The following work makes no attempt to compete directly with John Vargas Eyre's biography (1958), which benefitted from his being a former student of Armstrong's during the latter's final years at the Central Technical College. Whilst mine is a full biography, my concern has been to give considerably greater emphasis than Eyre did to Armstrong's remarkable career in retirement from 1911 to 1937, and less emphasis on the more technical details of Armstrong's contributions to the science of chemistry prior to 1911. I also pay less attention than Eyre did to his travels abroad, focusing more exclusively on his activities in England. Finally, by contrast to Eyre this biography pays close attention to the connections between scientific work and social relations, especially the question of gender and science.

A few words about my title: It is borrowed from the Canadian journalist, actor, playwright, and novelist Robertson Davies (1913–1995), whose novel *Fifth Business* (1970) I first read during my stay in Massey College, Toronto, where he was master in 1977. The novel, the first of his Deptford Trilogy, has stayed in my mind ever since and triggered the idea of seeing Armstrong (who was born a mile from Deptford) as an actor playing the role of "Fifth Business" in early twentieth-century science and culture. Following the title page, Davies explained his use of the phrase "fifth business," which he claimed had been coined by Thomas Overskou in an obscure Danish work on drama, *Den Danske Skueplads* [The Danish Theatre]. Davies wrote:

> Those roles which, being neither those of Hero nor Heroine, Confidante nor Villain, but which were nevertheless essential to bring about the Recognition or the dénouement, were called the Fifth Business in drama and opera companies organized according to the old style; the player who acted these parts was often referred to as Fifth Business.

Once Davies had become famous as Canada's leading author, critics searched for a copy of Overskou's work in Danish libraries to see what other insights it might contain about classical theatre and opera. The alleged quotation has never been found, and Davies eventually confessed that it was one of his literary jokes, a pure invention on his part. Joking apart, Davies's deep experience of acting at Stratford and writing plays himself meant that there was a genuine core of truth in the metaphor. The plots of classical theatrical and operatic works are often propelled comically or tragically by one of the "minor" characters—who turn out to be not minor after all.

Now that social historians of science frequently view scientists as actors, it seems apt to use the metaphor in interpreting Armstrong's long career. The following chapters will argue that despite his modest posthumous historical visibility, Armstrong was in fact a central figure in the development of the science of chemistry between 1885 and 1914, and not just in the British context. During this period, he and his students not only published important and wide-ranging research—indeed, often taking scientific "center stage"—but behind the scenes, Armstrong frequently played Fifth Business in many subsidiary plots and in a variety of fields. During his long retirement, no longer active in research, he firmly adopted this role and exercised considerable influence in his various passionate campaigns.

Taken as a whole, Armstrong's career provides a unique inside look by which we can better understand the history of British science, scientific institutions, science education, educationist theory, and social relations of science during the last third of the nineteenth and the first third of the twentieth century.

PART I

✷ 1 ✷
Becoming a Chemist

> I merely wish to claim, in all modesty, that biography should be the recognised province of the structural chemist: he alone can appreciate the complete interdependence of character and structure. The supreme interest of chemistry comes from the fact, that it is the study of character as affected by structure.
> —HEA 1931b, 238

Henry Edward Armstrong was born in the rural village of Lewisham, a mile from Deptford in Kent, on 6 May 1848, the first of seven children of Richard Armstrong (1827–1884), a scrivener in the French banking house of Deveaux in the City of London. He was baptized in Lewisham's church of St. Mary the Virgin two months later and raised as an Anglican. His father had eloped and married a distant cousin named Mary Ann Biddle (1824–1900) from Warwick a year previously. After his marriage, Richard had taken up a new profession as an agent for imported French groceries, and this international food business was to provide the growing Armstrong family with useful contacts within the City of London, which Henry would cultivate throughout his life.

Richard Armstrong was an omnivorous reader, and he encouraged this propensity in Henry, who became his favorite child. Mary Armstrong, on the other hand, seems to have had a difficult relationship with her eldest son. She disliked his diffidence toward the juvenile activities of his younger siblings and his total absorption in his own world of books and nature. She also had a difficult relationship with Henry's youngest brother, Louis (b. 1863), who, it seems, showed early signs of mental difficulties. Consequently, it appears that Mary Armstrong gave all her affection to the surviving daughters, Mary (b. 1850) and Harriet (b. 1855).

Henry Armstrong recalled being taken to London theatres as a child, where he saw performances by Henry Irving, Edmund Kean, and, later,

Ellen Terry. Many commuting city friends who had also set up households in Lewisham were entertained in the Armstrong family home. Some of them, like Hendericus Klassen, a Dutch merchant at the London Corn Exchange, or the wheat merchant Karl (Charles) Kekulé, the half-brother of the famous chemist, brought with them a repertoire of scientific gossip and comment. We may confidently assume that Charles Kekulé was in contact with August Wilhelm Hofmann, who had been directing the Royal College of Chemistry in London since emigrating from Germany in 1845. It is also probable that Hofmann's work, and the rapid progress of German chemistry generally, was a topic of conversation in the Armstrong household.

In the 1860s, Lewisham, now part of Greater London, was still a country town within easy reach of the city by the North Kent rail line that the South Eastern Railway had opened in 1849. Some of Armstrong's strongest memories as a child were of out-of-doors pursuits where his deep love of nature was founded. There, Émile-like, he discovered the Cosmos, scrambling over chalk and clay, exploring chalk-pit ponds, or, like Alfred Wallace and Henry Bates in Leicester a generation before, collecting butterflies with his future brother-in-law, Robert Adkin. He remembered breeding silkworms, keeping white mice, and fishing in the River Ravensbourne before it became polluted by petrol and oil. In middle age he would find a kindred spirit in the writings of John Ruskin, whose literary style, together with that of Thomas Carlyle, he adopted and adapted.

In an elegant memoir upon Armstrong's death, the botanist and physiologist Frederick Keeble, a lifelong friend, lyrically stressed the roles of nature and nurture in Armstrong's Lewisham upbringing:

> Perhaps it was only London calling—the Cockney voice he loved—that made Armstrong cling obstinately to the place where his childhood was spent; but it is more likely that he did so because Nature and Nurture—joint sponsors for man's behaviour—made him: Nature, by giving Armstrong country forebears and fashioning him in their image, sturdy, rugged, clear eyed, clear complexioned, with the countryman's gait, slow but tireless: Nurture by giving his boyhood a plot in the borders of the garden of England for a playground, so that his lifelong memories should be of gardens and rural scenes. (Keeble 1941, 230–31)

Armstrong's eldest son, Frank Armstrong, was also to recall that those who only knew his father as a London chemist,

> on going for a country walk with him, soon discovered a [new] side to his character. In the country the whole nature of the man seemed

to change and expand. Like a Cotswold village he became one with his surroundings. Those who were with him on such occasions—on country walks or in gardens—found out that Armstrong—agnostic and worshipper of knowledge as he was—meant absolutely what he said when he declared "men live by faith, not knowledge." (E. F. Armstrong 1941, 373)

Indeed, walking and hiking, fishing and sailing became spare-time passions throughout his long life.

Armstrong attended Colfe Grammar School in Lewisham, founded in 1652 and still oriented to the classics in Armstrong's day. He did, however, begin there to show an interest in the sciences. He later recalled reading John Pepper's *Boys' Play-Book of Science* (1860), to which he may have been led by his father taking him to see the exhibits and demonstrations at Pepper's Regent Street Polytechnic (Secord 2002; Brock 2004a). Pepper led him, in turn, to W. E. Statham's "Chemical Cabinet" (1842) and to the translation of Adolf Stöckhardt's *Chemistry* (1850).

> I recollect carrying out, as a lad, a fermentation and the terrible smell and taste of the product. I recollect the intense dissatisfaction I felt in just copying [Statham]. I could discover no method in what I read and was doing. The one book in which I found method was Trench's *Study of Words*. (HEA 1924b, 148)

We will return to Trench's influence in the next chapter. In addition to private reading, Henry's interest in science may also have been aided by a neighbor whose family became friends with the Armstrongs, the pharmacist Thomas Howard Lavers (1811–1893). Henry was destined to marry the Lavers' eldest daughter, Louisa.

Armstrong left Colfe's School when he reached the age of sixteen in 1864. A year later, after spending the winter and spring in Gibraltar (evidently for reasons of health) with his uncle Henry, who had become the chief warden of the convict settlement there, he started training in the cramped quarters of the Royal College of Chemistry (RCC) in Oxford Street, no doubt joining his father on the daily commute to London.

The Royal College of Chemistry

According to Armstrong's later account, his father had sent him to the RCC on "the advice of a shrewd engineering friend" (HEA 1920c, 4). The subject may have arisen in Richard Armstrong's conversations with City friends in which Hofmann was praised as someone whose teaching and research were doing good for Britain's industrial development. So Henry enrolled in the RCC for the summer term of 1865, in what turned out to be Hofmann's last months in London before he left in October for a professorship in Berlin. For the autumn term of 1865, his place as director of the college was taken by the Lancastrian chemist Edward Frankland.

A fellow student who joined classes the same day as Armstrong was Horace Taberer Brown (1848–1925), who became a lifelong friend. Brown's stepfather was a bank manager in the brewery town of Burton-on-Trent, where young Armstrong stayed during the summer vacation of 1865. This visit was his introduction to the brewing atmosphere and where he acquired "a nose and a taste [for beer] which I have never lost." It was Brown who introduced Armstrong to the fine German chemist Peter Griess, who had previously worked with Hofmann in London.

> Griess had then but recently taken the place at Allsopp's brewery of Dr Böttinger, father of the H[einrich] von Böttinger, who played so leading a part in the [Friedrich] Bayer [dyestuffs] firm at Elberfeld. After my return from Germany, I soon became intimate with Griess and through him with Heinrich Caro, who gained much inspiration of the greatest importance to the Azo-farben industry from his visits to Griess. (HEA 1927c, 186)

Here were two more worlds—brewing and dyeing—opening up to Armstrong, which were to play significant roles in his life. In 1890 he was elected an honorary member of the Institute of Brewing, a society whose local discussions he frequently attended and whose hop-field tours he frequently enjoyed (Jones 1937, 361).

Brown's "family circle was the most delightful and stimulating one that I have ever known; earnest but free and natural, without the slightest touch of highbrow gush" (HEA 1937f, 377). He and Horace had a common interest in nature. Unlike Brown, Armstrong was not versed in Virgil, but he claimed he was more widely read in the classics of Fennimore Cooper, Dumas, Marryat, Scott, Dickens, Thackeray, and Collins. In his last illness, Brown asked Armstrong to write his life story and left reams of helpful notes. Among these was Brown's "portrait" of Armstrong that the latter was proud to publish.

> Armstrong was a brilliant student, and even at that early age [1865] displayed an originality of mind and powers of independent criticism which marked him for future distinction. I owe much to Armstrong

during our lifelong friendship, and such success as I have attained in the scientific world is in great measure due to his counsel and stimulation at times when I most needed both. That we have not always seen eye to eye on every subject goes without saying, but should he survive me, of which there is a good chance, I should like him to know something of the gratitude that I feel towards him, and my intense satisfaction that with all our little differences no serious cloud has ever arisen between us. (HEA 1937f, 378)

After only six months, Brown left the RCC to become a junior brewer at Worthington's in Burton, but he returned to London in the summer of 1868 to Edward Frankland's private water laboratory, where he mastered the method of water analysis that Frankland and Armstrong had recently perfected. Brown was to apply the method at Worthington's, as well as in improving Burton-on-Trent's water supply and sewage system. He later joined the Guinness brewery in Dublin, conducted important research on fermentation, and was elected a fellow of the Royal Society (FRS) (HEA 1928h; HEA 1937f).

Armstrong was also enthused by the "impulsive energy" of one of the more advanced students at the RCC, Ernest Theophron Chapman, an occasional co-author of papers with Frankland, whose "flasks, usually big ones, were constantly boiling over; eventually he fulfilled our expectations [in 1872] by blowing himself to bits in an explosives factory in the Harz mountains" (quoted in Brock 2011a, 286). Armstrong saw little of Hofmann, who was leaving instruction largely in the hands of his able assistants, the "indefatigable" Herbert McLeod, the "skilled and exact analyst" William Valentin, and the Irishman Cornelius O'Sullivan. Another fellow student was the German Hermann Wichelhaus, who first suggested abbreviating Hofmann's neologism "quantivalence" to "valence" or "valency," and who later became professor of chemical technology at the University of Berlin. Everyone, Armstrong claimed, was working for working's sake. There were no external examinations, only an internal test at the end of each half year, "a sufficient test of our progress, and we took this in our stride never letting it worry us" (HEA 1920c, 4).

The Royal College of Chemistry, which had begun life as a private enterprise in 1845, had amalgamated with the government's Royal School of Mines in 1853 (Roberts 1976). It was there, in Jermyn Street, Piccadilly, that lectures were held, while the chemistry laboratory classes remained at the Oxford Street premises a half-mile away. Hofmann played the leading role in transferring to Britain the effective style of science instruction of his mentor, Justus Liebig; in doing so he led a British resurgence of chemistry, and especially organic chemistry, that Armstrong would later further advance. Under Hofmann's guidance the college had thus taken on aspects of a German university, at least for many of the students who, like Armstrong, were mature enough to exploit the opportunities for sampling the affiliated lecture courses in physics, geology, and biology offered in Jermyn Street. He later gave a telling capsule description of Hofmann.

> A Courtier and diplomatist of the first rank, gifted with great literary skill, he not only did much to encourage technical advance but also exercised a commanding social influence. Chemistry today is more in need of men of his type, of high social ability, than of mere technical expertise. (HEA 1927c, 185)

Armstrong sampled the courses given by Thomas Henry Huxley, John Tyndall, and Andrew Ramsay at the School of Mines in Jermyn Street, as well as Frankland's in chemistry at the RCC. Unaware, as a student, of Huxley's reputation, he "had no call to be a hero-worshipper" and consequently was unimpressed by Huxley's zoology teaching and even thought him a poor teacher for introductory classes. On the other hand, Huxley gave out opinions about things, but not "how to form opinions of your own." Only in his maturity did Armstrong learn of, and understand, Huxley's "pontifical importance in the teaching world." Armstrong was influenced by Huxley's polemical style of writing and learned from him how to be a propagandist for a particular cause.

The teacher whose lectures most impressed him, however, was Tyndall, the physicist at the Royal Institution (RI) as well as at the Royal School of Mines. Tyndall's "enthusiasm was as catching as Hofmann's, his demonstrations were marvellous." In contrast, he found Ramsay's geology lectures boring—though later in life, once he had read Charles Kingsley's exciting treatment of town geology (Kingsley 1872) and looked at Margate's cliffs with a medical friend named Arthur Rowe, he hastened to join the Geologists' Association and became an enthusiastic amateur geologist. By going on some of the association's field trips, he came to meet the Manchester geologist Charles Lapworth, whose son, Arthur, was to become one of Armstrong's best-known chemistry pupils.

Armstrong also read widely in the chemical literature of his day and was rewarded by his bookishness by being made the college's librarian. One final feature of his chemical education, and a curious one, was that while at the RCC he would often spend his Saturday mornings watching operations at St. Bartholomew's Hospital in the City. He took a keen interest in surgical operations and in the physiology and pathology that underlay the surgery. As he later confessed, "I have sometimes been surprised that I was not tempted to study medicine rather than chemistry: if I had then realized that it was no more nor less than a branch of chemistry, I should have done so; but there was little chemistry in medicine in those days" (HEA 1920c, 6). In fact, his career was to extend into aspects of biochemistry that became relevant to medical science.

Armstrong remained a registered student at the RCC for four semesters from 1865 to 1867. Frankland's employment of circle-and-dash (ball-and-stick) molecular formulae struck Armstrong as bold and useful, particularly when applied to the growing complexities of organic chemistry. Late in 1866, Frankland paid him the compliment of taking him into a private laboratory he ran in Notting Hill for the purpose of analyzing the quality of London's drinking water, which was then provided by a number of private companies. Given the minimum of help, Armstrong's task was to devise a new, accurate, and absolute method of determining the amount of organic matter (identified as polluting and dangerous) in water supplies.

> We devised the vacuum combustion process in which we were the first users, after [Thomas] Graham, of the Sprengel mercury-fall-pump, then a single tube. Not until three or four years after did [William] Crookes follow our example and use the pump, in an improved form in constructing his Radiometer. Meanwhile Frankland and [Norman] Lockyer were using it, in the spectroscopic work which gave Lockyer his start. We were at the beginning of things. (HEA 1920c, 6)[1]

Armstrong's technically difficult but accurate method of water analysis worked admirably and was published under Frankland's and Armstrong's joint names in 1868. It was his first publication (HEA 1868). He remarked later that "there was no trace of the slave driver" in Frankland. "Only those could work effectively with Frankland who needed little if any guidance; but at times he would step in and give invaluable advice or still more important assistance by coming to the worker's aid with his wonderful manipulative skill" (HEA 1920c, 6).

Frankland's skills had been largely self-acquired (C. Russell 1986). Disadvantaged by illegitimacy (which he kept a dark secret), shyness, and a poor ability as a lecturer, he had been apprenticed to a Lancaster pharmacist for five years before he entered Lyon Playfair's London laboratory in 1845. Through the influence of the German chemist Hermann Kolbe, who had joined Playfair's laboratory in the same year, Frankland spent three

---

1. See also Frankland to HEA, 8 January 1867, Royal Society [RS] archive, MM/10/90.

months at the University of Marburg in 1847 before joining Tyndall as a science teacher at the extraordinary Quaker school, Queenwood College, in rural Hampshire. There he had taught practical chemistry (analysis), botany, and geology, while Tyndall taught surveying, mathematics, and engineering physics. In 1848, Frankland returned to Marburg to study with Robert Bunsen, and after obtaining his doctorate there in 1849 he became the first professor of chemistry at the newly opened Owens College, Manchester, later the University of Manchester. Dissatisfied by life in the provinces and by the elementary nature of the courses required in Manchester, Frankland moved to London in 1857 to teach at St. Bartholomew's Hospital and the Royal Institution before succeeding to Hofmann's position at the RCC in 1865. According to Armstrong, who clearly adored Frankland (C. Russell 1996),

> Frankland was so thrown upon himself, he so developed the art of self-help, that he never learnt to order and use others sufficiently, which is the teacher's art; he kept counsel with himself. Frankland was a pattern worker to those who were privileged to work for him; they gained lessons for life... His lectures were clear, straightforward and logical and he took particular pains to illustrate them by well-thought-out, practical demonstrations. (HEA 1928b, 408)

But Frankland did not merely mold Armstrong's career by providing him with a solid chemical foundation, or with a love of the practical, or by giving him his first research problem, or by then packing him off to Germany. For Frankland was also interested in science education, as Armstrong came to observe.

The Frankland-Armstrong water analysis paper (HEA 1868) began with an historical account of the analysis of potable waters and how little accuracy there had been until Hofmann and John Blyth reported to the General Board of Health in 1856 (Hofmann and Blyth 1856; Bentley 1970). Working at the RCC, Hofmann and Blyth had noted that chemists needed to find a way of determining the amount of nitrogen in the organic matters dissolved in the waters, in order to have an accurate measure of sewage pollution. Ten years on, this was what Frankland and Armstrong proposed to achieve. In doing so they were highly critical of the usual methods for estimating the total solid constituents dissolved in water, the estimates of organic matter in water, and the estimates of the amount of oxygen required to oxidize and remove organic pollutants. The latter estimate would also involve improving the way the presence of nitrous and nitric acids and ammonia were estimated. The method Armstrong devised to achieve this involved adding freshly prepared dilute sulfuric acid to the sample of water; this mixture was then boiled in a Liebig condenser for a couple of minutes before the sample was evaporated to dryness. The residue was then transferred to a Liebig combustion train charged with copper oxide, which was also attached to a Sprengel mercury pump. The gases collected were a mixture of carbon dioxide, nitric oxide, and nitrogen, whose quantities were measured by an improved method of gas analysis already established by Frankland (Frankland 1868, 109–20).

Meanwhile, before their joint work was complete, Frankland's former Lancastrian pupil, J. Alfred Wanklyn, together with colleagues at the London Institution (LI), had published an entirely different method of determining the carbon and nitrogen content of natural waters. This publication had appeared as a monograph rather than as a paper presented at the Chemical Society (Wanklyn and Chapman 1868; Wanklyn, Chapman, and M. Smith 1868). In their method, nitrogenous materials were treated with potassium permanganate to form ammonia that was then estimated by titration. Frankland and Armstrong, however, argued that oxidation was never complete and therefore that it seriously underestimated the amount of nitrogenous pollutant. Wanklyn and his co-workers quickly answered that they did not consider it necessary to determine the exact amount of nitrogen in order to make a judgement on the amount of sewage contamination. This difference of opinion was to lead to a longstanding feud between Wanklyn and Frankland—a dispute that Armstrong was fortunate to escape (Hamlin 1990).

"Made in Germany"

In the autumn of 1867, on Frankland's recommendation, Armstrong traveled to Leipzig to study with the eminent organic chemist Hermann Kolbe. He spent five semesters there, and later reminisced:

> Kolbe's laboratory, in those days, afforded wonderful opportunities. About a dozen of us were doing advanced work in preparation for the degree—seeking independence. Each had his Arbeit—his definite problem—to view, as his chief aim in life: we were all proud of being called on to show that we could do something.... Whatever suggestion we made to Kolbe, he never discouraged us; his habit was the grasp the lapels of his coat, then to reply: "Try it, try it." We disputed with him continually before the blackboard, often for hours together, nearly always taking exception to his theoretical views—but without his being offended. And we constantly compared notes together. Each of us,

therefore, was interested in the solution of a whole series of problems. (Quoted in Hartley 1971, 220)

Because of his infamous diatribes and slanderous polemics in the 1870s against August Kekulé, Adolf Baeyer, Henry van 't Hoff, and others, Kolbe's justified reputation as a master chemist and teacher was eclipsed, while the reputations of those he attacked (in Armstrong's jaundiced view) "thrived on the publicity." Too often remembered for being on the wrong side, Kolbe's role in the history of nineteenth-century chemistry has received proper assessment by Alan Rocke (1993). Kolbe wrote clearly and precisely and abhorred looseness of language. He expressed his own views forcibly and without bowing his knee to authority. The similarity to Armstrong is obvious; the latter himself recognized this as a compliment. Kolbe was an outstanding teacher in the German tradition of Justus Liebig, Robert Bunsen, and Hofmann, and despite the unfashionable character of his theoretical views, his laboratories never lacked students or research workers. Kolbe did not merely reinforce his friend Frankland's tuition in experimental techniques but pushed Armstrong into the excitement of research in organic chemistry. Armstrong was left *selbständig*, able to walk alone (HEA 1924b, 149). In addition, Armstrong's propensity toward a sharp tongue and a critical attitude toward scientific theories, whether in attack or defense, was a legacy from Leipzig. Together, Frankland and Kolbe made Armstrong critical and a passionate believer in self-education through laboratory research.

Armstrong probably had only a reading knowledge of German when he arrived in Leipzig. Fortunately, Kolbe spoke English, having lived in London with Frankland for nearly two years. Their first meeting was curious; the two men, the forty-nine-year-old professor and the nineteen-year-old Armstrong, spent an afternoon nitrating phenol para-sulfonate by mixing phenol with sodium nitrate and sulfuric acid. Kolbe then suggested that Armstrong should investigate the mixture formed. The problem proved intractable, and it was only years later that Armstrong realized that he had been investigating a mixture of isomeric derivatives of phenol.

However, in going through the *Annalen der Chemie* he came across an abstract of Alexander Williamson's article on the compound formed between sulfuric acid anhydride and hydrogen chloride, SO_3ClH. He then recalled that, according to the theory of "types" that Hofmann and Frankland had taught him at the RCC, such a compound could be considered as analogous to hydrogen chloride, HCl. That being the case, sulfuric acid anhydride should also form a chloride. To his chagrin, however, he found that this had already been anticipated by one of Kolbe's former Russian students. On the other hand, the insight that two atoms of chlorine could easily be displaced from carbon tetrachloride suggested that sulfur might equally be displaced from carbon disulfide. Because such an attempt had already been made previously by Carl Claus, Armstrong was told that a further attempt would be a waste of time; nevertheless, he went ahead and succeeded in preparing carbon oxysulfide. The study of these oxysulfides formed the subject of his thesis (HEA 1870a) and second publication in English, which, with Frankland's help, he placed before the Royal Society (HEA 1870b).

Under the German system, Armstrong was to be orally examined in physics and mineralogy, as well as to face questions on his chemistry dissertation. Accordingly, besides attending Kolbe's lectures on organic and inorganic chemistry, he also audited those of Wilhelm Hankel in physics and Karl Friedrich Naumann in mineralogy. Both professors must have seemed old men to Armstrong; he certainly found their lectures "wooden and worthless," though he was to take up the subject of crystallography with enthusiasm in the early 1900s. He also voluntarily, and out of genuine interest, attended lecture courses in botany and physiology, the latter including vivisection demonstrations. The teacher was the eminent Carl Ludwig, whose lectures he found "marvellous." In addition, he audited courses on vegetable physiology and agricultural chemistry given by Wilhelm Knop, a pioneer of hydroponics. As if all these extras were not enough, Armstrong and a fellow student named Karl Knapp persuaded the professor of anatomy (F. F. Braun) to deliver a special course on anatomy on Sunday mornings. These lecture excursions took Armstrong away from the chemistry laboratory, much to Kolbe's disapproval. Nevertheless, master and pupil held each other in mutual respect.

Other pupils of Kolbe's whom Armstrong got to know were Gustav Hüfner (later distinguished for work on blood chemistry) and Ludwig Darmstaedter, while Carl Graebe (then a Privatdozent, soon thereafter professor in Königsberg) occupied an adjacent bench. He remained in touch with Darmstaedter for the next sixty years, watching with interest his career as a manufacturer of glycerin and lanolin who retired rich enough in 1906 to spend the remainder of his life as a collector of china and autograph letters of distinguished scientists (HEA 1926m). He also arrived in Leipzig in time to see Kolbe's Russian pupil Aleksandr Bazarov use a sealed tube to synthesize urea from carbon dioxide and ammonia (ammonium carbamate) at high temperatures, a process that was soon put to industrial use (Bazarov 1868).

In later life Armstrong painted a glowing picture of his time in Leipzig—his complete freedom to pursue his own lines of investigation, the sterling encouragement from Kolbe to try anything out to see where it led, and the opportunities given to challenge his supervisor's increasingly heterodox theoretical opinions without causing offense. In retrospect Armstrong recognized Kolbe as a German professor of the old school, "though with a wonderful sparkle of intelligence in his eyes and a most endearing personality when you learnt to know him—not the ogre he has since been painted" (HEA 1929g, 915). According to Armstrong, these were the golden days of *Lern-* and *Lehrfreiheit*, before commercialism and chemical industry made open discussion of ongoing research impossible.

This rosy picture, reported in the late 1920s, is contradicted by the occasional short letters he had, at the time, sent home to his father. Leipzig is situated in one of the less picturesque areas of Saxony, and unlike the beauties of Marburg, Giessen, or Heidelberg, it offered no riverside walks or hillside clambers. The summers were severely hot and the winters cold. Consequently, his letters home showed little enthusiasm for the town and countryside, apart from the many concerts and operatic performances he attended in that city of Bach. Vacations did, however, lessen the monotony of Leipzig life. In September 1867 he visited Dresden on the river Elbe, where the Deutsche Naturforscher-Versammlung was holding its annual meeting, and he was delighted by its architecture. At the Dresden conference he recalled meeting the chemists James Crafts, Rudolf Fittig, Charles Friedel, Peter Griess, Albert Ladenburg, Bernhard Tollens, and other young organic chemists who were just making their names. He recalled, too, the much older Fittig's astonishment that he, who appeared a mere stripling, had been the author of the huge joint paper on water analysis that had been abstracted for German readers from the Chemical Society's journal of 1868.

In April 1868, Armstrong left Leipzig to visit his laboratory friend Karl Knapp in Braunschweig. On returning he told his father that he had begun studying botany and how to use the microscope, as well as continuing Carl Ludwig's course on physiology. Curiously, however, unlike other British students trained in Germany, he did not seize opportunities for weekend and vacation explorations of the German countryside. The German cost of living was still cheap for Britons; even so, his father's allowance of twenty pounds a month for board and lodgings and other expenses proved a struggle.

In February 1869, when preparing for his doctoral examinations, he reported to his father in dismal tones on how his research had progressed "but slowly,"

> in fact with me very unsatisfactorily, fortune has not smiled on me and all my efforts as yet have been fruitless, but within the last two or three

days I have a hope and the odds are that I have at last been successful, at all events I shall probably know for certain by the end of the week. But still if this be the case I am afraid I shall not even then be able to complete my work in time to make my exam. at the end of this semestre [*sic*] and what I should like to know is whether I shall be able to work on and go through it the next [summer semester] as I should much like to do, that is to say providing that I get so far this term as to be sure of being able to do it then. Now that I have worked Chemistry so long I should not much like to cease studying it without taking my degree. I moot this question now and hope to have exactly your opinion and wishes on it as soon as possible as I shall regulate my work accordingly. (HEA to Richard Armstrong, 15 February 1869, quoted in Eyre 1958, 45).[2]

This voice does not sound like that of a mature twenty-one-year-old, nor like that of the confident and strident Armstrong of a decade later. His father replied, typically sympathetic, that his son could stay on—but also revealing that he was unwell. Perhaps this news provided the necessary spur, and perhaps we should not take Armstrong's claimed lack of progress too seriously. In May 1869, Armstrong reached the age of majority, though the celebrations were the family's in Lewisham rather than those of Armstrong in Leipzig. Presents were sent to Leipzig and a genuinely heartfelt letter of thanks was sent back to his parents.

May only all your good, too good wishes for my welfare and happiness come true. I am not capable my dear Papa of expressing all that gratefulness I feel towards you for your ever increasing kindness to me which I feel more and more every day; my only prayer is that with God's help I may during the years and years to come be able to prove to you through deeds what I now so lamely express by words. I feel it to be an utter impossibility for me to give you an idea of the immense obligation which I feel I lie under to you for all you have done for me. (HEA to Richard Armstrong, 9 May 1869, quoted in Eyre 1958, 48)

Despite his coming of age, Armstrong appears to have remained depressed because his research still dissatisfied him. By August 1869, when Kolbe was about to close the laboratory for the summer, an additional worry was his father's breakdown in health. Unable to continue his re-

2. This letter is one of forty-five such letters exchanged between father and son (AP2.134–46). Because the letters are in a fragile condition, I have quoted them from Eyre's printed transcriptions.

search project until the laboratory reopened in October, Armstrong frittered away his time in the excessively hot summer. He now suggested to his parents that he regretted choosing chemistry and wished that he had chosen a career in medicine. In a letter to his Lewisham family in November 1869, he expressed a hope that his chemical work would lead him "to do something in the physiological way as my inclinations are very great in that direction" (quoted in Eyre 1958, 50). It would be another thirty years before Armstrong realized this ambition by moving his research into biochemistry. In 1869 he again attended the annual meeting of the Naturforscher-Versammlung, which was held in Leipzig. There, he said later, he had met most of the German chemists of the time—though apparently not Kekulé.

At last, in December 1869, the problem of the sulfochloro derivatives of organic compounds began to take shape. To ensure his priority, he visited Hofmann in Berlin and read a preliminary paper at a meeting of the Deutsche Chemische Gesellschaft that Hofmann and Adolf Baeyer had founded in 1867 on the model of the London Chemical Society. For good measure, he also joined the German society, so that by the time of his death in 1937 he was its oldest member. In Berlin, he had time to admire the splendid laboratories that Hofmann had helped to design for himself, though he thought them less practical than the facilities that Kolbe had designed and recently opened at Leipzig. It was not until February 1870 that he felt in the position to deliver his thesis to Kolbe and his co-examiners. He confessed to his (by then worried) father that the thesis was difficult to write because his theoretical ideas were now "diametrically opposite" to those of Kolbe's and that he had to choose his words carefully. He was finally examined on 17 March 1870 and was passed immediately.

Why did Armstrong take so long to complete a piece of work, compared with other British students who studied in Germany? One answer might be that he was ill-advised by Frankland to choose Kolbe and Leipzig. Would he not have been better off going to Berlin to study with the eminent Hofmann? Or better still, with Bunsen at charming Heidelberg, where most British students were studying in the 1860s and 1870s (Brock 2013)? There was nothing to have stopped Armstrong from engaging in *Wanderjahre*, as most German students did. Why didn't he leave Leipzig after a semester or two and move to another German university to complete the doctorate? The paltry evidence suggests immaturity and a lack of self-confidence—evidence that is reinforced by his decision on returning to England to continue living with his parents until he married in 1877.

In a final letter to his father from Germany in February 1870, Armstrong recognized his fault.

Had I perhaps from the beginning of my stay here devoted myself to one special subject I might have long before this had matter enough for a dissertation, whereby I should have obtained comparatively little experience.... [but I] preferred to obtain as general a knowledge as possible, and have I am sure thoroughly succeeded in making myself at home in the practical part, not to speak of the theoretical. (HEA to Richard Armstrong, 6 February 1870, quoted in Eyre 1958, 52)

How should he apply his chemical knowledge after coming home? As he informed his father, he had the options of becoming an independent analytical chemist or consultant; working as an industrial chemist; or becoming an academic teacher if and when a university post became vacant. The latter was his preferred choice, and to that end he would have to publish as much as possible to bring his name prominently before fellow British chemists. He therefore needed a position that would allow him plenty of time for private research and even the chance to study practical physics in recognition of the advice that Frankland had given him.

In a final letter to his son, Richard Armstrong expressed annoyance at the time and expense of his education; but the letter also had an enclosure, a letter of introduction to August Kekulé, professor at the University of Bonn. It was written by Kekulé's London-based older half-brother, Charles Kekulé, the wealthy grain merchant well known to Richard Armstrong. Might this connection open doors for his son? The letter proved significant, for it is a curious fact that Armstrong appears not to have known prior to his arrival in Leipzig of Kekulé's groundbreaking proposal (1865) that the benzene molecule had a closed circular chain of carbon atoms. He only became aware of it because Kolbe had become a merciless critic of Kekulé's views.

On his way home, therefore, Armstrong paid a visit to Bonn and met the famous organic chemist, who with his theory of molecular structure, including the benzene ring, had transformed the understanding of organic chemistry. Kekulé showed him round the Bonn laboratories (which had also been designed and built originally for Hofmann, before Hofmann chose Berlin instead), had lunch with him and with Kekulé's assistant Theodor Zincke, and then had an afternoon walk in the *Siebengebirge* countryside with both men. Kekulé reminisced with Armstrong of the nearly two-year postdoctoral stint that he had enjoyed in London. "I have never forgotten the occasion," he later wrote. "We talked incessantly. Nothing in particular was said either of valency or of benzene but much of London—of [Alexander] Williamson, [John] Stenhouse, [William] Odling, Hugo Müller, men I did not then know, but with whom, later on,

I was to be intimately associated" (HEA 1929g, 914). But nothing came of the introduction—neither a paid position in Kekulé's laboratory nor the chance to become a Privatdozent in Bonn. Accordingly, in early April 1870 Armstrong arrived back at his parents' new home in Lewisham's High Street.

✷ 2 ✷

Cobbling a Career in London

> I thus at last realised that knowledge (Science) was one thing; the methodical use of knowledge (Scientific Method) another.
> —HEA 1933m

During the 1870s, Armstrong strove to fashion a proper career in metropolitan chemistry, at a time when "science" was still developing into the professional and societal role that it began more fully to enjoy in the twentieth century.[1] He would succeed, but his was hardly a simple path.

Upon his homecoming in 1870 he had changed completely in appearance, now full-bearded and wearing "foreign" clothes. He found his father worse for wear and subject to "nervous" attacks—a euphemism to cover his overindulgence in spirits. Richard's economic circumstances had nonetheless improved, and he had moved the family to a large house on the High Street with a spacious garden that extended to the River Ravensbourne. A former doctor's house, it contained an annex that had been formerly used as a dispensary and consulting room. Henry took over this annex as his own residence, while still taking his meals with the family.

The Lavers, near neighbors to the Armstrongs in their home above Thomas Lavers's pharmacy, found him rather straitlaced, leading the Lavers' adolescent and young-adult children to nickname him "Starch." Louisa Lavers was four years older than Henry, who probably suffered from shyness; they would not become engaged until 1875. According

---

1. In British English ca. 1880, "chemist" still evoked the proprietor of a pharmacy, and "scientist" was not yet fully accepted as designating a practitioner of "science." For a classic historical treatment of the place of science in society, see Ben-David 1971, and for the institutional background in Britain see Cardwell 1957. Desmond 1997 is an excellent parallel study of a prominent self-fashioning personality of the preceding generation, Thomas Henry Huxley.

to Frank Armstrong, writing four years after his father's death, before marriage—and for some time after—Henry was "extremely earnest and unsociable." Frank must have been told this by his mother, who evidently "had the greatest difficulty in getting him to go anywhere [because] his whole life was devoted to his work.... When he came home, he wrote letters until late into the night" (E. F. Armstrong 1941, 374). What Frank called "the latent social side" of his father's character was only slowly developed; eventually he became the most gregarious of men, an outstanding host who gave great dinner parties at his Lewisham home.

Words, Meanings, Method

During Armstrong's early years back in England, his views matured on how best to think through the meanings of words, and how to explore intellectual questions, including scientific ones. According to his testimony much later in life, his interest in scientific method[2] was originally literary—a surprising confession for someone who had little faith in the traditional literary education of the public schools.

> As a lad, I had read omnivorously and learnt not a little... I had, however, the definite feeling that something was wanting. I could not find any reference to "origins." I use the word advisedly, because the desire to know these things came in through using Trench's *Study of Words* at school. (HEA 1933m, quoted in Brock 1973, 61; R. Trench 1853)

The Irishman Richard Chenevix Trench (1807–1886), poet and professor of divinity at King's College, London, and from 1863 archbishop of Dublin, appears at first sight a curious hero for Armstrong, until it is remembered that Trench was the philologist who popularized the scientific study of languages in Great Britain and who founded the scheme for the *Oxford English Dictionary* in 1867. His *On the Study of Words*, first published in 1853, originated in lectures he gave to schoolmasters and pupil teachers in Winchester in 1845 "so that they shall learn to regard language as one of the chiefest organs of their own education and that of others" (R. Trench 1892, viii). The book had reached eighteen editions by the time of Trench's death and was already a typical school prize book before Armstrong had reached Colfe's School (M. Trench 1888). It was an intensely religious and moral work, and pre-Darwinian in the natural theology with which Trench

---

2. For a historical treatment of the development of scientific method from a different perspective—evolutionary theory—see Cowles 2020.

approached natural phenomena, but it was written in a rich prose whose beneficial influence on readers cannot be doubted. Although Trench discussed the origin of new words, including some scientific ones, the principal essay that engaged with, or was related to, the concept of method was concerned with the discovery of the relationship between words, especially homonyms.

Armstrong often mentioned the way that he felt Trench had made him "critical and anxious to get behind meanings." It was this influence, he said, which prevented him from liking T. H. Huxley's didactic approach. To be sure, the origin of Armstrong's lifelong emphasis on scientific method cannot be ascribed to Trench's influence alone; at best, Trench reinforced the lessons learned at home from a father who used words carefully. However, it surely helped lay the foundation of Armstrong's later vigorous prose style, often embellished by influences from Carlyle and Ruskin. As to method per se, Armstrong suggested that Trench's thorough questioning of evidence was given scientific meaning by Frankland's and Kolbe's research methods.

If we credit Armstrong's self-analysis, it was also given meaning ten years after his return from Germany by a singular legal experience involving patents for salicylic acid. This substance, found naturally in plants of the *Spirea* genus (hence "a-spirin"), was first synthesized by Kolbe in 1859 and shown by him (and others) to possess valuable properties as a food preservative, analgesic, and febrifuge. However, the cost of producing it synthetically was high until 1873, when Kolbe developed an inexpensive and elegant commercial method by carboxylation of sodium phenoxide (Rocke 1993, 304; Kauffman and Priebe 1978). This process was patented in February 1874 and awarded to Friedrich von Heyden, a pupil of a pupil of Kolbe's. However, in November 1877 the London druggists Messrs. Neustadt patented a virtually identical process whose sole difference was the use of anhydrous sodium phenoxide. The rival patent was worked by Neustadt's German partners, the well-established firm of Emmanuel Merck of Darmstadt. The aggrieved Heyden was granted an injunction against Neustadt in 1879; the latter then appealed. This case was one of the first of a great series of chemical patent lawsuits. At the Chancery hearing in February 1880, Armstrong found himself called as a scientific expert speaking for Heyden because he had personally prepared salicylic acid from carbon dioxide and phenol while working under Kolbe's direction in Leipzig. The appeal was granted in March 1880, the judgment crucially referring to "what the scientific witness, Dr Armstrong stated, that the discovery, the subject of the patent, took him entirely by surprise, although he had been a pupil of Kolbe's, and was conversant with what had been previously done by him in this matter" (*Times Law Report* 42 (1880), 300–03, quoted in HEA 1931d).

Armstrong was tremendously impressed by this case, the atmosphere of the Court, the dialectic, the way plaintiff and defendant marshalled their respective arguments, and the clear, incisive manner in which judgment was delivered. The experience seemed to him "the acme of scientific treatment" and made him reflect how often scientists failed this judicial model in their research and publications. Accordingly, he recommended that every scientific teacher should gain experience of trials in the Chancery Court, and he joked that legal proceedings had made him "the unpleasant critic I have since become" (HEA 1920c, 17). He soon found himself involved in other cases. He appeared as a witness against Bayer AG when it sued an English pharmaceutical firm for manufacturing "Aspirin"; and later he supported Ivan Levinstein when he was sued by the Badische Anilin- und Soda-Fabrik (BASF) in the case of a fast red dyestuff. This latter experience, he felt, had been the copingstone on his education. Scientific training had taught him to examine evidence and to ask questions about causes. Trench had caused him to worry about the meanings of words. And the patent actions had made him "alive to the need of a searching cross-examination and judicial consideration of every item for and against a proposition." Facts were one thing; "the methodical logical use of knowledge" was another.

In his training with Frankland and Kolbe, and in his own reading and thinking, Armstrong had come to understand the central importance of careful reasoning based on meticulously determined data. Such reasoning, he thought, was nothing more or less than what was called the scientific method. But Armstrong was not just a practicing scientist, he was also a teacher, and from first to last, Armstrong thought deeply about his pedagogy. His passionate campaign for the "heuristic" method of instruction, closely related to these ideas on scientific method, is discussed principally in chapters 8 and 12 below.

First Appointments

Home from Germany, Armstrong joined the Chemical Society at the first opportunity. He might well have taken up a career as a chemical consultant or, given that he was co-author of what had become the official government method of water analysis, he might surely have found employment with one of the water companies. However, he preferred the prospect of being a teacher and researcher like his mentors Frankland and Kolbe. He was therefore fortunate to have Frankland's immediate support in finding him an academic position. Frankland himself had held a part-time appointment teaching chemistry to medical students at St. Bartholomew's Hospital in the City of London between 1857 and 1864. When his successor there, Augustus Matthiessen, needed an assistant to help teach the small numbers of medical students who wished to take University of London examinations, Frankland was in a good position to get Armstrong the appointment. Armstrong was to coach students at "Barts" for their first MB examinations (equivalent to today's A-levels in the UK) for the next twelve years, "with no little gain in experience" of students and of the influence of examinations on the way they learned.

When he first joined the hospital, a new laboratory was being installed, and he and Matthiessen planned to collaborate on research that Matthiessen had published on the constitution of "the opium bases," that is, the alkaloids. Although his reputation was based upon his electrical research (which he conducted largely in his home laboratory), Matthiessen had for some years been investigating alkaloids with a succession of St. Bart's able assistants, including George Carey Foster and Charles Alder Wright. In 1869, just prior to Armstrong's appointment, Matthiessen had prepared apomorphine, a decomposition product of morphine that rapidly found use as an emetic and a putative cure for alcoholism. However, there were to be no joint publications, for Matthiessen committed suicide at the beginning of October 1870. Armstrong had, it seems, begun to collaborate with Matthiessen on the sulfonation of alkaloids; he wrote this work up and published it (HEA 1871). He did not continue research on alkaloids, but he closely followed Continental research on their chemistry and structures (HEA 1887e). He blamed lack of alkaloid research by British workers on the anti-vivisection Act of 1876 that inhibited British physiological chemists.

Armstrong was still only twenty-two years old and was not sufficiently experienced to replace Matthiessen and take on the entire teaching of chemistry to the hospital's medical students. Matthiessen's place was taken instead by one of Robert Bunsen's many British pupils, William James Russell (1830–1909). Russell had much more experience than Armstrong, having been trained by Alexander Williamson at University College, by Frankland at Owens College Manchester, and by Bunsen in Heidelberg, as well as having had two years of teaching medical students at St. Mary's Hospital in London before succeeding Matthiessen. He was to be elected a fellow of the Royal Society in 1872, four years before Armstrong. The two men got on well enough but went their separate ways regarding research, and never collaborated (J. Brown and Thornton 1955; Foster 1910–1911).

The St. Bartholomew's position brought in an income of only £50 a year plus a small amount from student fees. At the end of 1870, Armstrong added another part-time position that paid another £50 per annum. This new position was as the chair of chemistry at the private London Institution (hereafter referred to as LI) in Finsbury Circus, City of London, where he succeeded Frankland's pupil and later rival in water analysis, Alfred Wanklyn (Cutler 1976; Kurzer 2001). Wanklyn only used the post to conduct his own private research on organic chemistry and had never made the institution's primitive laboratory available to subscribers. With so much talk and publicity on education in the late 1860s, the LI's managers had different ideas. So, when the professorship was re-advertised following Wanklyn's resignation, it was made clear that the incumbent would be expected to involve "a project for establishing a practical chemistry course in the laboratory of the London Institution."[3] The post would allow the professor to request fees from students taking these practical classes. This income would be in addition to his annual £50 honorarium, but Armstrong presumably had to provide chemicals and shared apparatus when necessary. His teaching hours were from 6:00 to 8:00 p.m. Mondays to Fridays, leaving his days free apart from the hours he was required to lecture to medical students at the hospital.

In old age, Armstrong reminisced that he owed the LI position to the recommendations not just of Frankland, but also of William Odling, Warren de la Rue, and John Cargill Brough, all of whom he had gotten to know through the "B-Club," a dining and drinking club named after Section B (for chemistry) of the British Association for the Advancement of Science (BAAS) (HEA 1933i). He also recalled many years later that the laboratory was "little more than a coal hole," and yet there he laid the foundations of a school of research with nineteen papers centered on the chemistry of camphor (HEA 1920c, 15). Through Warren de la Rue, he met the German chemist Hugo Müller (1833–1915), a pupil of Friedrich Wöhler and Justus Liebig, the latter having recommended him to work for Warren de la Rue in his paper-and-ink factory in London. Müller had also been an intimate friend of Kekulé's when the latter was working in London in 1853–1855. Armstrong and Müller became lifelong friends, sharing a passion not just for chemistry but for natural history and nature generally. Müller was also foreign secretary of the Chemical Society during part of Armstrong's long period as secretary.

The LI gave Armstrong a new perspective. Its first professor had been William Robert Grove (1811–1896). Appointed to the LI in 1841, Grove had used the laboratory to develop an improved platinum-zinc battery. Armstrong eagerly read Grove's publications and found his thoughts turning

3. LI managers' minutes, 21 December 1870 (quoted in Cutler 1976, 172).

to the role of electricity in chemical reactions generally. His work at the LI began well, and the evening classes were well attended from the start. The managers were pleased, noting, "The high character of the laboratory as a centre of scientific research [and] a school of chemistry has been maintained."[4]

Unfortunately, the initial successful start was not sustained; evening classes disappeared in 1874, to be replaced by day classes. Armstrong used many of his students as research assistants, and an impressive run of publications ensued in both English and German. After 1874, however, the Institution's annual report ceased to mention Armstrong's classes. In effect, like his predecessor Wanklyn, Armstrong ran the laboratory as if it were his own private laboratory, even though it was hardly fit for serious research into organic chemistry. It had altered little since its erection in 1819 and was incapable of meeting the demands of organic chemistry of that day. It therefore says much for Armstrong's laboratory skills that he was able to generate publications between 1871 and 1875 using such impoverished facilities.

Armstrong also gave regular lectures during the LI's annual seasons, and he continued Brough's sequence of holiday lectures for children (Brough had been the LI's librarian and died young in 1872). By all accounts, Armstrong never became an impressive public speaker. He lacked a good projecting voice, though what he had to say was interesting, often arresting, and usually provocative (Gibson and Hilditch 1948, 623). Armstrong quickly found that the LI was ill equipped in chemical apparatus to perform lecture demonstrations. In response to Armstrong's appeal for new instruments, the LI managers confirmed the absence of chemical apparatus and the presence of virtually valueless apparatus for demonstrating mechanics and electrostatics. The latter were sold off and the money was used to purchase modern chemical equipment. But when Armstrong left the position, the managers decided not to proceed with further improvements, or even to replace Armstrong with another chemist.

While engaged at the LI, Armstrong understandably continued to look for new opportunities that offered a higher income and better laboratory facilities. Henry Roscoe, professor of chemistry at Owens College Manchester, seems to have watched out for academic openings for him. Roscoe mentioned Armstrong's name as a suitable candidate for a chair at the newly established Royal India Engineering College near Egham in Surrey; but it was Herbert McLeod, Armstrong's friend from RCC days,

---

4. LI annual report, 1871 (quoted in Cutler 1976, 144).

who was appointed.[5] Roscoe also invited him to apply for a vacancy he had engineered at Owens College, but Armstrong ignored the offer—he was too comfortable in London—and the position was filled by Bunsen's German student Carl Schorlemmer.[6]

Armstrong also actively sought new opportunities himself. Soon after he had secured the post at the LI, he applied for a vacant position of demonstrator at King's College London but was unsuccessful. In 1874 he was a candidate for a professorship at Yorkshire College in Leeds, but the post went to the Roscoe- and Bunsen-trained Thomas Edward Thorpe, who was later to become Armstrong's colleague in South Kensington and a personal friend. The following year Armstrong was a candidate for the Jacksonian Chair at Cambridge. He was the youngest of the six shortlisted candidates and was well supported by testimonials from Frankland, Roscoe, Benjamin Brodie Jr., Heinrich Debus, Henry Watts, and Kolbe, as well as being most favored by the resident professor of chemistry, George Liveing. The post was awarded to the Scot, James Dewar, who had been teaching chemistry at the University of Edinburgh. As Liveing and Roscoe told Armstrong privately, the medical dons on the Cambridge selection panel had naturally favored Dewar because he had published papers on physiological medicine. Dewar later became a close friend of Armstrong's when Dewar took on the additional and more public role as professor of chemistry at the Royal Institution in London from 1877 onwards.

Armstrong also declined some promising opportunities. In 1876, William Tilden, whom Armstrong had gotten to know through the British Association, moved from teaching boys at Clifton College in Bristol to a chair at Mason's College, Birmingham, and offered Armstrong the opportunity to succeed him in Bristol. The post offered £400 annually as well as far better laboratory conditions for research. Armstrong visited Bristol and was attracted by the neighboring countryside and the school's facilities; but his fiancée, Louisa, pointed out that much of the increased salary would get spent on travelling back to London to see their respective families. Her views proved decisive, and Armstrong declined the offer. There were also family ties to consider—meaning his father's alcoholism and declining health. Then, in 1877, the Fullerian Chair of Chemistry at the Royal Institution became vacant, and Armstrong immediately applied. In the event, as mentioned, the post went to Dewar, who skillfully arranged that his Jacksonian lectures at Cambridge took place in one term, thus allowing him to spend most of his time at in London (Rowlinson 2012).

---

5. Roscoe to HEA, 23 January 1871, AP2.498.
6. Roscoe to HEA, 23 and 29 October 1872, AP2.499–501.

In the same year, 1877, emboldened by the fact that the Librarian of the LI had had his salary raised from £50 to £100 the previous year, Armstrong suggested he deserved the same. He was told that the institution could not afford it. Then in December 1883 he received a letter from the managers that they would cease to pay his £50 honorarium (and a smaller stipend to his assistant) after Christmas of that year. The LI hoped, nevertheless, that he would continue as an honorary professor. This behavior is not quite as bad as it sounds, for the managers were well aware that Armstrong was going to become a better-paid instructor at nearby Finsbury College from January 1884—as will be described in the next chapter. Armstrong agreed to stay on at the LI in an honorary capacity, mainly so that he could complete some research that could not be done at Finsbury until its laboratories were fully up and running. He finally resigned the LI professorship in December 1884. The LI did not (could not afford to) replace him (Cutler 1976, 177–80).

Housed though it was in a beautiful and elegant building, the LI was never as successful as its similarly named rival in Albemarle Street, the Royal Institution, and it was in continuous financial distress from the time of its foundation in 1805 (Cutler 1976; Kurzer 2001). It limped into the twentieth century but gave up its premises to the University of London in 1912. Armstrong lived to see its fine building demolished and its even finer library dispersed in 1936.

But the LI provided Armstrong, the last of its professors, with income and a small laboratory that enabled him to start a line of research students whose work he reported in English and German journals. By 1873 he had also wheedled his way onto the council of the Chemical Society and was beginning to shape the society's business and publications. He planned to earn additional income by writing a textbook of organic chemistry while still residing in his parents' house in Lewisham. His *Introduction to Organic Chemistry* appeared in 1874, with a second enlarged edition in 1880 (HEA 1874a; AP1.14). He claimed that it was "one of the earliest attempts to introduce order into the subject" (HEA 1920c, 15). Benjamin Brodie, professor of chemistry at Oxford and a die-hard sceptic of atomism, congratulated him on avoiding "the various forms of 'pictorial chemistry' [that] has found favour ... and that you have adhered to the simple equatorial method of expressing results."[7] Armstrong was to remain a conservative chemist all his life. The physicist Oliver Lodge admonished him in 1908 for "trying to be too conservative" over the introduction of

---

7. Brodie to HEA, 26 February 1874, AP2.164.

physical chemistry, while admitting that "some conservatism on the part of a chemical leader is useful and desirable."[8]

Through the kindness of Henry Roscoe, Armstrong received an invitation to give one of the Manchester Science Lectures for the People in December 1875. He chose to speak on the chemistry of food, with appropriate demonstrations (HEA 1875c). This instance is the first indication of what was to become a lifelong interest, though it was only during his long retirement that he was able to publicize his interest and concerns about nutrition. As his reputation grew, other income streams opened up, particularly in examining for degrees. An external examinership at the University of Cambridge, where George Downing Liveing was professor, led to a friendship with Henry Fenton (1854–1929), Liveing's assistant. Fenton was a man to Armstrong's heart since he deployed a kind of heuristic method in instructing Cambridge students. Armstrong recalled that his son Robin, who became a physician, was deeply impressed by Fenton's teaching and had lauded him "as the one lecturer [at Cambridge] worth hearing" (HEA 1929c; Hutchinson and Mills 1929). Armstrong remained full of admiration for Fenton's discovery of dihydroxymaleic acid in 1896 and wrote of it as "one of the most masterly pieces of experimental work ever done" (Barbusinski 2009).

Marriage

By the late 1870s, Armstrong was confident enough that the income stream from St. Bartholomew's Hospital and the LI, plus consultancy fees brought to him via City contacts made through the LI, as well as literary commissions, provided a sufficiently robust income for him to support a wife and family. Accordingly, in 1877 he married Frances Louisa Lavers (1844–1935), the Lewisham pharmacist's daughter with whom he had become acquainted since his father's move to their neighborhood in 1870.

Henry and Louisa's first child, Edward Frankland ("Frank") Armstrong, was born in 1879, and a year later another son arrived, Henry Clifford ("Clifford") Armstrong. In 1882 the couple acquired their own home at 38 Limes Grove, Lewisham. Despite living close to his parents, Henry and Louisa naturally began to cultivate their own social circle, including his old school friend Robert Adkin (1849–1935), who had joined his father's tobacco firm in London and who had married Armstrong's sister Harriet Eugenie in 1874. In 1885 the bonds between the Armstrong and Adkin families were drawn still closer by the marriage of Robert Adkin's

---

8. Lodge to HEA, 21 September 1908, AP1.266.

brother Joseph Fletcher Adkin to Armstrong's remaining sister, Edith Annie Armstrong. This natural process of becoming independent from his parents caused something of a rift with Henry's mother. This maternal rift caused further difficulties between Armstrong's parents. They were devastated when their remaining homebound son, Armstrong's younger brother, Louis Ernest Armstrong, a commercial clerk in Lewisham, committed suicide by gunshot in October 1883 at the age of twenty. Louis had suffered a head injury when playing football, which led to mental derangement, and it seems that he had also suffered depression by negative comparisons to his more successful older brother. The inquest, at which Henry (and not his father) gave evidence, received national press coverage, to the embarrassment of the family. Henry never referred to this tragedy in his later writings.

At this point Richard Armstrong's business affairs began to go downhill; he was drinking heavily, and within a year of his son's suicide he also died. Henry Armstrong, poised to start a new career at Finsbury College, was left to pick up the pieces, which included serious debts. The settlement of his father's estate took many months and did nothing to improve his relationship with his mother, who survived until the spring of 1900, cared for by her married daughter, Edith Adkin.

Armstrong spent the first fourteen years of his long career at the London Institution. The students and audiences at the LI were very different in temperament from the young medical students at St. Bartholomew's Hospital. There he was not tied to an examination syllabus, a release that left him free to devise methods of teaching that he would develop in a relevant manner for the practical chemical trades followed by his students. Gradually he developed ways to interest his auditors and encourage them to tackle problems experimentally. As he grew in confidence as a teacher and leader of research, he also grew in social confidence, and under his wife's guidance and encouragement, he became the engaging chemist renowned for his friendliness, biting wit, and superlative knowledge of his subject.

3

Finsbury College

> [Finsbury] is the only Technical College of its standing which definitely
> refuses to spend its energies upon preparing its students for any outside
> examinations. It strictly adheres to the aim of training students directly
> for the industries or professions, mechanical, electrical or chemical,
> which they propose to enter.
> —Streatfeild 1912, 374

It was during Armstrong's tenure at the London Institution that the wealthy City Guilds of London (the so-called Livery Companies, that is, trade associations) began to be criticized for the neglect of their historical role as technical educators. The Guilds felt compelled to respond to the findings of successive royal commissions on the state of British education, the complaints of industrialists concerning the poor educational attainments of the workforce, the urging of leading Guildsmen, and the writings of leading scientists like Thomas Henry Huxley, that the maintenance of British industrial supremacy in the new era of electricity and synthetic chemicals was dependent upon the active promotion of technical education—especially in the political and commercial center of the United Kingdom, metropolitan London.

On the advice of the Society of Arts and a number of educational experts, the City and Guilds (as it became popularly known) decided to aid the cause of technical education by endowing a university-level teaching and research engineering institution within the City of London (London's central business and financial district), beginning in 1878. The overarching administrative structure would be known as the City and Guilds of London Institute (not to be confused with the London Institution). It became the model for other technical colleges around the country.

For financial reasons and because of the unavailability of a suitable plot of land within the City, the Central College of the new institute (also known as the Central Technical College or the City and Guilds College)—despite the word "central"—had to be erected in 1885 west of the City, namely in South Kensington. Central College was destined to be absorbed by the new Imperial College in 1907. Its elegant building was demolished by Imperial College in 1962 and was replaced by "buildings designed by architects with different ideals" (Harte 1986, 187).

But even before the completion of the Central College building, a different City and Guilds entity was born, Finsbury College.

The Science Buildings Movement

The rise of City and Guilds of London Institute, along with the associated creation of Central and Finsbury Colleges, was an outgrowth of a larger movement. Like the 1960s, the 1870s were a golden age for new academic buildings, not only in Great Britain but also in Europe generally, and especially in Germany, the country whose rivalry British scientists most feared (Johnson 1985). Between 1873 and 1874 the Cavendish Physical Laboratory was opened at the University of Cambridge to the design of W. M. Fawcett, with fittings devised by James Clerk Maxwell, the first Cavendish Professor, based upon the experience of the earlier Clarendon Laboratory at Oxford and of William Thomson's laboratory in Glasgow. In 1874, laboratories were opened at the new Royal Naval College at Greenwich, science colleges were opened at Bristol and Leeds, and George Carey Foster began to teach practical physics at University College London (UCL), to be followed three years later by W. G. Adams at King's College in the Strand. In 1878, the architect Alfred Waterhouse's Gothic-styled University College opened in Liverpool, with its extraordinary tiered chemical laboratory and lecture hall designed by its professor of chemistry, James Campbell Brown. Sheffield responded with laboratories at Firth College in 1879 and Mason's College in Birmingham.

The best source of information on this frenzied British activity is the large, illustrated volume entitled *Technical School and College Buildings*, composed by the English architect Edward Cookworthy Robins (1830–1918) (Robins 1885 and 1887). Although other professional architects, such as Waterhouse and Aston Webb, designed many more prestigious academic buildings than Robins, he gave much more detailed thought to the subject of design in buildings and interiors for scientific and technological education and research than his better-known peers did. Waterhouse's record is certainly impressive in quantity and in its geographical distribution: not just the City and Guilds' Central College in South Kensington (1881–1885), but also Gonville and Caius College, Cambridge (1868–1871);

Owens College, Manchester (1870–1871); the Yorkshire College at Leeds (1878); Liverpool's University College (1878); Girton College, Cambridge (1879–1881); and the magnificent Natural History Museum in South Kensington (1873–1881). However, Waterhouse designed precisely as his clients directed, and he dressed the buildings as they could afford. His science buildings effectively owed more to the quality of the technical advice he received from the scientists who were to use his buildings—notably Henry Roscoe at Manchester, Edward Thorpe at Leeds, Campbell Brown at Liverpool, and Richard Owen at the Natural History Museum.

Robins, on the other hand, was a close friend of Armstrong as well as of the engineers (Armstrong's future colleagues) William Edward Ayrton (1840–1919) and John Perry (1850–1920), and he reflected deeply and personally on the ideal environment for technical education.[1] As an executive member of the City and Guilds of London Institute, Robins was able to play an influential role in formulating its policy on technical education in the 1870s and 1880s. His book on technical school design continued to be used and cited by architects until the 1920s; only then was it being replaced as the paragon of laboratory and workshop design (Clay 1902; T. Russell 1903; Munby 1921).

As with most Victorian architects in the second half of the nineteenth century, most of Robins's bread-and-butter commissions came from religion and education, as churches expanded in the wake of the ominous 1852 religious census and as the middle classes increasingly determined to have their children fitted educationally for an industrial and commercial age. Robins designed dozens of London churches and dozens of elementary schools all over London. With the one exception of the flamboyantly Gothic Merchant Venturers' Technical School in Bristol (1882), none of his early commissions was for a scientific or technical institution.

What, then, gave Robins the authority to deliver addresses on technical education and the specific problem of designing buildings for this specialized kind of secondary education? The answer is to be found in another central concern of Victorian Britain, public health and sanitary engineering, for a large proportion of the special technical problems of designing laboratories and workshops were variations on the twin subjects central to the concern of public health experts: drains and ventilation. It was no

---

1. For Robins's extensive correspondence with Armstrong in the 1880s, see AP.1.378–430. The shifting meaning of "technical education" in Britain during this period has been teased out by Donnelly (1989). General surveys of the subject are provided by Cardwell (1957) and Brock (1996).

accident that Robins was a founding member of the Sanitary Institute of Great Britain in 1877.

During the cholera outbreak of 1852–1853, Robins was inspired by George Godwin's editorials in the *Builder* to take an interest in health and sanitation (Wohl 1983). He then played a leading part in the voluntarist District Board of Health, which had been set up in the Regent Square district of the parish of St. Pancras. Robins's views on the sources of infectious diseases (miasmatic, as befitted the era) were publicized in a pamphlet, *A Practical View of the Sanitary Question* (1854), and he joined Edwin Chadwick's metropolitan Health of Towns Association, whose propaganda purpose was "to diffuse among the people the valuable information elicited by recent inquiries, and the advancement of science, as to the physical and moral evils that result from the present defective sewerage, drainage, supply of water, air, light, and the construction of dwelling houses" (Chadwick 1847).

The Finsbury College Story

In 1875, Robins was appointed surveyor to the Worshipful Company of Dyers, one of the ancient Guilds of the City of London, and he was its Prime Warden in the crucial year of 1879, when the Guilds began to invest in technical education (Lang 1978; H. Gay 2000). Robins became a member of the important subcommittee C (Buildings) of the City and Guilds, together with Richard Wormell, Sir Frederick Bromwell, Sir Sydney Waterlow, and Sir John Watney. Robins's unpublished correspondence with Armstrong shows him to have been a zealous advocate of technical education. Such an educational program was necessary, Robins believed, not merely for the sake of manufacturers and artisans, but also for professional groups such as the architectural community to which he belonged. "Technical education," he wrote,

> is the complement and crown of all utilitarian education as contrasted with literary culture only. . . . It is pure science carried to its legitimate issue, that is to say, in varied applications to human requirements; it is the practical applications of scientific principles to special objects and purposes, and it is as necessary to professional men as it is acknowledged to be to manufacturers and artisans. (Robins 1887, 2)

In October 1879, Armstrong was appointed to teach at what was to become Finsbury College, on the basis of his experience at the London Institution (whose proprietors were all City men, with several being Guildsmen).

His colleague was to be the previously mentioned Ayrton, a pioneer of electrical engineering. These classes were the first significant outcomes of the creation of what became the City and Guilds of London Institute in November 1878 (E. F. Armstrong 1938a). But City and Guilds was just an administrative structure; where could the actual technical classes be held?

So as not to appear dilatory while the Central College building plan matured, the City and Guilds decided to sponsor a "Trades School" of evening classes in the large Cowper Street Middle Class School for Boys, in central London's Finsbury district (about a half-mile north of Finsbury Circus, just outside the boundary of the City of London). The headmaster, Richard Wormell, had long been a friend of technical instruction (Brock 2004b). Later, Wormell noted how Armstrong's heuristic method could be applied in teaching mathematics (Wormell 1900; Price 1994, 49). Wormell agreed to lease parts of the school premises after 4:00 p.m., and he also agreed to give up space in the large school playground to allow a new building to be erected. This move was strongly supported by Robins since it meant this technical college would be close to the Bishopsgate and Moorgate Street stations of the underground system; moreover, he was happy to see the proposed Central College placed in South Kensington in close proximity to the new Kensington Museum (the later Victoria and Albert Museum).

Anxious to make an immediate start, in February 1879, at the suggestion of Trueman Wood and Richard Wormell, the institute advertised two lectureships for evening classes in applied chemistry and physics at £150 per annum. In practice, the two lecturers were expected to teach the principles of chemical and electrical trades. There were fifteen applicants for the chemistry position. Armstrong applied on 26 September and outlined his proposed course as follows:

> As the lectures are to be of a character suitable to Artisans and Apprentices, I consider that it will be necessary to commence at the beginning and to teach general chemistry, always taking care, however, to dwell on its technical bearings. I would propose that the first course should be on the elementary principles of chemistry, and that it should include discussion of familiar phenomena such as combustion, and of well-known substances such as water, air, etc. (E. F. Armstrong 1938a; AP1.105–06)

Two further courses would provide more advanced discussions of inorganic chemistry, including chemical elements and compounds of technical importance. And another two courses would be on organic chemistry, since knowledge of the subject informed many manufacturing operations, as well as being important for understanding foodstuffs and life itself. Armstrong claimed that there was not a single "competent" teacher offering evening classes on this subject. Finally, he outlined a course of lectures on the chemical principles of brewing and coal gas manufacture. All of the proposed lectures would also involve experimental demonstrations. His appointment was strongly backed by testimonials from Frankland at the Royal College of Chemistry and Alexander Williamson at University College, the two most important and influential chemistry teachers in London. Many of the leaders of the City and Guilds were also subscribers to the London Institution and would have been well aware of Armstrong's activities there.

Armstrong was duly offered the post on 10 October 1879. And so it came about that on 1 November 1879, a large audience of city dignitaries and local artisans came to the Cowper Street School's assembly hall to listen to an address on "the improvement science can effect on trades and the condition of workmen." The speaker was not Armstrong but his colleague the electrical engineer Ayrton. The meeting inaugurated a series of evening classes on "the practical applications of electricity and magnetism (electric bells, light, telegraph, electric fire and burglar alarms)" which were to be held in the school's basement laboratory. Complementing these classes, in the school's attic, Armstrong (already well-known in the district for his industrial chemistry lectures at the London Institution) began to lecture and supervise laboratory work on "the first principles of chemistry." Ayrton's and Armstrong's courses, priced at five shillings per term, finally began in January 1880.

For the previous five years, Ayrton had been professor of natural philosophy and telegraph engineering at the Royal Engineering College in Tokyo (today's Tokyo Institute of Technology), where he had designed his own palatial physics laboratory (Brock 1981b; Brock 1996); he and Armstrong quickly saw that the City and Guilds plans for their respective laboratories were grossly inadequate. Not only would the proposed building in the Cowper Street School playground be too small to meet the likely student demand, it would be unworthy in comparison with the facilities available in other countries, the upstart Japan being the obvious example. The Cowper Street School's permanent architect was Edward Clifton (1817–1889), a former railway surveyor who had done much work in the City of London with William Tite (Anon. 1889a). Clifton had designed Gresham House as well as the East India House, before erecting the Cowper Street School in 1869. On Lyon Playfair's recommendation and headmaster Wormell's prompting, Clifton had incorporated laboratories and workshops for mechanics and chemistry. Consequently, it was

Clifton who was initially given the task of designing an inexpensive three- or four-story laboratory complex for the City and Guilds, to supplement the school's facilities.

When Ayrton and Armstrong saw the plans, they quickly dismissed them as unfit. Both men sent detailed criticisms and recommendations to the planning committee. Inter alia, Armstrong demanded specific machinery, such as a large gas engine which would serve both Ayrton's need to power various machines (including a dynamo) and Armstrong's need to conduct experiments in electrolysis and electroplating. His concluding comments are worth quoting at length:

> The proportions are modest in comparison with those for existing laboratories. They are still more so, in view of the fact that there is scarcely a laboratory in London where thorough systematic instruction, freed from the narrowing influence of the Science and Art Department examinations, is given in the evening, and scarcely anywhere a laboratory where instruction is given with special reference to the technical applications of chemical knowledge.... At the present time, it is impossible for anyone of ordinary average ability to become a competent educated chemist, not a mere analyst, by attending the English schools alone, and with very few exceptions all the English chemists of repute have received their highest education in Germany. On this account, nearly all posts of importance in our chemical works are filled by skilled German chemists. What is even worse, we have almost lost our position as manufacturers of articles such as the aniline colours and artificial alizarin, which require the highest chemical knowledge for their production. These are now principally manufactured in France and Germany, and as these industries were originally established by English enterprise and energy, it is a matter of great regret such should be the case. For these reasons, I regard the establishment of a Central [College], at which the highest class of education will be imparted, as being of the first importance; but on the other hand, a technical School such as that we have described [i.e., Finsbury] is equally necessary for the instruction of artisans and apprentices, and of lads who have just left the primary schools.... Great additional advantage would probably accrue from the combination of the two schools under one government, so that the courses of instruction could be arranged to harmonise; but if such a combination cannot be effected, it appears to me most undesirable that such a school should be established on hired ground, and thus be more or less subject to control not directly vested in the Guilds. (E. F. Armstrong 1938a, 19)

Ayrton similarly made it clear in his comments that Finsbury College (as it came to be called) would not compete with the proposed Central College. Finsbury's newly proposed day classes would be for lads of fourteen who had left their elementary schools prior to becoming apprentices; the evening students would be artisans and apprentices already following a trade. The Central College, by contrast, would seek its students among the sons of manufacturers who were destined to become managers of works, and among men who desired to be trained as scientific instructors (a class of men having the avocation referred to further on). The Central would also receive the best of Finsbury's students who required further experience of higher education.

Well over a hundred students—bankers, builders, engineers, insurance clerks, chemists, druggists, and others—attended Ayrton and Armstrong's evening classes in the first few months they were offered. Among the many students who attended the classes in the first few months was the twenty-two-year-old Arthur Robert Ling (1861–1937), the son of a local pharmacist who seemed destined to follow in his father's footsteps. Armstrong saw potential talent and encouraged Ling to investigate the halogen derivatives of the nitrophenols and quinones. Ling thereby became Armstrong's first evening-class research student. Ling's publications led to employment with the London Beetroot Sugar Association and a lifelong interest in the chemistry of starch (HEA 1937a; HEA 1930a). His last appointment was as professor of malting and brewing at the University of Birmingham in 1920.

Armed with this kind of evidence of local untapped talent, and with the support of Robins (who was acting in his capacity as the Dyers' Company architect and surveyor), Ayrton and Armstrong were able to persuade the City and Guilds to erect a much larger and better-equipped building in part of the school's playground on Tabernacle Row (now Leonard Street). The Drapers' Company gave £31,000 for the purpose. As Ayrton later acknowledged, it was because Robins "strenuously exerted himself to further technical education in Finsbury, that the various electrical, physical and mechanical laboratories now in Leonard Street, Finsbury, came into existence" (Ayrton 1887; Ayrton 1892, 8). In fact, it was Robins's report to the Guilds on 11 December 1880, in which he over-optimistically argued that a middle-grade technical school for engineering and applied art could be erected in the Cowper Street School playground for £12,000, that persuaded the Guilds to proceed (City and Guilds 1880). Much of his influence undoubtedly stemmed from the fact that he was able to argue that the fittings of a metropolitan technical school ought to be as good as those that the Merchant Venturers of Bristol had asked Robins to design there contemporaneously.

The four-storied plain brick building designed by Edward Clifton contained thirty-two rooms within a space of only 160 by 90 feet (Brock 1989). It was fitted with electric lighting powered by a dynamo in the basement, driven by a large stationary steam engine that also provided motive power to various machines throughout the building by belt drives. The basement also included laboratories for metallurgy, mechanics, and plumbing. The ground floor was entered from the street by a classical portico with a stone frieze bearing the figures of Newton, Wheatstone, Faraday, and Liebig, and it included workshops for the testing of electrical instruments and brewing chemistry, a drawing office, as well as Ayrton's and Perry's offices. On the first floor there were special rooms for teaching physics and mechanics supplemented by two lecture theatres, each seating two hundred students. The second floor was entirely devoted to chemistry, the largest of the two laboratories catering for ninety-six students, each place furnished with storage space for two students (so that the effective capacity was for 192 students working the day and evening shifts). Here Armstrong had his office.

Robins solved one of the perennial problems of chemistry laboratories by running fume hoods along the length of the benches. These conduits aimed to control the emission of fumes by a downdraft chimney connected in the basement to the furnace chimney in the basement. Victorian country houses were often successfully ventilated in this way, but at Finsbury it proved a troublesome and inefficient solution to the problem of ventilation. It was soon replaced by the introduction of electric fans placed on the roof—a direct application of contemporary mining technology. A warm-air ducted heating system also proved troublesome because of the dust it brought into the laboratories (particularly the balance room). This system had to be replaced by a system of steam radiators in 1909, long after Armstrong's departure. Other technical colleges and schools built before 1900 gained much know-how from these failures.

Unplastered walls served to emphasize a workshop or factory atmosphere of hard work, for students were expected to be in their places by 8:30 a.m. and remain there until 5.30 p.m., with a forty-five-minute lunch break. Noticeably absent from the building were reception and committee rooms; a library or refreshment room; or a common room for staff and students. Indeed, by oversight a lady's toilet was forgotten and had to be placed on the roof.

Although Clifton was given responsibility for erecting the building, it is clear that Robins was principally responsible for fitting it out after conversations and briefings from Ayrton and Armstrong. He also made personal visits to Mason's College in Birmingham, as well as to Owens College in

Manchester and the new Manchester Grammar School in September 1880. He further studied Hofmann's designs for the chemical laboratories at the University of Berlin (1865) and the facilities that Ayrton had enjoyed at the Engineering College in Tokyo. Costs inevitably escalated, the final cost being £38,000 (structure £23,000; fittings £15,000). This amount was a considerable expense given that the City and Guilds were committed to erecting two colleges instead of one. The decision to build both the Finsbury Technical College and the Central College was approved on 4 March 1881.

The foundation stone of what was in fact England's first technical college was laid at Finsbury by Queen Victoria's youngest son, Leopold (hence Leopold Street), on 10 May 1881. Curiously, by oversight, neither Ayrton nor Armstrong, nor any of the existing students, were invited to the ceremony.[2] However, the two professors were generously granted leave (with expenses paid) to inspect the fittings in chemical and physical laboratories in Germany in December 1881 and January 1882. This trip was probably Armstrong's first return visit to Germany since he left in the summer of 1870. John Castell Evans, who had been made Armstrong's laboratory assistant in 1881 and about whom we will say more below, was left in charge of the course running at Cowper Street School. The Finsbury building was not ready for use until 19 February 1883, because legal, labor, and cash-flow problems caused delays. Teaching, therefore, continued in the Cowper Street School premises with the first hundred students who had begun their courses in October 1882. The introductory address on "technical instruction" at the opening of the building was given by the newly appointed director, Philip Magnus (1847–1933)—though Armstrong later claimed that he had written it for Magnus to deliver (P. Magnus 1883).

Three further significant steps were taken during this interim period. First, at Ayrton's and Armstrong's suggestion, it was decided in March 1880 to admit day-release students from local factories and (with Wormell's enthusiastic support) from the Cowper Street School itself. Second, as already mentioned, in December 1881 Ayrton, Armstrong, and Robins went on an extensive tour of the continent to examine laboratory fittings. The tour provided Robins with much of the information and plans that he included in his *Technical Schools*. It also gave Robins's views extra weight with the City and Guilds subcommittee. Expenditure on fittings did cause some ill feeling, both between Armstrong (speaking for chemistry) and Ayrton (pushing for electrical hardware) as well as within the subcommittee,

---

2. William Phillips Sawyer (clerk to the Drapers' Company) to HEA, 20 May 1881, AP1.459.

whose members wondered whether their professors were spending money for their own research satisfaction rather than strictly for necessary teaching purposes. Although Robins acted as umpire, the possibility that expenditure on fittings would get out of hand was undoubtedly a major factor in the decision, made over Ayrton's and Armstrong's heads, to appoint Philip Magnus as temporary director of the new college in 1882 (Foden 1962; Foden 1970).

The third step was the appointment in 1882 of Ayrton's former colleague in Japan, John Perry, as professor of mechanical engineering, with additional responsibility for mathematical instruction. Perry, an unconventional but brilliant Irish engineer and teacher, had previously collaborated with Ayrton in Japan on telegraph problems and on the design of electrical measuring instruments. Together they also pioneered the use of graphical methods for recording instrumental readings (Brock and Price 1980). Hence the new building, originally designed to house two principal disciplines, was forced to do duty for three, with consequent severe pressure on teaching space. Lecture theatres had to be shared and the college had to make do for much of its existence without a waiting room, committee room, library, or restaurant. For these amenities, the adjacent Cowper Street School facilities had to be borrowed.

Additional evening courses were also added in 1882 when a Department of Applied Art, directed by A. F. Brophy, was taken over from the City School of Art, and a building trades class was also absorbed from the Artisans' Institute in St. Martin's Lane. Applied art continued to be taught in the basement of the Cowper Street School until 1891, when it was moved into temporary premises in Leonard Street until Finsbury College's extension was ready in 1906. (The department was finally closed in 1912.) The building trades course was ended as early as 1899, when it was transferred to the Shoreditch Technical Institute under the auspices of the London County Council's Technical Education Board.

The minimum school leaving age was fourteen until 1918, and there was no compulsion to provide secondary education until the passage of the Education Act of 1902. Finsbury College, by its admission of day students (in addition to the evening class students), therefore introduced a new kind of post-elementary school—the technical school—into the English education system. The college accepted students from the age of fourteen, though its records suggest that the average age for entry was seventeen. Although some of its students went on to the City and Guilds Central College in South Kensington for more advanced training to become captains of industry, Finsbury's more modest and successful intent was to produce well-educated foremen, sub-managers, and the men (and occasional women) who planned to take immediate posts in industrial works. Nevertheless, a number of distinguished chemists and engineers received their initial training at Finsbury: H. A. Crompton, W. J. Pope, G. T. Morgan, and John Read (future professors of chemistry at Bedford College, Cambridge, Birmingham, and St. Andrew's universities, respectively).

One of the first women students was Hertha Marks, who married the widowed William Ayrton in 1885. She narrowly missed the distinction of becoming the first female fellow of the Royal Society in 1902 for her research on the electric arc, because of the supposed legal difficulties in defining a "fellow" to include the female sex (Mason 1991a and 1991b). Of course, not all these alumni were taught by Armstrong, and most were students at Finsbury after he moved to "Albertopolis" (South Kensington) in 1885, but they all experienced the way that he taught chemistry, insofar as his teaching methods were continued at the college by his successor, Raphael Meldola.

In 1899. the Charity Commissioners suggested that Finsbury College might amalgamate with the new Northampton Polytechnic (now the City University), which they were proposing to build less than a mile away; however, the threatened loss of the Livery Companies' capital investment in Finsbury prevented this merger from happening. However, by 1920 the neighboring Northampton Polytechnic had become a thriving rate-supported institution with a large, up-to-date engineering school. Despite much agitation and protest, to which Armstrong added his voice, Finsbury College closed its doors in 1926. Its staff and students had been much depleted during the First World War, and its annual expenditure greatly exceeded its income from fees. Moreover, the building was now surrounded by better-endowed competitors whose efficiency had been modelled on it. Armstrong remarked, "The Finsbury Technical College was proclaimed a success but its methods are on the dust heap" (HEA 1926p, 728).

What had the seven-thousand-odd engineers and technicians trained at Finsbury gained from their education there? Its teachers, former students, and contemporary observers all spoke of a special "Finsbury method" or "Finsbury plan." Five features of the method can be identified.

First, the teaching in all three departmental areas (chemical, electrical, and mechanical) was analytic (or, in Armstrong's terminology, heuristic) rather than synthetic, deductive, or didactic. A fundamental principle was that students should be taught to think for themselves. Second, practice in the laboratory or workshop was more important than the lectures. All three professors—Ayrton, Armstrong, and Perry—developed printed instructions to accompany experiments and machinery tests.

These instructions raised questions the students had to answer through their own manipulations. Moreover, experiments and tests were usually quantitative in character, thus raising the issue of functional relationships between variables and the need to develop mathematical skills of analysis. Through Armstrong, "learning by doing" was developed into an educational movement for reforming and revitalizing science teaching in schools and colleges.

Third, the development of a fresh "practical" mathematics syllabus recognized that the mathematical skills needed by scientists and engineers were different from those of potential mathematicians—the skills professed by those who studied analytical mathematics at Cambridge. Perry developed *The Calculus for Engineers*, which emphasized decimals with approximations, the use of logarithmic tables and the slide rule, or in algebra the use of formulae without derivation, and functions studied by means of graphical or squared paper (Perry 1897; Brock and Price 1980). Practical methods in mensuration and geometry were also encouraged, as was numerical trigonometry. Such techniques would have passed Armstrong by since, as we shall see, he was no friend of physical chemistry in which mathematical skills were needed. However, he undoubtedly saw the wisdom of Perry's approach to the training of engineers. All these mathematical elements were variously mixed, while a close correlation was kept with the students' science lectures and laboratory work.

Fourth, all students, whatever branch of engineering they planned to enter, took a common first-year course that included chemistry, mathematics, mechanics, engineering drawing, and electrical and mechanical engineering, as well as French or German. Specialization was deferred until the second year. Armstrong was always proud that the introduction of an engineering element into his chemical course had given Finsbury and Central College students a special value in chemical works.

Finally, although students were selected by an entrance examination consisting of papers in English and mathematics (or, alternatively, they had to have passed London Matriculation, the examination qualifying a candidate to study at the University of London), this examination was imposed mainly to stimulate quality instruction in local schools. The Finsbury philosophy was firmly set against any form of outside examination. This policy had the advantage that syllabuses could be very quickly altered as technology improved or changed. For example, when during the early Edwardian period the demand for qualified electricians diminished, Finsbury was able to train more engineers capable of dealing with the new technology of the internal combustion engine. This kind of curriculum change could not have been implemented quickly if, say, the college had been tied to a University of London examination system. Both Ayrton and Armstrong were opposed to external examinations.

> To teach so as merely to fit students to pass certain examinations is, no doubt, as a rule, a demoralising practice; but when, as in the present instance, both the ethical education and the technological examinations themselves are happily in the hands of one governing body—the City & Guilds of London Institute—there ought, I think, to be no difficulty in adapting the one to suit the other, with good results. (E. F. Armstrong 1938a, 27)

A mark of the originality of the Finsbury Plan is that so many of its features now seem commonplace. Finsbury succeeded in its intended task of training technicians for Britain's "second industrial revolution," a revolution based not upon steam power alone but on electricity and the exploitation of synthetic chemicals. And through its factory-like architecture, its distinctive curriculum, and its philosophy of the laboratory and workshop, it exerted a profound influence on the development of British technical education. In the words of one of its alumni: "Finsbury blazed a trail which many have followed" (E. Walker 1933, 37; HEA 1934e; AP1.A7). That the college closed down in the 1920s could legitimately be considered not a mark of failure but a sign that its work as trailblazer was complete.

Observe, Experiment, Conclude

Finsbury's success is all the more curious because neither Armstrong nor Ayrton spent more than eighteen months in the completed college building. But that period, together with the teaching they had done in the Cowper Street temporary premises, was sufficient to ensure that the systems they put in place were continued by their successors. The contemporary national Department of Science and Art chemistry syllabuses began with the preparation and properties of metals and their salts, what Armstrong disapprovingly described as "test-tube chemistry." By contrast, at Finsbury Armstrong designed a course that started with the elementary principles and observations of familiar phenomena such as combustion, proceeded to pure and applied inorganic and organic chemistry, and concluded with a thorough outline of industrial processes. The whole scheme was illustrated by experiments performed by the students themselves, not demonstrated by Armstrong.

The Finsbury College practice-oriented curriculum and method of deferred specialization was designed to make students think, and it formed the basis of Armstrong's future zeal and propaganda for heurism. Thus, it will be worth examining the first-year chemistry course in detail. The course was developed jointly with Armstrong's young assistant, John Castell Evans (1844–1909), who had been appointed as senior demonstrator in 1880. In 1892, after Armstrong had moved to the Central College, Evans published outlines of this course as *A New Course of Experimental Chemistry* (J. Evans 1892). Evans had spent his Welsh youth as a schoolteacher before studying at the RCC with Frankland, who recommended him to Armstrong. He was also a Baptist preacher. As Evans and Armstrong made clear in several reports to the British Association Committee on Chemistry Teaching, the guiding principle for the Finsbury course was

> not to furnish the student with the knowledge of chemistry, but to help him to acquire that knowledge in the most thorough and scientific manner; to give him as little teaching as possible but good honest learning on the part of the student, and systematic guiding and helping to learn on the part of the instructor; and it is therefore not a collection of facts, but of methods and means of discovering facts and of drawing conclusions from them. (J. Evans 1892, 2)

The way to learn chemistry and to train chemistry students, they held, was not to list and accumulate facts but to proceed "systematically from the known to the unknown" by asking questions. Experiments were merely "questions put to Nature," but because Nature is very open-ended, the tyro had to learn how to close the system by asking definite questions. Further questioning might be needed in order to define the information gleaned still more precisely. For Evans, chemical instruction and learning involved both tacit and explicit knowledge.

For example, if we begin with the observation of the phenomenon of combustion and inquire into its causes and effects, the air will have to be rigidly confined—so confined that we can actually see what happens, and a fuel chosen that can be ignited after the system is closed or confined. Hence the choice of phosphorus in a stoppered glass flask. Furthermore, in order to ensure an unambiguous answer to the student's inquiry (that is, to ensure no hidden variables) both flask and phosphorus have to be as clean and dry as possible. The student received printed instructions to ignite the phosphorus and then to describe what happened ("a great change . . . with the evolution of heat and light"), and also what happened when the flask was opened with its mouth under water ("the air has diminished in volume" either "by mere condensation or by the actual loss of some part"). The student then learned to quantify what had happened by roughly calibrating the volume of the flask into five equal parts and repeating the experiment and finding, of course, that roughly a fifth of the air had disappeared and that the four-fifths remaining would not burn more phosphorus.

Only at this point was the student told to call the four-fifths portion "nitrogen," for it was a general principle of the Finsbury scheme "not to employ any chemical *name* or *term* until [the student] has discovered by himself the *thing* or *process* represented by it" (C. Evans 1897, 3). In subsequent experiments the student sought to discover whether phosphorus's reaction with air was unique and was led to discover, by systematic tests, that other substances such as iron and copper would also reduce the quantity of air by four-fifths of its volume. By the eighth experiment, the student was ready to isolate the one-fifth portion of air and to name it "oxygen," and to move on from air to water and discover a new fuel called "hydrogen." At this point, the student was introduced to carbon, to the distinction between elements and compounds, and to the laws of chemical combination. But only by experiment no. 91 was a student ready for equivalent weights, the notion of atoms, the determination of molecular and atomic weights, and chemical formulae.

Once that stage was reached the course became more conventionally systematic, for the new pedagogic aid of the *natural* classification of the properties of elements, or periodic law, could be used to treat elements in families or groups, while at the same time practical work on the *artificial* classification of elements by the experimentally determined groupings and separations of qualitative analysis was pursued. However, in the spirit of "prove all things," these exercises were not mere test-tubing, since the analytical Group Tables had to be deduced from chemical reactions and not absorbed or copied from such well-known aids as C. R. Fresenius's *Introduction to Qualitative Analysis*. All of this practical work was accompanied by parallel lectures on what might be called the "physics of chemistry," namely weights and measures, calorimetry, gas laws, calculations, etc., as well as regular tutorial problem classes.

Here, then, was an innovative course of chemical instruction emphasizing the practical and the quantitative; here was the course upon which Armstrong was to draw on for his lecture to the International Conference on Education, "On the Teaching of Natural Science as Part of the Ordinary School Course" (HEA 1884), which launched the heuristic movement that so happily triggered the school laboratory movement.

Armstrong had been one of the earliest chemists in Britain to realize the usefulness of Mendeleev's periodic system in teaching inorganic chemistry in a systematic way. He had first drawn attention to it in his essay on inorganic chemistry for the ninth edition of the *Encyclopaedia Britannica* when it appeared in 1876 (HEA 1876e). He also wrote a successful *Introduction to the Study of Organic Chemistry* (HEA 1880d), which reached a fifth edition by 1888. A far more expansive treatment of organic chemistry had been published by William Allen Miller in 1857. Miller had died in 1870, and Armstrong, collaborating with Charles Groves, undertook a major revision—a virtual rewrite—of Miller's third edition (W. Miller 1867). It appeared in 1880, partly displacing his own introductory text, and for a few years its thousand pages represented the definitive treatment of organic chemistry on the English market (HEA 1880c).

The success of Armstrong's *Introduction to the Study of Organic Chemistry* prompted the publisher Macmillan to commission Armstrong in 1885 to prepare a laboratory manual on inorganic chemistry. The publisher hired him to write this manual in the full knowledge that "your views on the subject differed considerably from those hitherto prevalent, but we were given to understand that they were very likely in the course of time to be accepted by the most intelligent teachers."[3] Despite this flattering proposal, not surprisingly in view of his new commitments at Central College and his many duties at the Chemical Society, it came to nothing. Hints of what it might have contained, however, may be found in his essay on inorganic chemistry for the ninth (1876) and tenth (1902) editions of *Encyclopaedia Britannica* (HEA 1876e).

Although Armstrong left Finsbury College in 1885, the method and curriculum for first-year teaching that he and Castell Evans had established was continued by Armstrong's successor, Raphael Meldola. It appears ironic that having spent almost five years planning and fitting out their ideal laboratory and workshop spaces for teaching and research, both Ayrton and Armstrong left Finsbury and moved to the Central College in South Kensington. They had spent nearly five years in Cowper Street and barely more than a year in the completed Finsbury College building. But the spirit of their teaching lived on through their replacements, including physicist Sylvanus P. Thompson (who also became the college's principal). The latter continued at Finsbury until 1896, when he too transferred to an identical post at the Central College (E. Walker 1933). Philip Magnus, who had hitherto administered the City and Guilds' educational affairs from an office in Gresham College, also transferred to South Kensington to become the organizing director of the Central College in 1885. He soon

---

3. Macmillan to HEA, 30 April 1885, AP2.459.

experienced difficulties with some prominent Guildsmen who succeeded in downgrading his post to that of "educational adviser" in 1888. Within a year, however, after the Central's administration became chaotic, Magnus was reinstated, and he stayed on at South Kensington until his retirement in 1915 (Foden 1962; Foden 1970). Like Armstrong, Magnus was not an easy man to get on with, and the two men clashed on various occasions.

Before leaving Finsbury, Armstrong and Ayrton submitted a report on how the college had worked since their appointments in 1879. They both clearly recognized that they had been participating and leading an educational experiment. They thought it had by and large succeeded. Syllabus overload had been caught in time and ameliorated by relieving the timetable at the end of the session 1882–1883. Even so, in running an interdisciplinary first year, the chemical students in particular had found it difficult to complete or leave an experimental operation before being called away to attend a class in another subject in a different part of the building. Experience had also shown the wisdom of admitting part-time students, and they recommended the success of using students' own notebooks to monitor an individual's progress. For the future, they recommended that their successors should have assistants. Armstrong's first research assistant (Castell Evans was his demonstrator) was Alexander Kenneth Miller (1856–1945), who was appointed in December 1884 (A. Miller 1946).

Because it took several months to appoint replacements for Ayrton and Armstrong, both men continued part-time at Finsbury until May 1885. From then on, they were permanently at South Kensington. Meldola joined the chemistry department in April of that year.[4] Armstrong's letter of appointment to the chair of the Central College, an appointment worth £1000 per annum, was sent to Armstrong by Magnus on 30 April 1884. A curious and irritating feature of the appointment was that it was renewable annually. In principle this meant that Armstrong (and the other staff) had to reapply for their positions each year. It was for this reason that Oliver Lodge decided not to be a candidate for the physics chair. He upbraided Armstrong "that you are doing your scientific brothers harm and lowering the dignity of your position by accepting office under such conditions."[5] As we shall see, this point was to become a significant issue in 1911.

---

4. Philip Magnus to HEA, 28 April and 1 May 1885, AP1.A/7. Tilden had been invited to apply but deferred to HEA, 24 March 1884, AP2.562.
5. Lodge to HEA, 7 May 1884, AP2.421–22. In his autobiography, Lodge claimed he was actually appointed to the post without his consent! (See Lodge 1931, 105.)

Chemical Research at the London Institution and Finsbury College

We have noted that Armstrong's position at the London Institution was abolished in December 1883, though he continued to be recognized as an honorary professor until December 1884, when the Finsbury College building was complete.[6] As a result, he had access to the London Institution laboratory for the eight years from 1871 to 1883. The research conducted there was published in both German and English; it was usually designated "Communications from the Laboratory of the London Institution" and numbered from I to XVII. Much of the work involved basic studies of the action of inorganic acids on a variety of organic sulfonic acids; these studies were entirely descriptive, but they included a substantial fifty-page investigation of the gases evolved when various metals were dissolved in nitric acid. This work was done with Joseph John Acworth (1853–1927), a London Institution pupil who subsequently took his German doctorate with Otto Fischer at the University of Erlangen in 1890 before founding his own photographic business involving the manufacture of dry plates.

It is notable that Armstrong's early papers on organic chemistry avoided structural formulae and used mainly empirical formulae, a result of his training under Kolbe in Leipzig. The so-called "type" formulae, mainly of the hydrogen or water types, were also used sparingly. It was not until 1875 that a Kekulé benzene hexagon appeared in an Armstrong paper published by the Chemical Society (HEA 1875b). This paper, on isomers in phenols, as he admitted, was written after he had read Wilhelm Körner's "remarkable paper on the study of isomerism in the aromatic series," which provided a compelling evidence-based absolute method to determine just where substituents are located on the benzene ring structure (HEA 1876a). (We will treat more of this matter in chapter 5.) Toward the end of the period, Armstrong and his London Institution pupils turned their attention to the constitution of camphor, and here tentative structural formulae were essential reasoning tools.

Because of the laboratory's destruction by fire in 1883, research during the year 1884 was limited to the availability of the Cowper Street School extension laboratory when evening classes permitted. In those years, 1871 to 1884, Armstrong published himself, or with collaborating pupils, over fifty papers, an article on inorganic chemistry for *Encyclopaedia Britannica*, three editions of his textbook on introductory organic chemistry, and his wholesale revi-

---

6. Raphael Meldola, unaware that the London Institution's proprietors did not plan to continue the professorship, had hoped to succeed Armstrong. See Meldola to HEA, 2 May 1884, AP1.297.

sion and expansion of W. A. Miller's hitherto outdated text on organic chemistry with Charles Groves. In addition, he began reviewing the occasional book for Norman Lockyer's weekly *Nature*. His chief collaborators were the Australian chemist Frederick Brown, George Harrow (HEA 1876b; 1876c; 1876d), and Alexander K. Miller (HEA 1883a; 1883b). These assistants were paid £150 per annum. Frederick Brown (1852–1922) had gone straight from school in Essex to study chemistry at the University of Strasbourg before returning to England to work with Matthiessen at St. Bartholomew's Hospital. Following Matthiessen's suicide, Armstrong invited him to help with investigations of sulfonic acids at the London Institution. Brown also engaged in independent work on fractional distillation in Frankland's laboratory at the RCC before taking a doctorate with Kolbe (on Armstrong's advice) at the University of Leipzig. In 1883 he emigrated to Auckland, New Zealand, as foundation professor of chemistry and physics at the new university (Anon. 1922). Armstrong did not see him again until 1914, when Brown planned to retire to England; however, he found the war conditions in England so depressing that he returned to New Zealand in 1918.

In addition, Armstrong collaborated at a distance with Edward Thorpe, whom he had met at the Bradford meeting of the BAAS in 1873, on the isomers of cresol (hydroxytoluene). Their two reports to the BAAS were presented at the Belfast and Bristol meetings in 1874 and 1875 (HEA 1874b; HEA 1875a). The latter report was notable in that Armstrong had become aware of Hans Hübner's work on substitution within the benzene ring. Armstrong illustrated the second report by Hübner's graphical representation of the locations on the ring (by then called ortho, meta, and para positions) of the hydroxyl and methyl groups. Apparently having been convinced by Körner and Hübner, Armstrong had thus begun to recognize the definite laws or rules for substitution within the benzene ring by 1875.

British technical education owed a great deal to the dedicated work that architect Edward Robins did for the City and Guilds and for technical education generally. He and Armstrong became close friends, as indicated by the fact that Armstrong christened his second son, Robert Robins Armstrong, in his honor; the child adopted the given name "Robin" and made his career as a physician. Edward Robins himself had to abandon his profession when he suffered a stroke in 1890. Incapacitated, he retired to Worthing, where he spent the last twenty-eight years of his life. Whether Robins was able to follow the later development of the Finsbury and Central Colleges is not known. While Robins remained incapacitated, we will see in the next two chapters how Armstrong was to use his work at Finsbury and the Central College to further his fame as a chemist and educationist.

⁕ 4 ⁕
The Central Chemist

> The most remarkable feature of the [Chemistry Department of the Central Technical College] is perhaps the great variety of subjects in which work of real value was done in a department which was always small in numbers, but for that very reason received an intensive cultivation which would have been impossible in a larger school or department.
> —Anon. 1916

Chemists have commonly named useful chemical reactions and syntheses after the person who first deployed or discovered them. There is no "Armstrong rearrangement," "Armstrong's law," "Armstrong synthesis," or "Armstrong reaction," though in industrial contexts one occasionally comes across "Armstrong's acid," naphthalene-1,5-disulfonic acid (Senning 2007, 30). As we have noted, Armstrong is remembered today primarily for his campaigns to improve the teaching of chemistry and science generally by the deployment of the heuristic method of instruction. However, Armstrong's chemical work was extensive and important.

Précis of a Career in Chemistry

Historians of chemistry such as Colin Russell have drawn attention to Armstrong's theory of residual affinity, while Alec Dolby has investigated his hostility toward physical chemistry and, in particular, toward the theory of electrolytic dissociation (C. Russell 1971; Dolby 1976). Right up to the time of his retirement in 1911, Armstrong remained a prolific organic chemist as well as a major figure in the academic and social life of the London Chemical Society. In honor of his forthcoming retirement, his past and present students, led by William Pope and Maurice Solomon, arranged a banquet in his honor at the massive Hotel Cecil in the Strand on 13 May 1911. Armstrong was tricked into attending by the subterfuge that the invitation was just to dine with a few friends. It turned out to be a splendid affair with 250 guests, made up from students and "men eminent in all branches of scientific industry." No women (not even Armstrong's wife) were present, though his four sons were invited. In proposing the principal toast, Pope said that it would not be an exaggeration "to say that no living chemist in this country has exercised a more generally stimulating influence on the progress of chemical research than our guest" (Crookes 1911; AP2.A8). Pope's panegyric was seconded by Solomon, who movingly extolled the value of Armstrong's first-year course at Central College in teaching students "the ability to think and act upon one's own initiative" (Crookes 1911, 255).

In reply, Armstrong initially expressed being overcome and wishing to go home so that he could collect his thoughts before returning to thank everyone. But if he did not have a prepared speech to hand, he soon got into his stride with reminiscences of training under Hofmann and Frankland and beginning research within six months of starting at the Royal College of Chemistry. Kolbe, the "arch heretic," had then taken him in hand, imbibing to him "some of my awkward tendency to express my opinions" (Crookes 1911, 256). He was rightly proud of the fact that, at the Central, he had trained five fellows of the Royal Society, three of whom held chairs at that date.[1] The Central experience had shown that it was no longer necessary to go to Germany for chemical training, to which he added that he was also proud of having made crystallography an essential element of teaching at the Central.

When Armstrong is remembered by contemporary chemists it is often for his proposal that Kekulé's alternating single- and double-bond ring structure of benzene was unsatisfactory and that a "centric" formula would be an improvement (HEA 1887b), but emphasis on this one pregnant suggestion undervalues his significance. As we shall see, Armstrong was a central figure in the development of organic chemistry between 1885 and 1914, more so than one might expect of a Fifth Business character. It is a verbal irony that this centrality coincided with his position as professor at the Central College of the City and Guilds of London Institute in South Kensington, as well as his long-term position as secretary of the Chemical Society. In addition, he had a powerful voice at the Royal Society, the

---

1. Armstrong was thinking of William Wynne (graduated 1896), Frederick Kipping (1897), William Pope (1902), Martin Forster (1905), and Arthur Lapworth (1910). They were followed after this time by Martin Lowry (1914), E. F. Armstrong (1920), and J. C. Philip (1921).

Royal Institution, and the British Association for the Advancement of Science.

Since Armstrong's chemical researches have been well described and analyzed by John Vargas Eyre in his biography (1958), and also by Armstrong's principal obituarists E. H. Rodd (1940) and F. W. Keeble (1941), it will not be necessary here to repeat their work. Generally speaking, Armstrong's work focused on seven principal areas of aromatic and heterocyclic chemistry: the structure of benzene and its substituents; the origin of color; the structure of terpenes and camphor; naphthalene and its derivatives; crystallography; the nature of solutions and the phenomenon of osmosis; and the chemistry of enzymes. Almost all of his published research was broken down into multiple numbered series of papers, each series forming a solid block of research on a particular topic.[2] Underlining all the research activity in these branches lay Armstrong's fundamental quest to understand the nature of chemical change, a common motivation of chemists in the days before electronic theories of the mechanisms of reactions began in the 1920s. More details on Armstrong's chemical research will be provided in the next chapter.

But before going any further, we should say something about the building in which the work was done and what exactly Armstrong taught there.

The City and Guilds Central Technical College

As we have seen, Finsbury College was intended as a stopgap in the larger ambition of the City and Guilds of London Institute of creating a model technical college, the Central Technical College (HEA 1916g; H. Gay and Griffith 2017). Its ambition was long thwarted by the lack of available (and suitable) land within the boundaries of the City of London. Eventually, because the 1851 Exhibition commissioners wanted to encourage scientific education along the lines of Prince Albert's vision for the land to the south of Hyde Park, the City and Guilds accepted an offer of a site adjacent to the Horticulture Society's gardens in South Kensington. The foundation stone was laid by the Prince of Wales in July 1881, and the still-unfinished

---

2. Among the topics were: "Communications to the Chemical Society from the London Institution," eighteen papers (1871–1875); "Isomeric Change in the Naphthalene Series," four papers (1887–1890); "The Origin of Colour," eleven papers (1888–1897); "On the Constitution of the Tri-Derivatives of Naphthalene," fifteen papers (1890–1896); "Studies on Enzyme Action," twenty-four papers (1903–1925); "Studies of the Processes Operative in Solutions," twenty-five papers (1906–1913); "The Origin of Osmotic Effects," four papers (1906–1911); and "Morphological Studies of Benzene Derivatives," six papers (1912–1914).

building was opened in 1884 to provide the site for an International Health Exhibition. Four professorships, each worth £1000 per annum for chemistry, engineering, mechanics with mathematics, and physics, were attached to the new college, and Armstrong, William C. Unwin, Olaus Henrici, and William Ayrton were duly appointed. Armstrong and Ayrton's appointments led to their replacements at Finsbury College by Raphael Meldola and Silvanus Thompson.[3]

The Central's splendid red brick and terracotta-fronted building dominating Exhibition Road was designed by Alfred Waterhouse, who took advice from a committee of experts that included Frederick Abel, William Perkin Sr., Henry Roscoe, and E. C. Robins. But because Armstrong and the other professors were not appointed until the building was nearly complete, Ayrton and Armstrong had to demand last-minute alterations and additional laboratory equipment and fittings. Armstrong and Ayrton were given permission to inspect German laboratories and assess their fittings in order to ascertain the state of the art in interior laboratory design. It was during this trip that Armstrong first met Emil Fischer, who had recently moved from a chair at Heidelberg to one in Erlangen. He was greatly impressed by Fischer's work in structural chemistry, so much so that he would later send his son Frank to Fischer after the latter had succeeded Hofmann in Berlin.

The original estimate of building and fittings costs had been £65,000, but by the time the building was ready for teaching in the summer of 1885, the total expense had risen to £100,000. Such an escalation of costs made the City and Guilds extremely parsimonious in the following years. The early years of the Central were also difficult because many of the City's Guildsmen had opposed building the college in "Albertopolis," as they felt it was too far from the City.[4]

To the right of the splendid marble staircase leading to two upper floors were two large lecture theatres, one for physics and engineering, the other for chemistry. On the upper floors, a portion housed Armstrong's students while the remainder of the building (including the basement) was given over to engineering. The second floor contained three laboratories for

---

3. Armstrong was not absolved from attendance at Finsbury until the summer term of 1885. See HEA to Magnus, 18 April and 1 May 1885, AP2.73–74.

4. The Central's building was demolished in 1962 to make way for Imperial College's expansion plans. As an example of the squeeze on expenditure, although Waterhouse had provided lift shafts, no lifts were installed until 1904. However, the delay was also caused by disagreement between Ayrton and Unwin over whether electric or hydraulic lifts would best serve both transport and the teaching of elevator engineering. Armstrong is reputed to have said he would just be grateful for a hypothetical lift.

advanced students of chemistry, together with preparation rooms and a small lecture room. Although roughly similar to his designs for Finsbury College, Armstrong insisted on more, and larger, fume hoods. Reagent shelves were also placed at right angles to the length of the benches, thus freeing up more space for apparatus (Robins 1887, 141, plate 46). Each of the eight benches for his research students had its own sink and glazed hood. All the first-year teaching of both chemists and engineers was placed on the third floor, which contained a laboratory for some two dozen students and a large balance room. The laboratory was lit by huge skylights and two large windows to the south. There were professorial offices along the corridors of the ground floor, which also housed a library serving all the staff and students (Robins 1887, 72–74, plates 11–13; Sheppard 1975, 238–42).

Although now possessed of probably the best and most up-to-date laboratories in the United Kingdom, and enjoying a munificent salary of £1000 per annum, Armstrong soon found that there were too many "managers" of the college. Not only was he responsible to the Guilds themselves, but he also found himself continually crossed by the chief administrator, Philip Magnus. The latter had transferred from Finsbury College to become secretary and organizing director of the City and Guilds, with his office in the Central building. Although chiefly responsible for organizing the City and Guilds' examinations, he was also responsible for budget control. Further annoyances came from engineering professor Unwin, who acted as dean until 1896, after which time the position was rotated between the four professors.

These admittedly rather trivial administrative annoyances prompted Armstrong to seriously consider transferring to the Royal College of Science when Edward Thorpe left that college in 1894. Despite his high reputation as a research chemist by this time, or perhaps because of this reputation, Armstrong let it be known that he would only move if the chair at the RCS were offered to him without competition. Perhaps he was hoping that the British government would follow the German model and "call" him to the post? Given the likely competition for the post, he did not apply. As a letter to Frederick Abel makes clear (AP1.43), Armstrong feared humiliation if he made an official application for the post but was not chosen, and he also implied that his relationship with Thorpe had not been amicable. Thorpe's successor was none other than Armstrong's Birmingham friend William Tilden. A mark of their deep friendship was that Tilden christened his first son Philip Armstrong Tilden (P. Tilden 1954).

Following the model set by Finsbury College, students were selected by an entrance examination that assessed candidates' ability at written English and mathematics. There were sometimes complaints that the examination discouraged potentially worthy applicants, but Armstrong pointed out that the professors had discretion to admit candidates who had technically failed parts of these entrance tests.

Courses in mechanical, electrical, and chemical engineering began in January 1885 with the intention of providing a higher level of instruction and training than was possible at Finsbury College for students who would probably attain responsible positions in industry or become teachers in technical colleges. Both Ayrton and Armstrong initially devised broad four-year courses, but these had to be trimmed back to three years on the orders of the City and Guilds. There was a common first-year course comprising engineering, mathematics, physics, and chemistry that in Armstrong's case was basically the heuristic chemistry course he had devised for Finsbury; it covered the chemistry of the atmosphere, the earth, chalk, sea water, and the conditions under which iron rusts. Many of his chemistry and engineering students hated the course at the time, claiming that at the end of the first year they knew no more chemistry than they had when they left school. Later in life, however, they came to realize "that they were being trained to observe and think, and even to devise experiments to provide answers to questions and, above all, to take nothing on trust, even from a professor" (Jones 1937, 362).

For their second- and third-year courses, students had the option of specializing in mechanical or electrical engineering (in which case they ceased attending Armstrong's chemistry courses) or in chemistry. The three-year chemistry course always proved less popular, since some 95 percent of the Central's students intended careers in electrical, civil, or mechanical engineering. As a former pupil observed later:

> Armstrong never had a vast amount of raw material for making chemists, distinguished or otherwise. It was probably the less brilliant chemical students and the men in other departments who owed most to him. The brilliant chemists would have arrived at distinction even without an Armstrong to help them, although they will be the first to testify to what they owed him. (Jones 1937, 362)

This fact, together with the Central's cumbersome administrative system and coupled with public allegations that the Central was a white elephant, occasionally made Armstrong's work life frustrating (Donnelly 1987).

By 1903, the number of students reading applied chemistry (the term "chemical engineering" was disapproved by the administration) was static compared with the numbers taking electrical and mechanical engineering. Although Armstrong refused to train up students for London University examinations, leaving it to students' own initiative if they wished to do so, it became apparent that chemistry students often failed the London BSc degree, whereas mechanical and electrical engineering students normally passed. Armstrong's answer to this apparent anomaly was that industrialists failed to encourage students toward careers in chemical factories and that this failure put students off from taking his courses. It also happens that other London colleges (University and King's Colleges in particular) charged lower fees for students who studied chemistry. As for failures at the external London BSc examination, they were due to his department's emphasis on practical as opposed to bookwork. A typical examination question set by Armstrong confirms that he was not interested in bookwork knowledge but in whether the candidates could think for themselves:

> How would you show that water which has been freely exposed to the atmosphere contains "air" in solution? What apparatus would you employ in order to extract and collect a quantity of dissolved air? (H. Gay and Griffith 2017, 94)

As for training engineers, the heart of the problem was probably that would-be mechanical and electrical engineers resented having to study chemistry in their first year when "they were least receptive of aught else but the practical side of engineering." Failing to interest the majority of first-year students, Armstrong came to develop a bias against the mentality of engineers generally (E. F. Armstrong 1941, 373).

In terms of research output, honors gained by his former students, and their successful employment, Armstrong's department soon came to enjoy more prestige than any other chemistry department in the University of London. Indeed, it could be said that the Central was the very center of chemical research in England between 1885 and 1911. Unable to accomplish his ambition alone, Armstrong persuaded the Guilds to support his teaching with salaried assistants such as Gerald Moody (1864–1943), his former demonstrator at Finsbury College. Moody proved a great success in teaching what was essentially the Finsbury heuristic course to first-year students at the Central, and in inspiring some to continue the study of chemistry into their second and third years.

As Gay and Griffith (2017) have noted, like his professorial colleagues in engineering and mathematics—but to an even greater extent—Armstrong seized the opportunity of his appointment to build the kind of department he wanted. In all but name, he created a replica of a German institute of chemistry. Although the number of students who opted to continue chemical studies into their second and third years was inevitably limited by the fact that most students had joined the Central to become mechanical, civil, or electrical engineers, there were sufficient numbers who continued to take the chemistry courses with a view to entering chemical industry. This field was not chemical engineering in the later sense of the term, in which specific problems of running plants and scaling up chemical reactions to bulk production were studied. But Armstrong did get students to study many topics in applied chemistry, such as coal and wood distillation, water analysis and purification, corrosion, and brewing. This approach proved the nucleus for the future Chemical Technology Department that would replace Armstrong's chemistry department following his departure.

With his City contacts, gained while teaching at the London Institution and Finsbury College, Armstrong was also adept at obtaining scholarships from the Guilds so that his most promising students might continue research in a fourth year. The Salters' and Clothworkers' Companies were particularly supportive in this respect. Such fourth-year students were to have very successful careers in industries such as cement manufacture, photographic chemicals, drug manufactories, and, especially, the brewing industry, where Armstrong had very supportive contacts. Others, like Arthur Lapworth, Frederick Kipping, Martin Lowry, William Pope, and Martin Forster, became eminent academic chemists who usually gained election to the Royal Society. Lapworth, it is said, proved particularly independently minded, in the style of Kolbe. According to Robert Robinson, Lapworth refused to allow Armstrong to read his doctoral thesis to London University before submission, telling Armstrong that "it is my work and I have no intention of changing it" (Robinson 1976, 69; Nye 1993, 44). Even so, Armstrong plainly influenced Lapworth's theory of the development of the instability of organic structures during chemical reactions.

A Visit to America

In 1897, Armstrong was given two months' leave of absence from the Central to take a trip to America to lecture on agricultural chemistry on behalf of the Lawes Agricultural Trust (he had been the Chemical Society's representative on the trust since 1889). During his absence, and without telling the college's administration or seeking its permission, he appointed Edward Howard Tripp to look after the laboratories. Tripp had taken a doctorate with Theodor Zincke at the University of Marburg that year and was seeking employment in England. Born and educated in Lewisham, it is possible that Armstrong had known Tripp as a boy and even encouraged him to go to Marburg to complete his chemical education.

When the dean heard about this arrangement, he had the locks of the building changed; Tripp was locked out, on the grounds that he was not an employee of the college. On his return from America, Armstrong was severely reprimanded.

Armstrong's visit to the United States can be followed from the report that he delivered to the Lawes Agricultural Trust in December 1897.[5] His predecessors in the biennial lecture series had included Robert Warington Jr. and Joseph Henry Gilbert in 1893. Armstrong's obligation was to give public lectures about the groundbreaking experimental agricultural research that John Bennet Lawes and Gilbert had been carrying out on the Rothamsted estate in Harpenden since the 1840s. He arrived in Minneapolis on 12 July to address a three-day meeting of the Association of Agricultural Colleges and Experiment Stations. He had prepared his illustrated lantern-slide talks with abstracts during the voyage. In his report to the Rothamsted trustees, he pointed out that the membership of the association was not confined to agriculturists but was mostly made up of the heads and principals of the land grant colleges that individual American states had founded under the terms of the Morrill Act of 1862. These colleges were, in fact, proper universities, meaning that in Armstrong's view the Lawes Trust would be wasting its time in offering highly technical lectures on agriculture. Armstrong found that Dr. Alfred Charles True (1853–1929), the director of the Office of Agricultural Experiment Stations, and Dr. Harvey Wiley (1844–1930), the chief of the federal government's Bureau of Chemistry in Washington, were of the same opinion.

Armstrong suggested that the trust might do better in future to follow the lines of the annual Harveian Orations that the Royal College of Physicians offered—that is, to present general talks on the progress of science and agriculture, rather than specialized lectures on current Rothamsted research. Alternatively, he noted that in the US there was an Association of Official Agricultural Chemists that met annually; but although, in principle, this association offered a better venue for Rothamsted's publicity drive, Armstrong doubted that time would be found for such British speakers at its meetings. It is clear from his report that Armstrong thought that it was not really worth lecturing to the Americans on agricultural advances now that nearly every state's land grant university had added an agricultural experiment station along the lines of Rothamsted. The trust must have taken this advice to heart since the lecture series did not continue after Bernard Dyer's tour in 1900.

5. HEA to Sir John Evans, chairman of the trust, 8 December 1897, RRL archives, LAT 2.5.

Besides introducing Armstrong to America, perhaps the greatest benefit he had from the experience was to cultivate the permanent friendship of Harvey Wiley. They were to share in campaigning for the improved quality of food and the eradication of artificial foodstuffs in human diet in both America and Britain.

Writing for Nature

In 1879, Armstrong began writing for Norman Lockyer's weekly journal *Nature*; he was to become a renowned contributor to the periodical for the rest of his life. Most of these contributions were made as letters to the editor, but gradually they extended to substantial book reviews from 1886 onwards. His first book review was a sharp criticism of a reissue of a textbook on *Chemical Physics* by Harvard University chemist Josiah Cooke (HEA 1886a). An equally sharp critique of a textbook by Cambridge chemist Matthew Moncrieff Pattison Muir (HEA 1888a; Muir 1887 and 1888), led Armstrong privately to make a half-serious suggestion to Raphael Meldola, his successor at Finsbury College: "Don't you think we might apply the letter ψ [psi] to such pseudo chemists as Muir?"[6]

Armstrong's first letter to *Nature* concerned the evidence that chlorine could be "dissociated" into its elements (HEA 1879 and 1880a; Brock 1973). In 1873, Lockyer had suggested that the high temperatures of the stars and nebulae might cause the terrestrial elements to dissociate into simpler substances and thereby give rise to the complex spectra observed by terrestrial spectroscopists. The observations appeared to offer support for the "calculus of chemical operations" that Oxford chemist Sir Benjamin Collins Brodie had proposed as an alternative to Dalton's atomic theory in June 1867. It seems from Armstrong's remarks that he was quite prepared, like many of his contemporaries, to believe that the elements of Lavoisier, Dalton and, more recently, Mendeleev, were composed from simpler materials (HEA 1880b and 1887f; Brock 1973).

Armstrong regularly attended meetings of the peripatetic British Association for the Advancement of Science each year in August or September. Sometimes he commented in *Nature* on papers that he had heard and was led into controversy with the speaker (HEA 1893b, 149–50; Smithells 1893, 198). By 1890, however, he felt that the association's procedures had become tired and, after fifty years, needed modernization. His letter to *Nature* (HEA 1890b) expressing these views elicited a response from

---

6. HEA to Meldola, 8 February 1888, Meldola papers, Newham Museum Service, Passmore Edwards Museum, Stratford, London.

his friend Tilden, who accused Armstrong of exaggeration (although he agreed that he did not relish spending a holiday period in stuffy rooms).

Armstrong returned to the subject two years later: not only were the procedures at the BAAS meetings not in tune with the times, but "the 'tripper' element had become too predominant, and that the credit of science would suffer if a large number of persons be permitted, year after year, to take a pleasant holiday, under the pretence of advancing science, while the number of true workers whose reputation alone upholds the claim of the Association to public recognition is but small" (HEA 1892b). The reading of papers had become "a solemn and dreary farce played to almost bare benches; and it is only in exceptional cases—such as Section A affords—that a small and devoted body of true believers worship at an inner shrine without regard to the general public, and are thus able among themselves to do work of high value to science." He urged the association to change course to cater for two functions: the advancement of science among professional men of science, and the popularization of science. The former was best done by making sectional and intersectional meetings the key feature, based around a controversial theme, and to abandon random papers. To promote this aim, papers should be pre-circulated to enable detailed discussion on the day, and these should be fully reported in the association's annual reports. Armstrong's comments possibly had some effect, since at the meeting in Nottingham in 1893, Section B held two intersectional discussions on bacteriology, led by Percy Frankland, and another on coal-mine explosions, led by Harold Dixon (Brock 1981a).

When Welsh journalist Ernest Williams published his sensational warning that Germany was overtaking Britain as a trading nation, Lockyer commissioned Armstrong to write an editorial for *Nature* in spring 1897 that ran over two issues (Williams 1896a and 1896b; HEA 1897b, 411).[7] The editorial urged the need to organize the scientific community to ensure that British industry and business be able to compete in a world market that included not only Germany but also now America and Japan; it addressed themes that were to dominate Armstrong's writing for the remainder of his life. The reason Germany had become such a successful trading nation, he asserted, was that it had developed a *scientific* system of education. Recalling Carlyle's warning in *Past and Present*, he wrote that

---

7. Williams's book, which arose from articles in the *National Review*, went into four editions. Armstrong observed that Germany was both a cultural and industrial leader; "the application of science to industry has brought the whole world into competition and only those who fully understand and can apply all the rules and every detail of the game can hope to succeed in it" (HEA 1897a).

Germany has but done her duty and lived according to the laws of Fact, and guided herself thereby; whilst we have followed the laws of Delusion, Imposture, and wilful and unwilful *Mistake* of Fact—for however unpleasant Carlyle's pessimism may be, it is impossible to deny the relevance of his conclusions to the present situation. (HEA 1897b, 409)

British industry needed to be reorganized along German lines, he felt, with science applied to industry and to education. In particular, the universities had to be reconstructed to allow a system of postgraduate education whose graduates would feed into the industrial system and prioritize research. "If English manufacturers will show their appreciation of science as the Germans do by giving employment *at fair* wage to men who have learnt to think for themselves as well as to work honestly and exactly, our schools will soon be filled to overflowing—genius will be attracted to them, and the tide of German competition will be easily stemmed in so far as chemists can stem it" (HEA 1899a). Armstrong's message went largely unheeded until the First World War.

In 1905, when the meeting of the British Association was to be held in South Africa, Armstrong was refused permission to attend by the City and Guilds, despite the fact that he had been elected vice president of the Chemistry Section (B). Going would have involved a month's absence from London, and the Guilds understandably did not see why they should pay a professor when he was not engaged in working for the Central College. On the other hand, he was allowed to attend the Winnipeg meeting of the British Association in 1909, when he was president of Section B for the second time.

The End of the Central

The Central Technical College (as it was officially renamed in 1893) was recognized by the University of London as an engineering training college in 1898. In 1906, the Haldane Committee produced a scheme whereby the independent work of the three colleges in South Kensington—the Royal School of Mines (which had moved to South Kensington from Jermyn Street in 1880), the Royal College of Science, and the Central College—could be streamlined and their courses coordinated. The result was the creation of the Imperial College of Science and Technology in 1907 and the abolition of the academic self-government that Armstrong and his fellow professors of physics, engineering, and mathematics had enjoyed. Duplication and overlapping of courses were to be avoided henceforth. The Central College therefore became Imperial College's Department of

Engineering. The end result was inevitable but nonetheless brutal when it came in 1911. None of the professors had long-term contracts, only annual ones. Armstrong was curtly informed that his contact would not be renewed for the academic session 1911–1912.[8] Armstrong's chemistry department was abolished, to be replaced by that of the Royal College of Science, whose precursor had been Armstrong's own school—Hofmann's and Frankland's Royal College of Chemistry.

Although aged sixty-three and within seven years of the usual academic retirement age, Armstrong was bitterly disappointed by these events. "The change was made without any of us being consulted," he complained.

> My course, which had proved to be of special service, was destroyed. Students were turned over to the tender mercies of the Royal College of Science, where they received the treatment meted out to students of professional chemistry. The Central engineering course lost its special value. Engineers are a sufficiently large and important class to deserve and receive special consideration they need; engineering is full of special problems for the chemist. (Eyre 1958, 163; Humphrey 1935, 105)

The closure was the destruction of what Armstrong liked to call "the Finsbury system" of training. There was no pension, though given the generous salary he had been receiving, he had probably made wise investment plans. His children were grown; apart from one daughter who still lived at home, they were independent; and he owned his own house. Nevertheless, opportunities to earn a little more from writing and consulting were to be welcomed. Despite many signs of bitterness, Armstrong's long years of retirement from 1912 to 1937 were to be, as we shall see, years of activity and enjoyment.

Fortunately, through the kindness of Tilden and Thorpe at the Royal College of Science, a small laboratory space was made available for Armstrong and his last couple of students to complete their work up until 1913. Delicate negotiations with authorities in the City and Guilds resulted in his being offered a half-salary "honorarium" of £500 per annum until 1913, when his last students would have completed their studies.[9] But an appeal in 1913 to engineer Sir John Wolfe Barry, president of the Old Centralians,

---

8. The impersonal letter came from A. L. Soper, assistant secretary of the delegacy of the City and Guilds College, dated 18 February 1911. In a separate, private letter of the same date, Soper apologized for having to send such a disagreeable letter (AP2.205–06).

9. John Wolfe Barry and Alfred L. Soper to Armstrong, March and April 1911, AP2.211–13.

asking whether he could have a room for further research at the City and Guilds College from 1914 onwards, came to nothing.[10] In principle, he could have taken a bench at the Royal Institution's Davy-Faraday Research Laboratory, as his former pupil Martin Forster did from 1913; but the administrative condition that researchers had to reapply for space each term, together with the outbreak of war in 1914, may have deterred him from taking such "temporary hotel accommodation" (quoted in K. Watson 2002).

Family and Lewisham Life

After Armstrong was appointed professor at the Central College, he became involved in promoting scientific interests in the Lewisham area. In February 1879, H. W. Jackson, a fellow of the Astronomical and Geological Societies whose main interest was in anthropology, aware that an impressive number of scientists and engineers lived in the district, founded the Lewisham and Blackheath Scientific Association. It was designed as a mutual improvement society whose purpose was "that those among its members having Scientific knowledge, shall, to the best of their power, enlighten their fellow members as to the aims, principles and methods of the particular Sciences with which they are specially acquainted" (Anon. 1879). The association planned summer outings to places of scientific interest, including a notable visit to Charles Darwin's home in the village of Down in July 1880, as well as geological and natural history fieldwork in the summer season. Not surprisingly, since Jackson lived in the same street as Armstrong at this time, Armstrong agreed to join the association's council under the first presidency of antiquarian Edward W. Brabrook. He remained on its council during the association's first decade and gave papers on a variety of his interests: starch and its function in plant life (1879), the manufacture of coal gas (1880), infectious diseases (1882), gas heating and cooking (1883), Kent's water supplies (1884), and the liquefaction of air (1886). He served as the association's president for 1887 and delivered an interesting address on "The Past and the Possible Future of Our Association" (HEA 1887a).

By then, although the association's membership stayed constant at around 130 members, attendances at meetings had, as so often happens in small organizations, dwindled to mere handfuls of mainly "elderly members." As Armstrong noted in his presidential address, the principle of the association, the relation between the expert improvers and "those willing to be improved," had broken down. Young people were not coming forward

10. Armstrong to Barry, 1 May 1913, and Barry's reply, 7 May 1913, AP1.114–15.

to benefit from joining the meetings, and, interestingly in view of the views he later expressed, he lamented the fact that women had not joined. Not surprisingly, in view of his interest in education, his address ended with suggestions as to how the association could benefit from liaising with the university extension movement's activities locally; but above all, how the curriculum of local (and national) schools might (and should) replace classical education with scientific instruction. "It is not yet recognised by any but the enlightened very few that the times are changed mainly because scientific discovery and the application of science to industry have revolutionised the civilised world," he declared (HEA 1887a, 23).

Although he did not use the term "heurism," he made it abundantly clear that local parents had to rise up in revolt against the "cattle show system" of education geared to competing for university prizes and insist that their sons and daughters be taught the language of science to feed their faculties of observation and finding out for themselves. He recommended that local parents read Herbert Spencer, Thomas Huxley, and Charles Kingsley. Unfortunately, Armstrong's hopes for the future of the association were short-lived, for there is no record of any further meetings after 1888, and it would appear that keener members drifted away to other organizations, such as the West Kent Natural History Society, which also met in Lewisham.

Armstrong's peroration of 1887 raises the question of how he and Louisa educated their own seven surviving children. Their children had all reached adulthood by the time Henry officially left the City and Guilds Central College in 1911. His eldest and favorite son, Edward Frankland (always known as Frank, 1878–1945), had been educated at the Central College and moved to Reading in 1906 as the chief chemist for the bakery and biscuit manufacturer Huntley and Palmer. On the strength of this appointment, Frank had married Ethel Mary Turpin, an insurance clerk from Kent, in 1907. Their first son, Kenneth, was born in 1909 and was destined to become an academic chemist. In Reading, where he had laboratory facilities, Frank had been able to continue collaborative work with his father on enzymes and to publish a comprehensive treatise on carbohydrates based upon the studies he had made with Emil Fischer in Berlin when he completed his postgraduate training as Fischer's assistant and demonstrator (E. F. Armstrong 1910). He would become one of Britain's leading industrial consultants and technical advisers (Gibson and Hilditch 1948). Eyre, who knew the Armstrong family intimately, suggested that Armstrong fell

> into the same error as did his own father in singling out the first-born son for his special concern and attention. [The other sons] were never

taken so completely within the orbit of their father's special care: it was always Frank's interests that were sedulously watched and cultivated. (Eyre 1958, 105)

By 1911, two of Henry and Louisa's other three sons, Henry Clifford (1880–1775, always known as Clifford) and Richard Robins (1886–1974, always known as Robin) were already in their twenties and thirties, while the youngest son, Harold Lavers Armstrong (1890–1984), was just completing his chemistry studies at the Central and was still living at home. In an angry letter to the *Times* in February 1910, Armstrong complained that Harold and his group of Central students had been refused leave to take a University of London BSc by research, even though Armstrong and Michael Foster had persuaded the university to grant such degrees some years before. At present, he complained:

> The London B.Sc. degree connotes cram and nothing more. My own youngest son is among those declined by the University. I am therefore forced, against my will, to send him to Germany. I will not submit him to the mental and moral degradation of the ordinary cram degree. I had hoped he would be able to boast that he had been "made in England" not in Germany like his father and his eldest brother [Frank] because in his time the conditions in London were so improved that it would be possible for him to be educated there. (HEA 1910c)

The refusal was made on the grounds that such a research degree was only permitted in exceptional circumstances under strict supervision and not as an alternative to the pass examination. He described how Frank, once admitted to a German university, had immediately engaged in research that had led to publications with "men so eminent as Profs. van't Hoff and Emil Fischer." It has not proved possible to verify that Harold did attend a German university, much less that he took a doctorate there; most likely the outbreak of hostilities with Germany in 1914 put paid to any such postgraduate study. Harold was destined to lead a varied life in England, America, and New Zealand. He filed a US patent for the destructive distillation of coal in 1921 (a topic that also greatly interested his father), and he eventually became a manager in the Hawker Aircraft Company. During the Second World War, he was a director general of aircraft production at the Ministry of Aircraft Production; afterward, he emigrated to California.

After war service, Clifford spent his career as a fuel technologist for the steel firm of John Brown in Sheffield. He was made OBE in 1944 for his services to Sheffield's fuel efficiency industrial concerns. He published a number of monographs on boiler management (e.g., H. C. Armstrong and C. V. Lewis 1935). Finally, by 1911, Robin Armstrong was beginning medical practice after studying medicine at Cambridge. As yet Robin and Clifford were still bachelors, but both would choose wives in 1917.

Two of the Armstrongs' daughters married. The eldest daughter, Edith Emilie (1882–1970) married the chemist and lawyer Stephen Miall (1872–1947) in January 1907. He was the son of Louis Compton Miall, FRS, professor of biology at the University of Sheffield. Stephen Miall shared the wit and wisdom of his father-in-law, and, as editor of the weekly *Chemistry and Industry* from 1923, he encouraged Armstrong in much of his writing in retirement. The second daughter, Annie (1883–1949), married Dugald William Lionel MacGregor (1877–1958) and emigrated to San Francisco in 1906 because her husband was in charge of the American office of Balfour, Guthrie, and Co. Finally, still in the family home in Lewisham, was the third and youngest child, Nora (1887–1972), who remained unmarried and looked after her parents in their old age. Throughout the Edwardian period and beyond, the Armstrongs entertained friends, pupils, and visiting scientists in their Lewisham home.

While this chapter has chiefly concerned Armstrong's role as a teacher at the City and Guilds' Central Technical College, it has also shown that he found time to widen his interests and become a commentator on and critic of contemporary science, both locally in his hometown of Lewisham and nationally. He began by way of book reviews, essays, and speeches at the annual meetings of the British Association. He also cultivated a wider interest in the role for chemistry in improving agriculture through his membership in the Lawes Agricultural Trust at Rothamsted. These wider activities were performed outside his working hours as a chemistry teacher and researcher, and, as we shall see in chapter 8, were further extended into a campaign for the improvement of science education generally. But before looking at the problem of science education, in chapter 5 we will look a little more closely at the work he and his students accomplished at the Central between 1885 and 1914, and in chapters 6 and 7 how he dominated the activities of the Chemical Society in the 1880s and 1890s. For it was the national and international reputation that he established at the Central and at the Chemical Society that gave him the authority to speak about education and the reform of teaching methods.

✳ 5 ✳
Chemical Research at the Central Technical College

> My laboratory was one of the best and most convenient in the country; the last thing I witnessed before going to Australia in June 1914, was its being hacked to pieces.
> —HEA 1920c, 18

Kekulé's "Marvellous Hieroglyph"

The gifted but cantankerous arch-conservative chemist Hermann Kolbe despised the brilliant theorist August Kekulé, whose ideas of detailed molecular structures, especially his theory of the hexagonal ring structure of benzene, provided what would become the basis of modern organic chemistry. Kolbe's vituperations, freely published during the 1870s in his own journal, often breached the bounds of propriety. How did Armstrong view the relative contributions of these two giants of chemistry, his mentor and his mentor's bête noire?

In his old age, Armstrong wrote a review of Richard Anschütz's definitive two-volume study of the life and works of Kekulé (1929). In the course of the review, he averred that Kolbe rather than Kekulé was "the real parent of the modern system of structure and formulae," that is, resolved structural formulae. But he also saw Kekulé as one of the greatest thinkers in the history of the science. The difference in their styles, he argued, was stark:

> The more so as Kekulé was so strikingly different from Kolbe: the one [Kekulé] the perfect aristocrat and man of affairs in manner, the other [Kolbe] almost *bourgeois* in appearance, a typical professor of the old school. (HEA 1929g, 915)

We saw in the first chapter that Armstrong had visited Kekulé just before his return home from Germany in 1870 and left Bonn "greatly impressed" by him. In 1894, he invited Kekulé to deliver the Faraday Lecture for the Chemical Society (Kekulé declined due to illness). Upon Kekulé's death two years later, Armstrong wrote: "[Kekulé's] theoretical conception of the benzene ring gave an impulse to the study of structural chemistry which has introduced order into the vast array of organic compounds... and has not, even yet, expended itself" (HEA 1896b, 301). Kekulé, he felt, was "a man of truly scientific habit of mind." Armstrong was present at the 1890 "Benzolfest," where Kekulé entertained the assembled colleagues and dignitaries with stories of inspirations from daydreams featuring dancing molecules and snakes, especially connected with his 1865 theory of benzene structure. Armstrong did not leave an account of the festivities, beyond noting with approval that Kekulé had there publicly declared Frankland's leadership in structural chemistry (Schiemenz 1993; Rocke 2010, 304).

Armstrong twice discussed what he thought Kekulé's real contribution to chemistry had been. He quite disagreed with the claim of Kekulé's obituarist, Francis Japp, that Kekulé's impact on chemistry through the benzene symbol had been immediate (Japp 1901). On the contrary, Armstrong stated that until the mid-1870s the structure of benzene received no special attention, and no one foresaw what an important role it would come to play in chemical industry and synthesis. Only after the end of the Franco-Prussian war and (Armstrong claimed) the demonstration by Kolbe of how easy it was to make salicylic acid from phenol did organic chemistry begin to be regarded as a paying subject (HEA 1929g, 916). Armstrong reported that when he had first gone to Leipzig to study under Kolbe for his doctorate in 1867, he had never heard of Kekulé's benzene formula. Only after much controversy, and the opening up of naphthalene and anthracene chemistry, to which he himself made notable contributions, did Kekulé's hexagon prove itself "to be the most marvellous hieroglyph ever devised, satisfying all reasonable minds" (HEA 1929g, 916).

In Armstrong's eyes, besides Kolbe, the key worker had been Kekulé's former assistant in Ghent, Wilhelm Körner, who in 1874 had devised an experimentally based absolute method to determine just where on the benzene ring various substituents (such as hydroxyl, methyl, or carboxyl groups) were located (Koerner 1874). For the first time chemists could actually write structures for the molecular groups that occupied the so-called ortho, meta, and para positions on the benzene ring. Armstrong noted with surprise that Körner's important paper was missed by the indexers of the *Royal Society Catalogue of Scientific Papers* despite the very long abstract (a virtual translation) by Armstrong himself that had appeared in the *Journal of the Chemical Society* in 1876 (HEA 1876a; 1929g, 916). This oversight was partly due to the fact that "Guglielmo Koerner's" pathbreaking work had appeared in a minor Italian journal, as a consequence of Körner's transfer to the University of Palermo.

All of Armstrong's earliest contributions were purely empirical—that is, they were merely descriptions of experimental procedures, calculations of empirical compositions, and illustrative formulae. No *structural* formulae were used in these early papers, nor was there any use made of "type" formulae or Kolbe's proprietary "radical" formulae. This state of affairs began to change only in 1874, after Armstrong read Körner's paper and then headed a British Association committee investigating the isomeric cresols and their derivatives. His collaborator was Thomas Edward Thorpe, a German-educated chemist who had recently been appointed to a chair at Yorkshire College in Leeds, a post that Armstrong had also applied for in June 1874 (AP1.255). In the absence of any relevant correspondence between the two chemists, it is not possible to glean how the collaboration came about; but, in any case, the resultant reports were all written by Armstrong, and it is evident that the investigation of cresols took place at Finsbury College and the London Institution.

Cresol or cresylic acid, a coal tar extract, is hydroxytoluene, $CH_3C_6H_4OH$; it is a homologue of phenol. In their first report produced for the Belfast meeting (HEA 1874b), Armstrong described how he prepared cresylic acid and showed that it was a mixture of the three known isomers of cresol. In the second report, given at Bristol in 1875, he revealed that the investigation had become an investigation of "the laws of substitution in the phenol series" (HEA 1875a). He noted that Kekulé's hexagonal structure of benzene predicted the existence of three isomeric di-substituents at the ortho [1,2], meta [1,3], and para [1,4] positions, but asserted that usually only ortho and para isomers were identifiable. He pictured this result graphically, in a Kekulé-style hexagonal structure, with dashed lines indicating directive bonds leading from the hydroxyl carbon to the ortho and para positions within the hexagon. The bond notation had been suggested by Hans Hübner, a former student of Kekulé's, in 1869.

Armstrong was to develop this directional formula into what became his "centric" formula. He was always opposed to Kekulé's suggestion of 1872 that the carbon atoms of the benzene ring mechanically oscillated between single and double bonds. Such an explanation was profoundly unsatisfactory, he thought, since it implied that the unsaturated benzene molecule ought to exhibit an olefinic character—which it didn't. He was delighted, therefore, when in 1887 the calorimetric and thermodynamic work of Danish chemist Julius Thomsen revealed that the heat of formation of benzene was typical of neither single nor double bonds. This result seemed to support the alternative idea of "residual affinity," a theory that Armstrong had developed in a paper for the Royal Society in 1886 (HEA 1886b; Thomsen 1882–1886; Kragh 2016). After abstracting Thomsen's important results for the *Philosophical Magazine* in February 1887 and emphasizing that they demolished the supposed olefinic character of benzene, he suggested a "centric" formula alternative to Kekulé's model.

> I venture to think that a symbol free from all the objections may be based on the assumption that of the 24 affinities of the six carbon atoms, 12 are engaged in the formation of the six-carbon ring and six in retaining the six hydrogen atoms, in the manner ordinarily supposed; while the remaining six react upon each other, acting towards a centre as it were, so that the "affinity" may be said to be uniformly and symmetrically distributed. I would, in fact, make use of the following symbol. (HEA 1887b, 108)

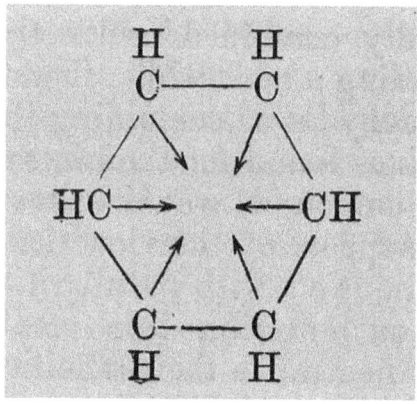

Armstrong emphasized that he did not suppose noncontiguous carbon atoms were directly connected; rather, in his opinion, all the carbon atoms exercised a mutual "influence" upon one another. The various carbon affinities acted toward the center of the ring in the sense of vectors, or what he called "resultants." He recognized that a similar centric formula had already been proposed by Lothar Meyer, but rightly emphasized their theoretical difference (Meyer 1883, 262). In Meyer's formula the six valence forces were free and unsatisfied; in Armstrong's model there was "an excess of [negative] affinity beyond what is required to maintain the C_6H_6 ring and I do not consider that each carbon atom can be supposed to have an affinity free" (HEA 1890d).

Although Armstrong did not immediately explain the advantages of his formula and symbol—for example, in terms of understanding aromatic substitution reactions—he often stressed its significance for representing the chemical *character* of the molecule, rather than its absolute structure. Because benzene did not behave like an unsaturated ethylene-like (olefinic) molecule, Kekulé's oscillating single-double bond symbol misrepresented that character and was, therefore, illegitimate. On the other hand, Armstrong was a practical chemical communicator. Because the centric formula was complicated to write and to print, he was more than content to use the unadorned hexagon for ordinary purposes. For Armstrong, all that an unadorned benzene hexagon implied was "a mere scheme in indicating orientation and also as the implication of the existence of a closed system or nucleus in the hydrocarbon and its derivatives" (HEA 1929g, 916).

These views were spelled out in greater detail in a long essay on chemistry for the tenth edition of the *Encyclopaedia Britannica* in 1902, by which time, as he noted sardonically, Adolf Baeyer's brief flirtation with the centric symbol had been "thrown over." As Colin Russell pointed out, Baeyer's centric formula was based upon the idea of a mechanical strain theory in which the six-carbon-ring structure was maintained by the resolution of affinity forces pulling carbon atoms inwards and prizing them apart (C. Russell 1971, 253; HEA 1902a, 726). It therefore had no connection with Armstrong's idea of residual affinities. Armstrong disdained the other possible option, diagonal affinities, instinctively objecting to affinities that could act across one another. In an important aside, he reminded readers that chemistry was difficult:

> The end the chemist has in view is to gain clear conceptions of the nature of the primary materials, i.e., the elements, with which he has to deal, and in understanding of the operations into which they enter. He is called upon, however, to study this behaviour in crowds and companies in circumstances of great difficulty, in order that he may form an opinion of their behaviour as individuals and depict their characters. (HEA 1902a, 746)

Armstrong stuck tenaciously to his centric model. In a British Association discussion of the laws of substitution in benzenoid compounds in 1899, he likened the mechanism of substitution to a game of musical chairs in which benzene flipped backwards and forwards from an olefinic to a centric form as chairs were removed and players were guided to safe seats by the desire of couples to sit together (HEA 1899c and 1899d). This analogy may sound inconsistent with his objection to Kekulé's oscillating single-double bond system, but apart from better explaining benzene's symmetrical properties, Armstrong believed that the experimental conditions of substitution reactions strongly implied that a catalytic process was involved in the mechanism. In the article "Chemistry," published in the tenth edition of the *Encyclopaedia Britannica* in 1902, he wrote:

> In the case of mono-derivatives, probably the centric mechanism breaks down initially, and a compound is formed by the union of the agent with contiguous carbon atoms of the hydrocarbon; this change is immediately followed by one in the reverse direction in which one of the radicles [sic] thus introduced into the molecule becomes separated together with an atom of hydrogen. (HEA 1902a, 727)

Armstrong (like several of his contemporaries) developed empirical rules concerning how a monosubstituted benzene nucleus directs the entry of a second substitution to either the ortho/para (1:2 and 1:4) or meta (1:3) positions. In 1887 he generalized the rule that in any monosubstituted benzene C_6H_5X, if X is monovalent, then the entering di-substituent will be ortho/para;[1] but if X contains polyvalent atoms, the orientation of the second substituent will be meta. The rule seemed to work generally apart from amines (like aniline) where, in strongly acidic conditions, di-substituents were directed meta rather than ortho/para (HEA 1887d; HEA 1900d). The reason for this exception only became apparent in the electronic era, with the idea of the formation of the $ArNH_3^+$ ion. Armstrong's rule was greatly extended by Alexander Crum Brown and J. Gibson's work in Edinburgh in 1892, and it was their general rule that prevailed in textbooks until the work of Christopher Ingold in the 1930s (Crum Brown and

---

[1]. The quantity of ortho-substitution was always lower than for para-substitution. This remained another unresolved puzzle until the advent of electronic theories.

Gibson 1892). Armstrong and others were searching for the mechanism that explained substitution. In Armstrong's case, he drew upon the idea of residual affinity and the formation of an intermediate addition compound that resolved by means of an intramolecular transformation that settled to form a symmetrical ortho-para di-substitution product. Armstrong never pursued the subject further, possibly because he was equally interested in why substitution sometimes took place in side chains rather than in the nucleus, and above all because he was equally interested in finding rules and explanations for substitutions in the naphthalene nucleus.

Armstrong's claims for residual affinity were further developed in Switzerland by Alfred Werner, who supposed that affinity was disposable (or potentially shared) within a molecular structure. Thus, within a closed hydrocarbon ring or an open chain, adjacent elements with a strong mutual affinity might drain affinity from other attached atoms, leading to alternating strong and weak bonding. This concept was fully developed by Werner's German student Bernhard J. Flurscheim (1874–1955), who had settled permanently in England by 1900. There is no evidence that he and Armstrong ever met or discussed the matter of benzene substitution, though Armstrong surely must have been aware that Flurscheim's model provided a rationale for Armstrong's rule.

Naphthalene Chemistry

The empirical formula of naphthalene was first determined by Michael Faraday in 1826 as the hydrocarbon $C_{10}H_8$ (here using modern symbols), but it was another forty years before Emil Erlenmeyer produced evidence for its structure as two fused benzene rings; the structure was soon thereafter confirmed by Carl Graebe. The elucidation of naphthalene's chemistry is necessarily complicated because of its double-ring system, and this elucidation was not helped by confusing ways of naming and denoting its derivatives. The impressive investigations of Armstrong and his collaborators revealed ways to determine the direction of substitutions within the naphthalene rings by oxidation to phthalic acid derivatives.

According to modern numbering conventions, "Armstrong's acid" is naphthalene-1,5-disulfonic acid. But the older notation, current in Armstrong's day, was more complicated, deploying Greek-derived prefixes rather than numbering. Substituents in the right-hand hexagon were named in clockwork fashion so that substituents 1–2 = ortho, 1–3 = meta, 1–4 = para, 1–5 = ana, 1–6 = epi, 1–7 = kata, 1–8 = peri, 2–6 = amphi, and 2–7 = pros. Only the prefixes "ortho," "meta," and "para" have survived in modern chemistry. The number of possible isomers was also a further complication in understanding naphthalene's chemistry. Monosubstitution simply creates one of two possible isomers, but the introduction of a second substituent may be either homonuclear (when the second substituent enters the same ring) or heteronuclear (when the second substituent enters the other ring). Consequently, di-substitution with one entering group means ten isomers are possible, and if the two substituents differ, a total of fourteen isomers are possible. Tri-substitution with three identical substituents also makes fourteen isomers possible. It gets ever more complicated from there.

Although Armstrong had used the tool of sulfonation to investigate the chemistry of naphthalene as early as 1871, and although he pursued occasional investigations with his assistant N. C. Graham (HEA 1881), naphthalene only became a definite subject for his research when he was at the Central College. There he was able to collaborate over an eleven-year period with William Palmer Wynne (1861–1950). Together, they synthesized all possible mono-, di-, tri-, and tetra-substitution products of naphthalene in order to confirm its double ring structure and complicated isomerisms. Wynne had become Armstrong's research assistant at the Central College in 1886. He soon became a close friend of Armstrong and of his Lewisham family. He taught at the Royal College of Science from 1891 until 1902, when he was elected to a chair at the Pharmaceutical Society. After only two years there he moved to Firth College in Sheffield in 1904 as professor of chemistry, whereupon his research ceased for several decades. It seems he devoted his time at Sheffield entirely to teaching and student affairs. Curiously, it was only after he retired from Sheffield in 1931 that Wynne took up research again. This new research included the reappraisal and detailed publication of all the joint work on naphthalene chemistry that he had carried out with Armstrong in the 1890s.

Wynne had no official position at the Central College after 1891, but he continued to work in Armstrong's private laboratory in the evening, on weekends, and during Royal College of Science vacations, using a master key that Armstrong had provided him. Armstrong had never sought the permission of the Central's administration for this arrangement, and when it finally came to the notice of director Philip Magnus in December 1897, Wynne was humiliatingly marched out of the building and made to hand in his key. Armstrong was furious and pleaded with the Central authorities. Negotiations went on for months that became years, and by the time an agreement was finally reached in March 1904, Wynne was in Sheffield. In this shambolic manner their important research collaboration on naphthalene derivatives was brought to a sudden end in 1897. Half-completed preparations and apparatus of Wynne's remained in a corner of Armstrong's laboratory until it was closed in 1914.

The collaboration with Wynne, which was aided by grants of materials from German dyestuff manufacturers, produced a total of forty items in the *Proceedings* of the Chemical Society. These items were deliberately short, concise reports of experimental procedures and conclusions, not full papers that deserved presentation in the Chemical Society's *Journal*. However, Armstrong did provide thorough reviews of his and Wynne's work in conjunction with sweeping reviews of continental work on naphthalene. These reviews appeared as eleven reports on naphthalene derivatives, written in conjunction with his RCS colleague William Tilden and published by the BAAS.

Between them, Armstrong and Wynne succeeded in preparing pure crystalline samples of all ten of the isomeric dichloronaphthalenes, and all fourteen of the trichloronaphthalenes—outstanding feats of preparative chemistry and structure determination. When he left London, Wynne apparently took the samples with him, but it was not until the 1930s that he repurified them and checked the findings. The work was finally summarized by Wynne in a definitive article on naphthalene derivatives that he compiled for Thorpe's *Dictionary of Applied Chemistry* in 1947.

Color Chemistry

Armstrong produced an influential theory of color in carbon compounds, which he first announced to the Chemical Society on 2 March 1888 (HEA 1888c). He began by reviewing previous attempts to explain color in terms of constitution and molecular structure, starting with the pioneering work of Carl Graebe and Carl Liebermann, and of Otto Witt (Graebe and Liebermann 1868; Witt 1876). The German chemists had drawn attention to the fact that since the majority of colored organic compounds were decolorized by reducing agents, such compounds must contain double or triple bonds. Witt had named color-conferring groups, such as NO_3, $N=N$, and $C=O$, "chromophores," and compounds that contained such groups he termed "chromogens." Moreover, noting that the intensity of color was often increased by adding salt-forming groups (for example, phenolic or amino compounds), he called such groups "auxochromes" (E. Watson 1918).

While Witt's observations and classification were of considerable practical significance to the dyestuffs chemist, the work was purely predictive and did not provide a detailed structural theory of color. The latter was Armstrong's goal. He began by drawing attention to the difference between saturated and unsaturated hydrocarbons. The former, built on single bonds between carbon atoms, were colorless even in the infrared and ultraviolet regions of the spectrum. The unsaturated hydrocarbons, on the other hand, were not only more reactive but also often displayed color in the infrared and ultraviolet. Building on Adolf Baeyer's strain theory and believing that "affinity has direction," Armstrong pictured double and triple bonds as curved affinities. Color therefore was simply evidence of strain within the molecule, analogous to electric current arising from what Faraday had called an "electrotonic state."

Armstrong then drew attention to the fact that the majority of dyestuffs whose structures were known contained a quinone group or might be supposed to produce one by tautomerism. For example, Theodor Zincke and Hans Bindewald had shown that the benzene-azo-α-naphthoquinone prepared from the coupling of α-naphthoquinone and phenylhydrazine was identical with that prepared from the reaction between α-naphthol and benzenediazonium chloride. Hence, there was as much justification for formulating azo dyes by the quinonoid type $O=C_6H_4=N.NHR$ as there was for the formulation $HN=C_6H_4=N.NHR$, "according to whether they were derived from phenols or amines" (HEA 1888c). And, in the case of the colorless para-rosaniline, Armstrong suggested that when converted into colored salts, water was eliminated and a quinonoid structure assumed. Armstrong never claimed that this quinonoid theory of color was original, but it did appear to him that it was of importance to group together facts and conclusions with regard to the character of colored compounds in general.

Armstrong was one of the earliest chemists to link research in progress together in a series of papers. Between 1882 and 1896, he published eleven consecutively numbered papers "On the Origin of Colour." We need not follow these further, except to note, from the third paper onwards, Armstrong's use of his quinonoid theory increasingly to explore and speculate about what he described as "isodynamic change," meaning intramolecular rearrangements, as well as about the phenomenon of fluorescence, which he saw as "a feeble manifestation of that which we ordinarily describe as colour." In 1892, a dispute with Walter Hartley, professor at Dublin's Royal College of Science, allowed Armstrong to elaborate how he believed the "quinonoid mechanism" conditioned color. Color, he supposed, demanded the presence of two "color centres" corresponding to the presence of two double bonds in the ortho and para positions. It was these bonds that cooperatively produced color by interacting with light waves. At some point Armstrong must have realized that he was in the realm of Helmholtzian physiological optics, for—supposing the distinction between visible and physical color was "inherent in the human optic mechanism"—perhaps color perception itself was dependent upon a quinonoid mechanism.

Armstrong's theory of chromogenesis received strong support from several British dyestuff chemists, including A. H. Green and R. J. Friswell, while it allowed Armstrong to keep his idea of a centric form of benzene and other aromatic compounds in the chemical literature, along with the concept of residual affinity, which he also applied to the case of inorganic complexes. However, from 1904 onwards, beginning with the preparation of the colorless quinone-imines by Richard Willstätter and others (and with further examples of non-quinonoid colored compounds that could not be explained in terms of molecular rearrangements), more and more evidence accumulated that Armstrong's theory was untenable. Edwin Watson critically reviewed the several attempts by Adolf Baeyer, Willstätter, Heinrich von Liebig, and others to modify Armstrong's quinonoid theory to account for these exceptions (E. Watson 1918). Besides finding these theories wanting chemically, Watson noted that most of the supporting experimental work was vitiated by its dependence on "selective absorption in the visible part of the spectrum." Indeed, Hartley's work showed that any meaningful theory of color would have to be based upon the quantitative examination of changes of absorption across the whole optical spectrum (Murrell 1964).

Armstrong did not participate in this development, his last research on color appearing in 1905 (HEA 1905c). On the other hand, a lecture to the Society of Dyers and Colourists in February 1925 demonstrates that he believed that the quinonoid theory was still basically sound (HEA 1925e). He also stuck tenaciously to the centric model of benzene. In 1909, for example, when discussing James Dewar's contributions to low-temperature research, he reasserted that "Kekulé's formula" (meaning the single- and double-bond hexagon) "no longer holds the field. At the present the tendency is to accept the centric formula as a more suitable symbol" (HEA 1908e). As we have seen, he also represented naphthalene and anthracene centrically. In all these cases, support came from Dewar's work on the structure of amorphous carbon and from that of Brühl on atomic refractive constants. Armstrong reiterated this point at the meeting of the British Association in Canada that same year (1909) in a long and entertaining presidential address.

> The structure of benzene has been the subject of much discussion.... I trust I shall not be accused of parental bias if I urge that the centric formula is the best expression of the *functional activity* of the hydrocarbon

benzene and its immediate derivatives. The attempts which have been made of late years to resuscitate the Kekulé oscillation hypothesis in one form or another appear to me to be devoid of practical significance. Any formula which represents benzene as ethenoid [olefinic] must be regarded as contrary to fact. But in considering the properties of benzenoid compounds, generally it is necessary to make use of the Kekulé conception as well as the centric expression. (HEA 1909c)

At the Kekulé "Benzolfest" in 1890, Hofmann delighted the assembled guests with an elaborate metaphor concerning the "benzene tree" and its fruits and blossoms.

> Understandably enough, there is no lack of industrious workers who busily strive to collect the harvest. Keen climbers have already climbed up to the third or fourth branch. Some of them we can see working at a dizzy height. Most of them, however, are on the bottom branches of the benzene tree. Some of them have already collected enough and are about to get down. Others still cannot separate themselves from the rich harvest, and yet another group are already quarrelling with their neighbours about the harvest. (Brock, Benfey, and Stark 1991)

Although Armstrong was often a quarrelsome man, and although he begged to differ over the structure of benzene, he seems to have remained content to explore many of the branches of the benzene tree rather than arguing or harvesting extensively from any one of them. While destined to lose the case for quinonoid color chemistry, Armstrong had the pleasure of believing that Kathleen Lonsdale's X-ray analysis of benzene in 1929 vindicated the centric formula (Lonsdale 1929). In his own chivalric metaphor, the eighty-five-year-old Armstrong could honestly say in 1933: "Beginning with the year in which Kekulé hung up the hexagon, and the shield of benzene, above the main gateway of our castle of chemistry, I have seen almost everything happen" (HEA 1933i).

Crystallography

Although chemistry had close associations with mineralogy at the turn of the nineteenth century, the study of crystals largely developed independently from chemistry. However, with the emergence of structural organic chemistry, some chemists began to use crystals as a measure of purification and identification via melting points. In his early days at the Central, Armstrong began to see crystallography as a valuable tool in the determination of the three-dimensional structure of molecules—stereochemistry. Armstrong's interest in the possibilities of crystallography probably stemmed from chemists' deepening awareness of stereochemical isomerism and from Henry van 't Hoff's and Joseph LeBel's demonstration in 1874 that cases of such isomerism could be explained by assuming that the carbon atom's four valence bonds were directed tetrahedrally. In 1891, Armstrong and his new student William Pope prepared sobrerol, an oxidized terpene, and showed how it formed left- (*laevo*) and right-handed (*dextro*) crystals when they were separated mechanically in the manner memorably devised by Pasteur with the tartaric acids (HEA 1891c). Pope was to become famous independently of Armstrong for his later work on the spatial configurations of both inorganic and organic molecules.[2]

In 1886, a year after the opening of the Central, Armstrong made crystallography a compulsory part of the curriculum and engaged Henry A. Miers (1858–1942) as a part-time lecturer. Although Armstrong himself did not participate directly in the determination of crystal structures, some fifteen papers were generated by Miers's and Armstrong's pupils, published in the *Zeitschrift für Krystallographie* between 1886 and 1895. Miers continued this activity until March 1896, when he moved to Oxford as professor of mineralogy.[3] Subsequently, in 1915, it gave Armstrong much pleasure when Miers was elected to Britain's first chair of crystallography at the University of Manchester.

Unfortunately, the Central's parlous finances did not permit a replacement for Miers, but by 1895 experienced senior students like Pope were able to give instruction in crystal measurements to Central students, and from 1901 this role was continued by Martin Lowry into the twentieth century. It would appear that Miers's lectures were also attended by mature outsiders who included William Barlow (1845–1934), a wealthy amateur geologist and mineralogist; Scots lawyer William P. Beale (1839–1922), who had a deep interest in mineralogy and who had made some private studies of chemistry with both Hofmann at the Royal College of Chemistry and Bunsen at the University of Heidelberg; and Alexander Herschel (1836–1907), who had taken up his father's interest in mineralogy on retiring from teaching physics in Newcastle. Armstrong encouraged Barlow in particular and was delighted when he developed a geometrical theory of

---

2. It was, in fact, Armstrong's female student Clare de Brereton Evans who, in 1897, first demonstrated the asymmetry of quinquevalent nitrogen. No acknowledgement to Armstrong was made (C. Evans 1897). Armstrong did not put his name to this work or, indeed, to any work by his female students.

3. Miers to HEA, 2 March 1896, AP1.309–11.

crystallography with Pope that remained a rival theory to X-ray analysis until the 1920s. The Pope-Barlow modelling system satisfied Armstrong's conviction that molecules had to have spatial configurations that, in turn, determined their visible crystal form.

With his student J. F. Briggs (HEA 1892c), Armstrong found that three distinct crystals could be separated out from para-chloroaniline when sulfonated. Briggs and Armstrong concluded that these crystals were isomers whose shape depended on the differing amounts of water of crystallization in each form. In 1895, Armstrong and Briggs succeeded in making the meta-acids of halogenated aniline by using fuming sulfuric acid in excess. This result implied that ortho- and meta-acids were formed by different mechanisms. There followed a study of how sulfonation occurred within benzenoid systems, investigated by means of crystallizing the complete series of sulfonyl chlorides and bromides prepared from the para-dichloro and para-bromo-benzenesulfonic acids (HEA 1900c). Armstrong's plan, as he related it, was to obtain definite conclusions that external form could be correlated with internal molecular structure (HEA 1910e). It came as a godsend, therefore, when Barlow and Pope began to publish work that suggested definite connections between crystalline form and molecular configuration. This work led Armstrong to a series of seven papers with his research students Colgate, Crothers, and Rodd that ended in 1914. Armstrong was satisfied that the research demonstrated "clear proof of the existence of a *benzene framework* in all the molecules" investigated.

> Structural formulae, however, are as much symbols of function as of the relative arrangement of the parts; that it should be possible to go further and define the position of these parts in space and the geometrical outcomes of their arrangement is a remarkable achievement in which chemists may well take pride. (HEA 1910e, 1584)

Alas, it was not to be, for within a decade, the Barlow-Pope crystal theory was to be replaced by the Braggs' X-ray modelling of crystals.

Terpene Chemistry

Armstrong and several of his pupils at the Central spent much time investigating the important family of related compounds called terpenes, especially focusing on the structure of camphor. By the 1870s it was known that terpenes were unsaturated hydrocarbons all possessing the formula $C_{10}H_{16}$, whose chief oxidation products were camphor, $C_{10}H_{16}O$, and camphoric acid, $C_8H_{14}(CO_2H)_2$. When either of these products was treated with sulfuric acid and other reagents, another hydrocarbon, cymene, was formed. Initial work between 1878 and 1883 at the London Institution and Finsbury College concerned the large number of isomeric changes that were involved in turpentine and camphor chemistry. Many other chemists were investigating terpenes, including William Tilden at Clifton College and Mason's College Birmingham, and Otto Wallach, whose extraordinary research on this subject at Bonn and Göttingen won him the Nobel Prize (1910).

Armstrong resented Wallach's entry into the field of terpene research; he excused the failure of his group at Central to rival Wallach's by the fact that he, like Tilden, was overwhelmed by also having to administer the Central laboratory. Further terpene research was done in the first decade of the twentieth century at the Central in collaboration with five capable advanced students: Frederick Kipping, Arthur Lapworth, William Pope, Martin Forster, and Martin Lowry. However, these studies merely confirmed and extended some of Wallach's and Julius Bredt's work (Partington 1964, 867–71; Eyre 1958, 236–43). The moral was, as Armstrong said in a memoir on Wallach in 1931:

> Wallach was able to accomplish his work because he was under conditions [at Göttingen] which were the outcome of centuries of loving care for the universities and a public belief in the value of education. Here, fifty or sixty years ago [i.e., ca. the 1870s], even Oxford and Cambridge were scarce known to natural science. Cambridge came fairly rapidly to the fore, but Oxford was slower. Meanwhile schools of university rank [like the Central] have been established in every considerable town in the country; perhaps some of us who have contributed to this end may prove to have done work of far more value than that on essential oils. (HEA 1931g, 602)

It is a measure of their importance that these terpene investigations launched the aforementioned pupils on distinguished careers as chemists and teachers; all five became fellows of the Royal Society.

Enzymes

Following the British Association meeting in Winnipeg in 1909, Armstrong had taken the opportunity to visit California where his daughter Annie (1883–1949) had settled with her husband, Dugald MacGregor (1877–1958), who was in charge of the San Francisco offices of Balfour, Guthrie, and Co. It was an opportunity to see grandchildren he had never met. It also turned out to be the chance to meet Berlin-educated biologist Jacques Loeb (1859–1924), the "apostle of the mechanistic conception of biology" (D. Fleming 1973, 445). Loeb had emigrated to America in 1891 and become professor of biology at the University of California at Berkeley. There the two men met and talked about enzymes and the tropic effects in organisms that Loeb was then studying.

> I left feeling that I had met a man of real worth and complete honesty of purpose. Ever afterwards, I gave special attention to his work. We are too little alive to the value of even brief personal intercourse, as creating bonds of sympathetic understanding and appreciation. (HEA 1927d)

Armstrong was to follow Loeb's later research on colloids, proteins, and enzymes, as well as his book *The Mechanistic Conception of Life* (1911), with the greatest interest.

Armstrong had been elected the Chemical Society's representative on the Council of the Lawes Agricultural Trust at Rothamsted in May 1889.[4] Loath to be a sleeping member of the committee, Armstrong embarked on fundamental research on plant growth and physiology in the early 1900s. An additional stimulus for this work was that his son Frank had returned from working on enzymes with Emil Fischer in Berlin, and father and son began to collaborate. Since he was representing the interests of the Chemical Society when sitting on the Rothamsted committee, one would have expected his research to have appeared in the society's *Transactions*; instead, they appeared under the auspices of the Royal Society. Although Armstrong remained a fellow of the Chemical Society all of his life, with a few exceptions he stopped publishing papers in its journal after 1903. His last paper in the Chemical Society's *Transactions*, written with Frank, was the first in a series of reports on enzyme research; but the rest of the series appeared in the *Proceedings of the Royal Society* (*PRS*) from 1904. When the *Proceedings* divided into an **A** and **B** series covering the physical and biological sciences respectively, the papers were passed to the appropriate secretary; consequently, some of the enzyme papers appeared in the physical proceedings and some in the biological. It seems likely that Armstrong wanted Frank to be elected FRS as rapidly as possible, and this desire may have driven the choice of *PRS* over the *Journal of the Chemical Society* (*JCS*) (which also printed *Transactions of the Chemical Society*). In fact, Frank's election was not achieved until 1920. Armstrong's enzyme papers with his other students and assistants were also printed by

---

4. Wyndham Dunstan to HEA, 16 May 1899, RRL archives, LAT 1.1.

the Royal Society. The most likely explanation is that Armstrong's enzyme interests were no longer directly chemical but were of greater biochemical and physiological interest.

However, there is evidence that he was interested in the hydrolytic processes involved in fermentation a good deal earlier when he criticized the use of the word "ferment" by physiologists (HEA 1890a). The series of enzyme studies eventually numbered twenty-four papers, beginning in 1903 and ending in 1925. The work established that enzyme action depended upon the formation of a complex (which might be labile and subject to rearrangement) that subsequently resolved into two or more substances, one of which was physiologically specific to an organism. The studies included physical measurements of the rates of such reactions, especially that of the hydrolysis of urea by urease and the character of emulsin (HEA 1912e). Besides general studies of enzymes as hydrolytic agents, the research diverged into the function of plant hormones in stimulating enzyme action and thereby regulating plant metabolism and the transportation of nutrients within plants (Eyre 1958, 251–59; HEA 1912d). Referees' reports were generally favorable, though often accompanied by remarks such as "The views of the author are somewhat speculative."[5]

Radioactivity and Luminosity

Although Armstrong was never a radiochemist, like all contemporary chemists he followed the investigations of Ernest Rutherford and Frederick Soddy, as well as the work of Marie and Pierre Curie, with the greatest interest. The Rutherford-Soddy idea of atomic disintegration was by no means accepted without controversy. For chemists generally, the idea of atoms (which themselves had been the subject of scepticism throughout the nineteenth century) disintegrating into smaller particles and in so doing transmuting into different elements was strange and controversial. For one group of chemists, led by Armstrong, a chemical—or more specifically, a photochemical—rather than a physical explanation of the Rutherford-Soddy results seemed far more likely. In Armstrong's eyes, the phenomenon was a particular case of his general electrolytic model of chemical change.

In 1903, Robert Strutt (later Lord Rayleigh) drew attention to the fact that because air had been shown by Crookes and others to conduct electricity, possibly the ionization was due to a feeble radioactivity emanating from the walls of the Crookes tubes. To his surprise, the effect was common

---

5. Alexander Scott, referee report, 16 June 1904, RS archives, RR/16/138.

to all the materials he tested. He drew the conclusion that "the observed ionisation of the air is not spontaneous at all, but due to Becquerel rays from the vessel" (Strutt 1903). Armstrong begged to differ: might the ionization not be due to a very ordinary chemical effect, equivalent to William J. Russell's observation that light was not the only thing to affect photographic plates (W. Russell 1897; HEA 1903a)? The exact nature of the phenomenon, in Armstrong's interpretation, was clear in a paper he had presented to the Royal Society a year previously, when he was vice president (HEA 1902e). His starting point was H. Brereton Baker's demonstration that chemical substances showed no interaction if they were in a state of high purity and dryness.

> The amount of *impurity* present in the gases [hydrogen and oxygen] being reduced to a minimum . . . change takes place at a very slow rate when heat is applied; and even when a considerable amount of water is present, the amount of associated impurity is too small to raise the conductivity—the rate of formation of conducting systems—to a point at which the rate of change would be such as to give rise to an explosive wave. (HEA 1902e, 102)

Appealing to his theory of reverse electrolysis, as he had done in 1893 when first drawing attention to the possible significance of Baker's experiments, together with the discussion he had offered in his presidential address in 1895, he turned to discharges in vacuum tubes. These tubes were invariably made from soft glass that contained contaminants, while the imbedded electrodes invariably housed occluded gases. He therefore contended that the discharges demonstrated by Crookes and others were caused by the presence of impurities. All the effects that Crookes had attributed to "radiant matter" (which by 1902 he now explained by means of "electrons") might well be due to occluded impurities that effected an electrolytic action.

> It has long seemed to me that luminosity and line spectra are the expression—the visible signs—of the changes attending the formation of molecules from their atoms, or, speaking generally, *that they are consequences of chemical changes*, as chemical change being one which involves an alteration of molecular composition, or it may be of molecular configuration, as it is conceivable that even changes involving but the formation of isodynamic (tautomeric) molecules—changes in molecular structure unattended with changes in molecular size—may give rise to such manifestations. (HEA 1902e, 103)

He went on to suggest how such molecular transformations might explain phosphorescence as due to "oscillatory changes in molecular structure." This line of thinking brought him to radium. The element's radioactivity and luminous effects might well be analogous to molecules like nitro-camphor undergoing isodynamic change. The argument, he contended, suggested that radioactivity was not necessarily an isolated phenomenon "but a concomitant of some chemical changes." Substances like cane sugar and saccharin had been observed to glow or flash when crushed, and several observers had photographed the evidence. The business of conductivity in air or gases (then explained by physicists as being due to ionization) needed much more careful assessment. Thus, C. T. R. Wilson's recent cloud chamber demonstrations were open to criticism because the gold leaf of the electroscope might have undergone chemical changes that would have affected the results and have nothing to do with ionization.

> I venture to think that until the phenomena of conductivity presented by gases have been studied not merely with the same, but even with far greater, care than has been devoted to the study of those attending gross chemical changes in gases, it is premature to conclude that gases undergo ionisation—using the word in its modern [i.e., not Faraday's] sense. I also think that the question whether mere molecules cannot form conducting systems has not yet received in any way the attention it deserves from those engaged in these inquiries. (HEA 1902e, 109)

Surprisingly, Armstrong offered no comment on J. J. Thomson's determination of the ratio of charge to mass of the so-called cathode rays in a Crookes tube that constituted the discovery of the first subatomic particle, later called the electron.

He returned to the subject a year later in a joint paper with his assistant, Martin Lowry, to make the connection between radioactivity and luminous effects even more explicit (HEA 1904e). The paper cited evidence they had found that "luminous manifestations"—that is, photochemical effects such as triboluminescence, fluorescence and phosphorescence—were the outcome of "oscillatory changes in molecular structure." On the basis of the tautomeric changes that were accompanied by luminous effects in many organic systems, they confidently asserted that radioactivity was "an exaggerated form of fluorescence in which radiations unnoticed by substances generally—capable of penetrating substances generally—become absorbed and rendered obvious." They stressed that this interpretation was a rational *chemical* explanation of physicists' findings.

Rutherford and Soddy's observation that thorium disintegrated into their "thorium-X" [radon] was due to the isodynamic nature of thorium and did not represent the disintegrations of thorium atoms (Trenn 1977, 127ff). Radioactivity, Armstrong and Lowry contended, was an example of a compound breaking down, not of an atom disintegrating.

The nature of radioactivity and, in particular, the strange properties of radium, were the subject of a major discussion at the York meeting of the British Association in August 1906. In one of his last appearances, Lord Kelvin declared that Soddy was wrong in asserting that radioactive phenomena implied that elements were transmutable; for Kelvin, the emanation of helium from radium was more an indication that radium contained helium, just as William Ramsay had found helium contained in cleveite minerals (Kelvin 1906). Armstrong agreed: physicists had not provided enough evidence that radium "transmuted" into helium. As he said at the York meeting, "physicists are strangely innocent workers; formulae and fashion appear to exercise an all-potent influence over them" (HEA 1906a). A return to a truly cautious approach—Kolbe's approach—was necessary. By all means one should speculate and try out "the most cracked-brain hypotheses" as a means of guiding inquiries, he thought, but one should not do so in order to court public popularity, as he thought the "radium school" had done.

In 1913, Armstrong had the opportunity to review Soddy's book on radium in the quarterly *Science Progress* (Soddy 1908–1912; HEA 1913a). He rightly opened his review with a warning that the public were being gulled into "a belief that radium has magic virtues which make it a cure for all sorts of evils and in setting an entirely fictitious virtue upon it—to serve commercial ends." He praised Soddy's account of radium science for its lucidity and eloquence (rare praise from Armstrong) and recommended it to the lay reader and to every school library. He begged to differ, however, with Soddy's claim that radioactivity formed a *new* science different from traditional physics and chemistry because radioactive substances appeared to maintain "a perennial supply of energy from year to year without stimulus and without exhaustion." This claim, Armstrong felt, misused the word "perennial." He disagreed with Soddy's irritation that radioactive researchers had been overly attacked for their innovative concepts; scientific workers, argued Armstrong, had to be conservative and could not accept something until it had satisfactory proof. The scientific, or more especially the chemical, community had been right to be sceptical of a substance that "without rhyme or reason, at a perfectly constant rate . . . gives out an amount of energy altogether extraordinary in comparison with that given out in any of the cases of chemical change known previously." Radioactivity, he imagined, was like a live shell packed with ammunition that erupted in miniature explosives at regular intervals without affecting the explosive shot left behind; and the radium shell, it was supposed, was derived from a still-more-complex ammunition shell of uranium. When radium was discovered, he continued, it seemed that the chemists' dream—that Mendeleev's periodic law implied that all the known elements were related in Proutian harmony—had been found correct. What then was an "element?"

Radium decomposed into helium and a so-called emanation, noted Armstrong; but that did not mean radium was a compound of these two materials, but rather that they were already present in radium in some active form. Radium was therefore not an element at all, and neither was helium. The chemical parallel was a substance such as nitrogen chloride which, when exploded, produced nitrogen and chlorine. Armstrong attributed the production of the gases to the nitrogen chloride undergoing resolution into its atoms, which recombined to form molecules of nitrogen and chlorine.

$$2NCl_3 = N + N + Cl + Cl + Cl + Cl + Cl + Cl = N_2 + 3Cl_2$$

By analogy, Armstrong supposed that the helium molecule consisted of several atoms of a "protohelium" that had such an intense affinity for one another, it made the molecule completely inert—an idea he had, in fact, proposed in 1895 when argon had been discovered by William Ramsay and Lord Rayleigh. By this analogy, uranium, thorium, and radium were comparable to the paraffin series $C_nH_{2n+2} = C_nH_{2n} + H_2$. In the case of the hydrocarbons, however, the intermediate products are unstable and easily isomerize—a weightless change of structure corresponding to the observation that when emanating alpha and beta rays, the radium sample's weight remained unaltered. How, then, to explain the constant rate at which radium decomposed? Armstrong's answer was that decomposition (disintegration) was a reversible reaction: the protohelium and the emanations reunited more frequently than they separated, and the radium did not explode because of the intense affinity of the protohelium entities.

The modern reader will have noticed that Armstrong's *chemical* explanation of radioactivity founders on the fact that the inert gases like helium are monatomic. Armstrong was undeterred: "I venture ... to suggest that it were time to discard the fiction that the gases of the argon family are monatomic molecules which has so long retarded progress" (HEA 1913b, 655). Armstrong's chemical explanation was too much for the physicist Sir Oliver Lodge (1913a, 672). While conservatism in science was a proper attitude up to a limit, Lodge felt, it was "transgressed when facts are ignored and hypotheses wildly manufactured in order to retain some old and superseded exclusive and negative generalisation" (Lodge 1913b, 197). Lodge easily trounced Armstrong's position. Radium had proved its status as an element spectroscopically and by its formation of regular salts and compounds. It was entirely reasonable to refer to an atom of radium even if the element underwent spontaneous fission: etymology ("a-tom" = in-divisible) was beside the point; meanings changed with new discoveries. But to find a tortuous analogy between radium's fission and that of ordinary chemical composition was ridiculous. Physicists had correctly recognized that radioactivity was a quite different phenomenon. Armstrong's hypothesis was extraordinary. Why did he "strain himself into this singular attitude of gratuitous hypothesis, instead of yielding gracefully to the logic of facts?" And the chief fact that exploded Armstrong's reasoning was that the inert gases were monatomic, *not* diatomic—and Lodge spent the remainder of his critique explaining how and why physical chemists had shown this fact by the kinetic theory of gases.

> If the argument does not appeal to Prof. Armstrong, physicists are not to blame; but the circumstance that it does not so appeal is largely responsible for the attitude which he has consistently taken up in connection with those unwelcome, or rather let us say indigestible, chemical discoveries which have been made by purely physical processes. (Lodge 1913b, 201)

This last thrust was a reference to Armstrong's attitude toward the theory of ionization, which will be discussed in chapter 9.

Even though Lodge had thoroughly dismantled his proposals, Armstrong could not stay silent, blustering that Lodge's advertisement in *Nature* had made him "shiver in my shoes" throughout the summer, and abandon a holiday to "arrange my worldly affairs" in advance of Lodge's expected attack. His draft reply was probably found unacceptable by Ronald Ross, the new editor of *Science Progress*, and appeared instead in a rival quarterly entitled *Bedrock*, which had been started by a group of four sceptical scientists in April 1912 as a rival journal for the layman published by Constable (HEA 1914b). As its impudent title suggested ("Sir Oliver Lodge, Intolerant, Infallible"), Armstrong's reply was more an ad hominem attack on Lodge than a reasoned explanation of his own speculations and hypothesis. Lodge was vulnerable to jousting, due to his tentative support for the physical mediumship of Eusapia Palladino in the 1890s, and for his speculations concerning "continuity" and a physical ether in his presidential address to the British Association at Birmingham in 1913.

Lodge had completely misunderstood his review, Armstrong declared. He had accepted Soddy's and other physicists' findings without reserve but offered an alternative atomistic and chemical explanation of radioactivity without bias. However, although the specific heats argument for the monatomicity of the inert gases was mathematically and kinetically sound, he refused to take it seriously. It was:

> merely proof that no energy is wasted in doing internal work within molecules. But this is precisely what my hypothesis requires, my postulate being simply that the constituent "atoms" of the Helium molecule are so firmly interlocked that they behave as one. (HEA 1914b, 418)

And he concluded his reply by extolling the importance and value of speculation in the development of scientific ideas, saying that Lodge had engaged in precisely such speculation in defending continuity against discreteness in his Birmingham speech. Interestingly, Soddy referred to Armstrong's suggestion that the inert gases were possible compounds in his Royal Institution lecture on the nature of elements but made no comment on its plausibility (Soddy 1917–1919).

Armstrong (and Lowry) offered no further contribution to a chemical explanation of radioactivity, and as the physical evidence provided by Rutherford, Soddy, Thomson, and Aston became stronger and stronger, Armstrong quietly accepted the physical atom of protons and electrons. This acceptance is clear in his final revision of the chemistry entry for the thirteenth edition of *Encyclopaedia Britannica* in 1926, which he reprinted in his *Essays on the Art and Principles of Chemistry* in 1927. There, he contented himself by merely stating that radium and the other radioactive elements, as well as the production of helium from them, remained mysterious.

Classification of the Elements

Armstrong had long accepted that the so-called elements were probably genetically related in some fundamental but as-yet-unknown way. He therefore welcomed Mendeleev's periodic law when he first became aware of it as a student in Leipzig. He drew attention to the periodic law in his essay on inorganic chemistry for the ninth edition of *Encyclopaedia Britannica* in 1876. There he firmly stated that:

> The establishment of the periodic law may truly be said to mark an era in chemical science, and we may anticipate that its application and extension will be fraught with the most important consequences. It reminds us how important above all things is the current determination of the fundamental constant of our science—the atomic weight of the elements, about which in many cases great uncertainty prevails; it is to be desired that this may not long remain the case. It also affords the strongest encouragement to the chemist to persevere in the search for new elements. (HEA 1876e, 544).

As the final sentence suggests, it appears likely that he had seen Lecoq de Boisbaudran's paper on gallium that year, in which the Frenchman identified Mendeleev's missing element, "eka-aluminum."

As secretary of the Chemical Society in June 1889, it also fell to Armstrong to read Mendeleev's Faraday Lecture, "The Periodic Law of the Chemical Elements," to the Chemical Society when the Russian chemist was unexpectedly called back to his homeland because of the sudden illness of his son. This occasion must have reinforced Armstrong's interest in Mendeleev's great generalization. However, as an organic chemist, Armstrong developed an idiosyncratic view of how the periodicity of the elements should be displayed. He devised a periodic table of his own in 1900, when he updated the article on chemistry for one of the supplementary volumes of *Encyclopaedia Britannica* that comprised its tenth edition (see next page). Because the entry did not appear in print until 1902, it was only then that he aired his ideas to the Royal Society as its vice president (HEA 1902a, 712; 1902b).

As an organic chemist, Armstrong could not but help notice the analogy between the close regularity of increasing atomic weights across horizontal periods of the periodic table and the way the unit difference in the paraffin series of homologues was CH_2—an analogy first brought forward decades earlier by Jean-Baptiste Dumas and Max Pettenkofer. Newland's octaves had set the tone for Mendeleev's arrangement, but homologous elements were usually sixteen units apart, not eight. Accordingly, Armstrong argued that the first horizontal period of the periodic system should begin with hydrogen (no. 1) and end with oxygen (no. 16) before the second period began with undiscovered entities (nos. 17 and 18), followed by fluorine (no. 18). The more complicated table produced thereby had sixteen vertical series, or families of similar elements, and a huge number of gaps. Against all contemporary evidence, he assumed that the recently discovered inert gases were diatomic like nitrogen, arguing that their individual atoms were probably extremely active, but satisfied their mutual affinities completely when coupled together, so becoming inert.

TABLE OF THE ELEMENTS.

1 H	2 He	3	4	5	6	7 Li	8	9 Be	10 Ne	11 B	12 C	13	14 N	15	16 O
17	18	19 F	20 A	21	22	23 Na	24 Mg	25	26	27 Al	28 Si	29	30	31 P	32 S
33	34	35 Cl	36	37	38	39 K	40 Ca	41	42 Kr	43 Sc	44 / 45 / 46 / 47 / 48 Ti	49	50	51 V	52 Cr
53	54	55 Mn	56 Fe / 57 / 58 / 59 {Co, Ni}	60	61	62	63 Cu / 64 / 65 Zn	66	67 X	68 / 69 / 70 Ga	71 / 72 Ge / 73	74	75 As	76	77 Se
78	79	80 Br	81	82	83	84 / 85 Rb	86 / 87 Sr	88	89	90 Yt	91 Zr / 92 / 93	94	95	96 Nb	97 Mo
98	99	100	101 / 102 Ru / 103 Rh / 104 / 105 Pd	106	107	108 Ag / 109 / 110 / 111 / 112 Cd	113	114 In	115 / 116 / 117 Vc	118 Sn / 119 / 120	121 Sb	122		123	124 Te
125	126	127 I	128	129	130	131 / 132 / 133 Cs	134 / 135 / 136 / 137 Ba	138 La / 140 Ce / 141 Pr / 144 Nd	[150 Sm	156 Gd	163 Tb	166 Er	173 Yb]	183 Ta	184 W
			191 Os / 192 / 193 Ir / 194 / 195 Pt	196	197	198 Au	199 / 200 Hg / ? R		[204 Tl	207 Pb	208 Bi]	232 Th			240 U

Speculation on such a subject will be justified if it but lead to further appreciation of the rhythm which undoubtedly underlies the relationships subsisting among the elements. That work in plenty is left for the chemist to do is certain. (HEA 1902b, 94)

Armstrong returned to the subject in January 1912, when he gave the Thomas Graham Memorial Lecture to the Royal Philosophical Society in Glasgow (HEA 1912b). He still argued that the inert gases were most likely diatomic, saying that "an inert element is a contradiction in terms," while admitting that this claim was "merely giving expression to my personal sentiment." Ironically, later that year, Francis Aston, working at the Cavendish Laboratory at Cambridge, would deploy a vacuum tube to demonstrate that neon possessed two different atomic mass numbers, 20 and 22.

Armstrong's speculative table fell completely flat, and there is no evidence that other chemists ever cited the Royal Society paper. Moreover, the advent of radiochemistry in the hands of Soddy and Rutherford, and of the electron by J. J. Thomson, had already undermined the basis of Armstrong's speculation. On the other hand, when coupled with the idea of Soddy's isotopes, the basis of Armstrong's "visionary" table does not seem so outrageous after all. Armstrong asserted his arrangement as "foreshadowing the possible passage from one to another [element] by an increment of a single unit of mass, which was subsequently to be justified by Aston's discovery and apparently also by Rutherford's cruel treatment of nitrogen, he is supposed to have knocked a spot of hydrogen off the atom" (HEA 1927r, xv–xvi).

Not surprisingly, he returned to the subject in his third contribution to *Encyclopaedia Britannica*, in its twelfth supplementary edition in 1922 (HEA 1927r, 621). In this case, however, he used standard atomic weights, but indicated how non-integers arose through isotopy and returned to the more acceptable layout of nine horizontal periods. He also now accepted that the noble gases were monatomic. In a further conciliatory mood, he gracefully acknowledged that "chemists are still further indebted to physicists for the aid they have recently received from them in classifying the elementary series and 'fixing' their possible number" (HEA 1927r, 23). As an afterthought, he observed that "as in human families there may be marked diversity of character within a family [of elements]." The sceptical chemist had been finally convinced.

This chapter has surveyed a major part of Armstrong's chemical research. Trained by an arch-sceptic, Hermann Kolbe, Armstrong quickly assimilated the structural chemistry pioneered by Frankland and Kekulé that Kolbe hated. His PhD topic involving the investigation of sulfonation of organic compounds proved fortunate, since the organic salts prepared could be easily purified and crystallized. This work enabled him to contribute to the classical methods of structure determination involving analysis of an unknown compound into known compounds, and the subsequent synthesis of the known into the previously unknown from which a tentative structure could be pictured.

An important side issue in such determinations in the aromatic series was the deduction of where a second substituent was to be placed on a mono-derivative of benzene. Such investigations led him to suggest a centric valence model for the structure of benzene rather than the one initiated by Kekulé. In this research, he was a leading player rather than a background "Fifth Business" character. Several of Armstrong's deductions—as applied to benzene, naphthalene, and anthracene derivatives—came to play a role in the gradual emergence of electronic ideas of structure that only fully emerged in the 1920s. Although he was a classically trained structural chemist, Armstrong developed a fully fledged explanation for chemical reactivity under the stimulus of reading Faraday's work on electrolysis and following the work of schoolteacher Herbert Brereton Baker on the inertness of chemical reactivity under extreme conditions of dryness. This electrolytic theory will be examined in later chapters, as will Armstrong's deep scepticism of Svante Arrhenius's and Wilhelm Ostwald's claim that solution chemistry was best explained in terms of the dissociation of solutes into ions.

Another source of ideas proved to be the crystallography that, at first, was simply employed as a tool of purification and identification, but which—following the work of his pupils Barlow and Pope—seemed to offer a further (in fact, a stereochemical) tool for understanding the spatial characteristics and reactivity of molecules. Although replaced in the 1920s by the Braggs' X-ray models, the Barlow-Pope model was enthusiastically employed by Armstrong in imagining the chemical world.

Finally, while reminding readers of his contributions to terpene and naphthalene chemistry that is now unrecognizably embedded in twenty-first-century chemistry, we note how, during the Edwardian era, his interests turned completely toward biochemistry and the actions of enzymes in living processes. Having always been interested in physiology and its medical implications, work on enzymes and hormones was to stimulate his interest in food, its purity, and its nutritious value during the long years of his retirement.

✳ 6 ✳

Running the Chemical Society

> Armstrong and the Chemical Society were practically the same thing: officially he was the Secretary, but in fact he was the Director, and when he was elected President, his rule was supreme.
> —Kipping 1938, 59

Armstrong joined the Chemical Society on 1 December 1870, within a few months of his return from Germany and appointment at St. Bartholomew's Hospital. Three years later he was elected a member of council, a position he retained virtually uninterruptedly for the rest of his life—a unique position for a fellow of the society and one never to be repeated. In 1875, while teaching at Finsbury College, he began his service as a reforming secretary of the society alongside William Henry Perkin Sr., who had discovered mauve in 1856 and who had been made a secretary in 1869. Perkin was about to retire from being an active manufacturer of dyestuffs in order to devote himself to pure research, such as the study of the hydroxyanthroquinones. However, he remained a joint secretary with Armstrong until 1883, when he became president and was replaced as co-secretary by John Millar Thomson (1849–1933), the head of the chemistry department at King's College London.

Armstrong stayed in this powerful position of secretary, serving nine successive presidents, including Frederick Augustus Abel, John Hall Gladstone, Henry Enfield Roscoe, Joseph Henry Gilbert, William Henry Perkin Jr., Hugo Müller, William Crookes, William James Russell, and Alexander Crum Brown. This feat was another unrepeated record of service in the history of the society. In fact, he only retired from the position of secretary in order to be elected president for the years 1893 to 1895. He then became a vice president, and, like all surviving vice presidents, he had the permanent right to attend meetings of council. This right was something Armstrong clearly cherished and frequently utilized until he was in his late eighties. It is true to say, therefore, that no other Chemical Society fellow, before or since, has served the administration of the society so long and so well.

Armstrong transformed the society in various ways. Although the total of 150 research papers he published in the society's journals between 1868 and 1903 did not exceed the output of some of his peers, it was his administrative and critical services that were remarkable. Although his researches after 1900 were usually published by the Royal Society, this fact did not mean that his interest in the Chemical Society was any the less.

The Society's Publications

In the first decade after its founding in 1841, the Chemical Society had issued three volumes of *Memoirs and Proceedings* in irregular fashion; these were replaced by sixteen volumes of a *Quarterly Journal* that were issued more regularly between 1847 and 1864. As the number of contributions increased, and in order to speed up publication, A. W. Hofmann persuaded the council to issue the journal monthly from the beginning of 1862; thus, when Armstrong joined the Chemical Society in 1870, he received what was then known simply as the monthly *Journal of the Chemical Society*. The decision as to which papers deserved publication was by then made by the editor, Henry Watts; he was advised by a publications committee, which had been created in 1848 and whose duty was to handle the refereeing of potential papers. The abstracting of papers in other British and foreign journals, hitherto included in the journal, had been abandoned in 1862. It was not until 1869 that the then-president, Alexander Williamson, persuaded the council to publish monthly digests of all the chemical literature, both British and foreign (Moore and Philip 1947, 447; HEA 1896c, 243).

The work of the abstractors (for which Armstrong volunteered for German papers on organic chemistry immediately upon joining the Chemical Society) was coordinated by A. J. Greenaway, the subeditor, for publication in the *Journal*. Abstractors were paid a small fee, so this work provided additional income for Armstrong. His success as an abstractor also led editor Watts to invite him to contribute to a new edition of his *Dictionary of Chemistry*. Armstrong came to admire Watts as "an incomparable editor [whose] literary advice was of the greatest value to all who worked with him and to the contributors to the Journal of the Society" (HEA 1920c, 15). The costs of the abstracting service were met by increasing the price of the society's subscriptions and by a generous grant from the British Association for the Advancement of Science. The system had soon got into difficulties, with abstracts falling into arrears and the burden of editing falling on Watts's shoulders.

A particular issue was whether any hint of criticism or scepticism should be allowed to play in an abstract. Benjamin Brodie, professor of chemistry at Oxford, was angered by one critique and, as a consequence, any hint of criticism in an abstract was ruled out of order. More general criticism of the society's publication policy of mixing the printed abstracts with original papers came mainly from William Crookes's weekly *Chemical News* and reached its peak at the anniversary meeting of the Chemical Society in 1877. As principal secretary, it fell to Armstrong to put things right by reforming the duties of the Publications Committee and the composition and makeup of the monthly journal.

This reform restored the balance and muted further criticism. Armstrong praised the Physical Society for starting an abstracting service for physics papers that would also be of great use to chemists. Having now established an efficient abstracting service, it was time, Armstrong suggested, to begin a complementary service of "monographs which will tell us exactly what is the state of knowledge in the particular little chapter of the subject of which each monograph treats" (HEA 1894c, 355). This prescient idea produced no fruit until 1905, when the society, under the presidency of Sir Henry Roscoe, launched its *Annual Reports*.

By 1900 it was plain that there was far too much, and very expensive, duplication of abstracting by complementary sciences with the *Chemisches Zentralblatt* in Germany and the similar efforts of the American Chemical Society (ACS). Armstrong was contacted by Henry N. Stokes, a chemist working for the American Geological Survey in Washington, suggesting that there ought to be just one English-language abstracting journal for chemistry, analogous to the German *Zentralblatt*. Stokes was aware that Arthur Noyes at MIT had begun issuing chemical abstracts in a house journal that had then been taken over by the ACS in 1897. It seemed to Stokes a waste of effort for both the London and American Chemical Societies to produce abstracts of the world's chemical research in English. Armstrong brought Stokes's letter to the attention of the council on 16 March 1899, and as a result a committee composed of Armstrong, Charles Groves (by then the editor of the *Journal*), and William Wynne were selected to begin negotiations with the American Chemical Society for a pan-Anglophone publication of abstracts.[1] Although negotiations

---

1. Chemical Society council minutes, 6 (16 March 1899), 180; 6 (29 March 1899), 182. All Chemical Society council minutes can be found at the RSC archives in London.

went on for some years concerning a cooperative venture, the ACS eventually decided to go it alone after 1902, when Noyes's cousin, William A. Noyes, became editor of the *Journal of the American Chemical Society*. The ACS's *Chemical Abstracts* began publication in 1907, at which juncture the British were relieved of publishing abstracts in the *Journal of the Chemical Society* (Watchurst 1974).

Armstrong sat on the Publications Committee for at least twenty-four years from 1875. In 1885, Armstrong persuaded it to launch a monthly separate report in the form of abstracts of papers that had been read to the society during the previous month. These *Proceedings* were also to be reprinted in collective form at the end of a session in the society's *Journal*. The *Proceedings* came to occupy an average of around 150 pages of each annual volume of the *Journal*. Indexing over a thousand pages of the *Journal* had become too much of an effort for its editor by the mid-1880s; in 1885, therefore, the society agreed to pay for a professional indexer. Ironically, in view of Armstrong's later intolerance of professional women, the task was given to a married female chemist, Margaret Dougal, from 1885 to 1909.[2]

Problems also sometimes arose over refereeing. In April 1888, Walter Hartley complained that one of his papers had been rejected as "theoretical" and not a record of experimental work. He demanded to know whether the council was "cognisant of such a rule" and if so, why had at least four recent papers of a theoretical nature appeared in the *Transactions* during the past year? Armstrong was instructed to inform Hartley that the council was unaware of any rule but that nevertheless, no irregularity had occurred. Even so, it was decided that in future, when a paper was rejected, authors would be sent a letter informing them that the paper had been deposited in the society's archives. Armstrong then placed a printed note on the back cover of the *Transactions* stating that to ensure priority in disputes, rejected papers would be placed in the archive.[3] In another, later complaint that involved Armstrong himself, he demanded to know why an article of his had been amended after he had approved the proof before printing. The editor, in fact his friend Wynne, had deleted an unnecessary critical reference to "ionism" that bore no relevance to the discussion. The then-secretary (Alexander Scott) explained that the

---

2. Margaret Doule Dougal, née Robertson (1858–1938), had conducted research on inorganic chemistry with Thomas Edward Thorpe at the Royal College of Science. After her husband died, she retired in 1909 and remarried (Creese 1991).

3. Chemical Society council minutes, 19 April, 17 May, and 21 June 1888, ff. 107, 109, 113. The RSC no longer houses any such archive.

deletion did not affect the sense of the article and so the council rejected Armstrong's complaint. He then walked out of the meeting![4] There was no further mention of the matter when Armstrong attended the next council meeting in January 1902.

Although the Publication Committee always had twenty-two members, often only a half-dozen turned up to meetings, thus allowing Armstrong considerable power in approving or disapproving of proffered papers, including his own. On one occasion in 1890, when both Armstrong and William Ramsay were present on the committee when their own papers were being discussed, Armstrong's paper was sent for refereeing, but Ramsay's was rejected. Perhaps this example suggests that the committee acted without fear or favor?[5] However, at the next meeting, when Armstrong was absent, Ramsay argued that in future two referees independent of the committee should be chosen. Armstrong objected but eventually agreed that such a process would be fairer.[6] Armstrong and the committee were also continually faced with the problem of getting contributors to shorten the length of papers. As president in 1901, Edward Thorpe (1901, 877) referred to the powers of the committee as necessarily "draconian" to condense the length of papers and to cut the excessive costs of printing the *Journal*. The most notorious example was that of Spencer Pickering, one of whose lengthy papers on solutions cost £250 to typeset. When his next, equally verbose paper was considered, Armstrong and Groves acted as arbitrators and rejected it.[7] As late as 1919, Armstrong was still urging a reduction in the length of papers as well as better regard for literary form and conciseness of expression (HEA 1919i, 10).

The duties of the society's secretaries were quite demanding, even allowing for the fact that they were shared with W. H. Perkin until 1883 and that the more routine correspondence fell on the shoulders of an in-house clerk. In May 1878, Armstrong also joined the Library Committee, which was chaired by Henry Watts, who also edited the society's *Journal* and *Transactions*.[8] And by the 1890s, he was also serving on the Research Fund Committee, the Index Committee, and the Publications Committee—

---

4. Chemical Society council minutes, 19 December 1901, f. 73. The paper concerned was "The Part Played by Residual Affinity in the Formation of Substitution Derivatives." No paper with that title was ever published.

5. Chemical Society Publication Committee minutes, 27 February 1890, RSC archives.

6. Chemical Society Publication Committee minutes, 8 May 1890, RSC archives.

7. Chemical Society council minutes, 21 May 1891.

8. Watts was librarian from 1860 to 1884 and editor from 1850 to 1884.

and somehow combining all of this work with a busy life as teacher and researcher at the Central.

The Publications Committee role proved important when preeminent German chemist August Wilhelm von Hofmann died in Berlin in 1892. Armstrong joined a committee composed of Frederick Abel, Hugo Müller, William Odling, and Perkin to plan a memorial day of lectures on Hofmann, and Armstrong offered to produce an analysis and review of Hofmann's research. In November 1892, Armstrong outlined the proposed program, consisting of papers by Lyon Playfair, Abel, Perkin, and himself.[9] It was Armstrong who masterminded the splendid daylong meeting at which Hofmann's memory was commemorated, and it was he who edited the papers delivered and published in the *Journal* in 1896. The four memorial lectures were later republished to form the backbone of the society's first volume of memorial lectures, printed in 1901—again edited for the press by Armstrong (HEA 1893c; Chemical Society 1901).

In May 1891, Armstrong (wearing his hat as secretary of the Chemical Society) co-wrote his first letter to the *Times* with his colleague William Ayrton (acting as president of the Physical Society). It was to be the first of many such letters during his long life. In this case, the two colleagues were protesting against the government's possible plan to create an additional art gallery at South Kensington, which would thwart scientific expansion on the site (HEA 1891b). In practice, the possible conflict between "art" and "science" was solved by the creation of the Victoria and Albert Museum in the Brompton Road by 1908 and the Science Museum in Exhibition Road from 1914 onwards (Sheppard 1975).

There were also obituaries to write. Here, most notably, was Armstrong's promise to write the obituary of his revered teacher Sir Edward Frankland following his death in August 1899. A year later, the agreed submission date of 25 October 1900, having passed, he was asked by the president (Wyndham Dunstan) what had become of the obituary. Armstrong replied:

> The task has proved a far more difficult one than I had expected owing to the importance of the issues which his work raises. This includes organometallic compounds, valency, combustion, water supply among other subjects. I have given a great deal of time to the essay already, but like other people, I have had my own official work to do and needed some rest, and am not yet near the end of my task. Valency especially

---

9. Chemical Society council minutes, 16 June 1892, f. 197; 17 November 1892, f. 210.

has given me a great deal to think about and I am hoping to put forward a useful discussion of this subject which will bring it up to date.[10]

He finally delivered a short lecture on Frankland's life on 31 October (HEA 1901d), and the council generously gave him a new deadline of 31 January 1901, for the final copy to be published in the society's *Journal*. It soon transpired that he wanted a further extension until 31 October 1901, following the long vacation.[11] That deadline also passed with no obituary in sight. Dunstan, treading delicately, then asked Armstrong's colleague, Wynne, if he could find out what the problem was. Wynne replied at the end of April 1902 that it "is expected to be completed soon," one of the problems being that photographs of Frankland had not yet come to hand from the family.[12]

Another year went by with no finalized obituary from Armstrong, prompting the council—led by the Royal College of Science chemists Thorpe and Tilden—to propose in February 1903 that a stern letter should be written to Armstrong demanding that the extended text of the memorial lecture he had delivered on 31 October 1901 should be immediately forwarded for publication.[13] Once again, Armstrong filibustered and wrote a response accusing the council of bullying.[14] The council tabled the letter without comment, probably concluding that Armstrong had embarked on a book-length biography rather than the required obituary. In June 1904, Wynne (who was about to be appointed to the chemistry chair at Sheffield) tried some further gentle nudging.[15] Once again, this effort had no effect. Curiously, it was not until 1905 that Frankland's son Percy offered a portrait of his father for inclusion in Armstrong's memorial lecture, only for the council to reject it. Presumably, the Chemical Society had abandoned hope that Armstrong would ever deliver the expanded lecture for printing.

The actual problem all along was not procrastination; it was surely that Armstrong had discovered (or deduced) that Frankland's birth had been illegitimate, making the writing of any such biography in Victorian times challenging without doing damage to Frankland's posthumous moral standing. That would have been the last thing Armstrong wanted. The full story of Frankland's parentage, no longer at risk of controversy or negative

10. Chemical Society council minutes, 15 November 1900, f. 268.
11. Chemical Society council minutes, 20 June 1901, f. 44.
12. Chemical Society council minutes, 30 April 1902, f. 104.
13. Chemical Society council minutes, 18 February 1903, f. 161.
14. Chemical Society council minutes, 18 March 1903, ff. 168–69.
15. W. P. Wynne to HEA, 12 June 1904, AP2.590.

construction, was only revealed eighty years later by Colin Russell's superb research (C. Russell 1986).

The Society's Semicentennial Jubilee

In December 1890, the council noted the approaching semicentennial of the society. Armstrong reported plans to hire rooms at the Society of Arts in February 1891 for a celebratory meeting, followed by a banquet in the City of London's Goldsmiths' Hall. Following the successful event, which large numbers of foreign members attended, it fell to Armstrong to compile a monograph commemorating the jubilee. Armstrong was clearly overburdened with work, since the volume was only finally published in 1896, five years after the jubilee (HEA 1896c). Armstrong had a notable, and valuable, success in persuading Robert Warington Jr. to exhibit and donate a collection of his father's memorabilia of the society's founding in 1841.

Armstrong had planned to compose a review of the society's scientific work from 1841 to 1891, part of which had been set in type by 1896. Since it was not ready for inclusion in the jubilee volume, the council ordered that the type be kept. Instead, William Odling wrote a short and unsatisfactory review of the development of chemical theory since 1841 (Odling 1896, 26). Armstrong promised to complete the survey in time for the annual meeting in 1897, the year he was elected a member of the Athenaeum.[16] However, nothing more was heard of the project and a separate historical publication never appeared.

However, some of the material Armstrong had prepared was probably incorporated into his first presidential address to the society on 22 March 1894, when he offered a survey of the society's publications. He reported that by that date, 2500 copies of the *Transactions of the Chemical Society* had been circulated to fellows and other societies. He also claimed that "during the past 15 years, few papers have been communicated to the Society which I have not read in manuscript." This claim led him to ask why, having reached its peak as an organization, did the society have competitors? "Chemical literature," he argued, "is fast becoming unmanageable and uncontrollable from its very vastness" (HEA 1894c; Meadows 1974). The number of papers published continued to increase worldwide and new journals were continually being created. The only way chemists could avoid becoming swamped by the literature was if they all agreed to publish

---

16. Chemical Society council minutes, 19 March 1896 and 21 January 1897 (quoted in Watchurst 1974, 44). For HEA's election to the Athenaeum, see AP1, Diplomas 5.

everything in just one language and one journal. This idea was a dream, he admitted, but a practical step would be if all chemical research were published by the Chemical Society and all the other rival English journals ceased publication (HEA 1894c, 345). He could not see, for example, why the Royal Society published papers on chemical subjects unless it was for the personal vanity of the author who was hoping to be elected FRS.

> It seems to me, therefore, that eventually one of two courses must be adopted in this country—either the societies engaged in doing similar work must become affiliated, or our Society must return to the practice of early days and publish lengthy abstracts of papers communicated to societies such as the Royal Societies of London and Edinburgh, in order to bring these papers properly under the notice of chemists generally. (HEA 1894c, 354)

Such an arrangement could also lead to an agreement to print volumes in a uniform size and paper quality that was comfortable to hold and read at home or on a train journey, as well as cutting back on prolixity. It would also lead to the cessation of the deceitful practice of publishing the same paper in different journals—a practice that Armstrong himself had been guilty of in his early years. He was concerned, too, about the increasing number of specialist journals, pointing to publications emanating from the Society of Chemical Industry (SCI), the Society of Public Analysts, the Society of Dyers and Colourists, and the Institute of Brewing.

As Moore and Philip noted in their short history of the society, until the end of the nineteenth century the society was effectively run by a self-elected, autocratic, and very much London-Oxbridge-centered council composed of senior chemists who were usually in their fifties (Moore and Philip 1947). This state of affairs led effectively to a perennially conservative attitude, exacerbated and preserved by the fact that when presidents completed their term of office, they became vice presidents who were then permitted to remain on the council until they died. Dissatisfaction amongst younger chemists, especially those working outside London and Oxbridge, elicited a kind of rebellion during the 1890s.

This campaign was led by regional fellows under the leadership of Arthur Harden, a biochemist from Manchester who had recently transferred to London to work at the Lister Institute, and a personal friend of Armstrong's. Harden's companion in arms was another Manchester chemist, Philip Hartog, who was beginning to shine as an administrator. Between them, in 1898 they collected 540 signatures of provincial fellows in a memorial to the council that pointed out that, although fellows outside London formed the majority of the membership, they had little say in the society's management. One way of alleviating this state of affairs, the writers suggested, was to introduce postal voting. If it proved necessary, the charter should be amended to allow this. On taking legal opinion, the council was advised that the charter would have to be amended to allow postal voting; but that a supplementary charter would not be granted unless the society could demonstrate that such a measure was more or less the unanimous wish of all the fellows.

On learning of this legal advice Harden and Hartog moved at the society's annual general meeting in 1898 that the fellows' views on a supplementary charter should be sought by a postal circular. The council then made the obvious reply that this kind of survey could not be carried out since postal balloting was not permitted under the charter; that the society was not allowed to use its funds for such a purpose; and that, in any case, no action would be possible unless the result was unanimous. The last straw in this piece of prevarication was that if Harden and Hartog, or any other individual fellow, would bear the cost of a posted questionnaire, they were free to do so (this response suggests the kind of maneuver for which Armstrong was well known) (Moore and Philip 1947, 85). The democratization of the society therefore was placed in stalemate until it reared its head again over the membership of women (see chapter 7). The rebellious demonstration by Harden and Hartog upset Armstrong, and he was still marked by the experience nearly thirty-five years later.

A Royal Society Interlude

As we have seen, Armstrong had been elected a fellow of the Royal Society in June 1876, though he never published any papers in the society's *Philosophical Transactions* and only began to publish work in the society's *Proceedings* in 1902. Much of his research was supported from a government grant, administered by the Royal Society, amounting to £1,775 between 1872 and 1914. His fellowship was frequently fraught by difficulties of his own making. For example, in 1889, the council had to reprimand him for delays in reporting on three papers that he had been asked to referee, and he often failed to report properly on how he had spent awards from the government grant that the Royal Society administered.[17] We might excuse

---

17. Herbert Rix (assistant secretary) to HEA, 9 October 1889, RS archives, NLB/3/786. Archibald Geikie and Joseph Larmor (secretaries) to HEA, 2 June 1904; 9 May 1907; and 14 June 1907, concerning grants dating back to 1888, RS archives, NLB/23/2/484, NLB/35/275, NLB/35/571.

such lapses as due to his having been a harassed individual with too many calls on his time, but they were also symptoms of a general disdain for authority.

As noted in the foregoing section, Armstrong was critical of the Royal Society's *Transactions* in his presidential address to the Chemical Society in 1894. Released from the secretaryship of the Chemical Society in 1895, he accepted election to the Council of the Royal Society in 1901, a role he continued in for only one year. During this time on council, he actually served the term as RS vice president. The role made him even more critical of the Royal Society, and in a parting shot in November 1902, he fired a highly critical but also constructive memorandum to the Royal Society's secretary, Irish physicist Joseph Larmor.[18] Armstrong claimed the society was losing its way and that unless changes were made it would lose its position as the most important body in the country for promoting natural knowledge. "At present," he claimed, the society "follows, but it does not lead." Moreover, its publications—meaning the irregularly printed *Philosophical Transactions* and the monthly *Proceedings*—were "somewhat trivial in character" in that they no longer "represent[ed] the high-water mark in all branches of science." He deplored the society's failure to print its publications with any degree of regularity and the sheer "complexity of the machinery of publication." The consequence was that "many authors [both FRS and non-FRS authors] prefer to publish through other channels." The time when the society could be proud of the *Philosophical Transactions* and the *Proceedings* lay only in the past.

Armstrong was not just negative; he also made some positive suggestions, the most important being that the Royal Society's position at the head of the British scientific estate made it an eminently suitable platform to discuss what he described as "borderland problems." He also suggested that the society could improve its publications by appointing editors who had sole command of what appeared in the *Transactions* and *Proceedings*, rather than following the cumbersome procedures hitherto adopted by the physical and biological secretaries. Doing so would produce faster publication times (as was the norm with the Chemical Society and with Crookes's commercial *Chemical News*) and avoid the use of committees and referees. As Aileen Fyfe has shown, Armstrong's suggestions had little effect in his lifetime, and it was not until the 1960s that the Royal Society radically altered its publication practices (Fyfe 2020).

These criticisms were not entirely new. In private correspondence, Armstrong and his close friend James Dewar had grumbled about what

---

18. 6 November 1902, RS archives, CMP/8.

they saw as the increasing bias of the Royal Society toward the biological as opposed to the physical sciences (Harrison 1988; Csiszar 2018, 260).[19] Dewar's letters were frequently filled with critical comments of the society's current president, Lord Lister. Not surprisingly, Armstrong was not pleased to have the *Proceedings* split into two series, A and B, thus separating papers in the biological and physical sciences. More to his taste was the creation of the Faraday Society in 1903 as an explicit site for interdisciplinary discussion of problems in physical chemistry—though he was soon upset by its takeover by mathematically inclined chemists and the "ionists." He did not continue with his membership.

In 1892, Armstrong was also briefly involved in a council debate on the candidacy of Sir Henry Howorth (1842–1923) for fellowship election to the Royal Society. Howorth, a Conservative politician, not only had no scientific credentials but had criticized the glacial theory in favor of a Noachian flood. But he had powerful friends on the council, and though Ray Lankester, George Romanes, and two other fellows supported Armstrong's opposition to his election, Howorth was duly made FRS in June 1893. This contretemps was part of a general debate that year concerning whether the council had too much power in determining elections.[20]

The Geneva Conference on Organic-Chemical Nomenclature

In August 1889, to coincide with the Universal Exposition held in Paris that year, the Société Chimique convened an International Chemical Congress. The French society suggested four topics for discussion: the analysis of foodstuffs, agricultural materials, pharmaceutical products, and how chemical nomenclature might be unified by international agreement amongst chemists. Unfortunately, because invitations to foreign delegates were issued late, only a handful of foreign chemists were able to join the meeting. Of the three hundred delegates who did attend, the vast majority were French; Armstrong was not among them because he prioritized his family's holiday (Anon. 1889b). The most important decision made at

---

19. See 17 January 1892, RS archives, MC/15/253; and 9 November 1893, RS archives, CMB/43.

20. The controversy began with an anonymous letter in the *Times* dated 1 December 1892, and continued on 24 and 27 December 1892 and 3 and 4 January 1893. Armstrong's memorandum with A. H. Green, R. Lankester, E. R. Poulson, and G. Romanes appeared on 31 May 1893 (p. 5) and suggested that the Welsh physicist J. Viriamu Jones be substituted for Howorth on the voting sheet. Howorth was elected by a majority of 4 to 1 on the grounds of archaeological publications. See Tilden to HEA, 4 June 1893, AP2.569; and RS archives, CMB/43.

the Congress was to appoint a committee to "consider the unification of chemical nomenclature" (HEA 1892a and 1892d; Crosland 1962; Hepler-Smith 2015). Initially a purely French committee of nine (including Marcellin Berthelot and Charles Friedel), it quickly decided to be international in scope by electing Carl Graebe (Switzerland), Friedrich Beilstein (Russia), Adolf Baeyer (Germany), Adolf Lieben (Austria), Emanuele Paternò (Italy), Per Cleve (Sweden), and Ira Remsen (USA). The sole English member was Armstrong, indicating his high international status in organic chemistry.

Two years and some forty meetings and worldwide correspondence passed before the committee was ready to announce its conclusions. This time, Friedel made sure that a truly representative group of chemists would discuss the proposals at a meeting held in Geneva beginning on Easter Monday, 18 April 1892. Some thirty-four chemists attended, and Armstrong was joined by John Hall Gladstone and William Ramsay to represent British interests. Friedel, who had dominated the preparations since 1889, was unanimously elected president for the discussions, which lasted five days.

Friedel drew up the final report with its series of recommendations. The report began with the proposal that saturated hydrocarbons must have names ending in -ane and concluded with recommendations concerning the naming of azo and diazo compounds. While the recommendations were to be considered as open to further discussion amongst the worldwide chemical community, there was unanimity that while chemists were free to call compounds colloquially by whatever best suited local languages, it was absolutely necessary to also adopt a systematic name that "could at once be translated into the corresponding formula" and used in creating an international official register of chemical compounds, such as the one that Beilstein had already prepared.

Regarding Armstrong's role in the proceedings, it is clear that he was able to enforce nomenclature principles that had already been adopted by abstractors of the Chemical Society—rules such as the one that particular terminations of the names of chemical substances should indicate their function within a molecule. Interestingly, Armstrong also foresaw that English, rather than French or German, was more than likely to become the lingua franca of chemists worldwide; hence it was important that English chemists take the lead in devising an acceptable nomenclature in the publication of papers. In his report for *Nature*, he also stressed how convivial the meeting had been. The delegates had been uniformly congenial, united in their mission "to explain the enterprise to chemists generally" and to establish that structural formulae were the basis for the names of compounds (HEA 1892a, 57; Hepler-Smith 2015, 23). The editors of the *Chemische Berichte* soon found problems with the Geneva nomenclature rules even though Beilstein happily adopted them for the third edition of his *Handbuch der organischen Chemie* between 1892 and 1906.

The Noble Gas Saga

Although the Geneva conference delegates had been a congenial group in which Armstrong and Ramsay had played their roles, their relationship soon deteriorated. In 1894, the third Baron Rayleigh, John William Strutt, began his celebrated collaboration with William Ramsay that led to the discovery of argon in the atmosphere, the first "noble gas." Rayleigh's colleague at the Royal Institution, where he was professor of natural philosophy, was James Dewar—a difficult man whose one and only deep scientific friendship was with Armstrong. It seemed incredible to many chemists, including Armstrong, that following the analysis of air in the eighteenth century there could still be an undiscovered constituent. During his presidency of the Chemical Society in 1894–1895, Armstrong frequently voiced Dewar's criticism that if the unknown gas existed, Dewar would have found it in his technically brilliant liquefaction of air. He shared Dewar's view that the new gas was probably an allotrope of nitrogen—itself an inert gas. By January 1895, however, Armstrong had grudgingly admitted the existence of argon, though he strongly doubted that it was monatomic. Equally significantly, he attributed the discovery to Rayleigh rather than Ramsay, who was one of the new breed of physical chemists that he distrusted.

The matter did not end there. In March 1897, Dewar was proposed as president of the Chemical Society, and Ramsay was forwarded as a rival candidate by a clique of younger chemists led by the Cambridge chemist Pattison Muir. Armstrong asserted privately that Muir's opinions were not to be trusted. After a campaign bristling with dissensions, Dewar was elected and succeeded in making the peace (HEA 1897b). Ramsay's turn would come later, but Armstrong never forgave Ramsay for accepting a candidature when it was "Dewar's turn." Armstrong is usually considered to have been the anonymous author who poured scorn on Ramsay's work on the noble gases in the pages of Crookes's *Chemical News* under the pseudonym *Suum Cuique* ("may all get their due"), but this identity has never been confirmed. Against this assumption, neither before nor afterward did Armstrong ever shield his views under a pseudonym; he was always direct.[21] The maverick Irish American metallurgist, Stephen Emmens, who gulled William Crookes into half-believing that he might have transmuted base metal into silver, proposed the Manchester physicist Arthur Schuster as the anonymous author (Emmens 1899, 34).[22]

A quarter of a century later, when commenting on J. J. Thomson's review of the fourth Baron Rayleigh's biography of his father, Armstrong observed of the biographer's account of the discovery of argon:

> The account given is only partial, in no way a complete presentation of the episode: among others we would like to have heard the views of [George] Gordon, the discoverer's devoted laboratory servitor. Lord Rayleigh [the fourth Baron Rayleigh] was but young [aged 19] at the time of the achievement and cannot have been aware of the state of feeling among chemists—nor, probably was his father; he will not have known in what reverence we held his father's work. I suppose I was behind the scenes as much as anyone, the more as I was president of the Chemical Society, was thoroughly acquainted with Ramsay and his ways and, as is well known, an intimate friend of James Dewar. (HEA 1925a, 47; Thomson 1923)

It was evidently at Armstrong's insistence that the Chemical Society awarded Rayleigh the Faraday medal in March 1895

> in recognition of the important service he has rendered to Chemistry by his discovery of argon. The Chemical Society advisedly took the view that it was *his* discovery. There was the strongest possible feeling among chemists that his name alone should have been associated with the discovery. (HEA 1925a, 47)

In other words, the feeling was that Ramsay had muscled his way into research that properly belonged to Rayleigh. Nevertheless, added Armstrong, "on the same occasion, I had the pleasure of calling upon Ramsay to make public his startling discovery of helium in cleveite, which was

---

21. He did use pseudonyms in the 1930s, but he did so for comic effect; everyone knew the writer was Armstrong.

22. Ramsay's pupil and collaborator Morris W. Travers was convinced that the writer was Armstrong (Travers 1928; Brock 2008, 362).

thereupon confirmed by Crookes" (HEA 1925a, 47).[23] Ramsay's former colleague Morris W. Travers was annoyed by Armstrong's remarks. "It is ... with positive pain that many chemists have read Prof. Armstrong's letter upon this subject. What does he mean by 'behind the scenes' and 'Ramsay and his ways'?" (Travers 1925). Travers pointed out that in his speech on receiving the Chemical Society's medal, Rayleigh had not only expressed embarrassment because Ramsay was not also receiving the award, but also commented:

> In some quarters there had been a tendency to represent that antagonism existed between chemists and physicists in the matter, though such although had never entered [Rayleigh's] head. Professor Ramsay was a chemist by profession, while he himself had dabbled in chemistry from an early age, and had followed its development with very keen interest. (Travers 1925, 122)

Travers then recalled that, as Ramsay's assistant from 1894, he knew "as a fact" that when Ramsay told Rayleigh of his preparation of argon, he had placed the discovery at Rayleigh's disposal, "but Lord Rayleigh was equally willing to allow Ramsay to go forward with the work alone." Travers's point was that the discovery was shared, and neither Ramsay nor Rayleigh had never had any regrets about sharing the honors. Travers then offered a piece of damning evidence known to him alone:

> Even during the preliminary stage, there were attempts to disturb the friendly relations between the two discoverers. A well-known chemist [meaning Armstrong] called on Ramsay, and after being shown everything, after the manner of Ramsay, went home and wrote to Lord Rayleigh, telling him that he must place no reliance on Ramsay's work. Lord Rayleigh sent the letter on to Ramsay, with a brief comment, which Ramsay passed on to the author [Travers]. Later the "suum cuique" letters in the *Chemical News* showed chemists that there were people in their ranks capable of the most unworthy actions. (Travers 1925, 122)

All told, Armstrong did not come out well in this dispute.

---

23. In 1895, HEA ensured that Rayleigh became an honorary fellow of the Chemical Society. See Edward Frankland to HEA, 27 February 1895, RS archives, MM/10/99. The augmented edition of Strutt's biography of his father (Strutt 1968, 417, 420) contains further material on the relations between Armstrong, Dewar, and Lord Rayleigh.

Social Activity in the Chemical Society

Armstrong had what now would be regarded as a clear bias against women chemists and women generally. But he loved society. The last, but by no means the least, contribution made by Armstrong to the Chemical Society was his chairmanship of the Chemical Club from 1873 onwards (E. F. Armstrong 1941, 374; Moore and Philip 1947, 87). The club had been created in November 1872 and limited to forty chemists elected by ballot. Meetings were held monthly following the reading of papers at the society. The club continued the traditions of the B-Club and Red Lions Club, as well as the more selective Royal Society Club (H. Gay and J. Gay 1997). Its first secretary was Alexander Pedler, who passed the job on to Armstrong the following February because Pedler had taken a post in India. The dinner arrangements were made by Mrs. Armstrong, who was no doubt seen by her husband as fulfilling her hereditary role. Forster commented that this informal club was "almost amoebal in the simplicity of its operation and was incalculable value in promoting harmony and founding friendships among the people whose main activity was to advance the study of chemistry, and whose consequent devotion to the laboratory tended to hamper their social instincts." Frank Armstrong also commented on how

> the older men were met in level terms by the younger, and many enduring friendships were formed. Here also it was possible to entertain appropriately distinguished foreign visitors to our shores. (E. F. Armstrong 1941, 375)

By the 1880s, the club regularly hosted an annual dinner for the current president. The W. H. Perkin Dinner, arranged by the Armstrongs for 23 April 1884, was a particularly elaborate affair involving entertainment by the Grenadier Guards.

Armstrong was also involved in an offshoot of the Chemical Club that involved beer drinking in pubs around Piccadilly, where chemists would congregate after a Chemical Society meeting and before the last train home from one of London's main stations. The group finally settled in the Gambrinus restaurant in the basement of the Café Royal, and its members—mainly German-educated chemists—referred to themselves as the "Gambrinus." Again, according to the testimony of Forster, a past president condemned the group as the "pothouse brigade," but in its defense Forster recorded that the principal activity over a pint of beer was "the advocacy—or condemnation—of new chemical hypotheses." Armstrong would undoubtedly have been in his element in such convivial surroundings that allowed him to condemn "the ionic hypothesis."

Finally, Armstrong ran a chemists' dining group of (mainly) former pupils called "the Catalysts," which met at the Florence Restaurant in Piccadilly. A surviving menu, annotated by chemical engineer Herbert A. Humphrey for 1 November 1918, suggests that the name was chosen to "mean a fire-brand of a peaceful nature, who after his labours takes his rest in an inn." Armstrong apparently divided chemists into three groups: analytic, catalytic, and paralytic![24]

Armstrong was a highly proactive secretary of the Chemical Society, very much in the "Fifth Business" line. Not content with carrying out his administrative duties to the letter, he made efforts to improve its publications, to stimulate discussion at its meetings, and to improve its social activities. That these responsibilities, when combined with his teaching and laboratory research, sometimes overwhelmed him, is clear from his failure to deliver invited obituaries to a deadline and from his delay in editing the society's memorial lectures on Hofmann or the society's jubilee volume. His experience of running the Chemical Society also made him critical of the Royal Society. An FRS since 1876, as his administrative commitments to the Chemical Society diminished, he began to be highly critical of the way the Royal Society was governed. Just as the Chemical Society had needed reforming, so did Britain's senior scientific society, he felt. He believed in the unity of the chemical professions and disapproved of the attempts by some of the society's members to blackball potential members who practiced chemistry commercially, but he nevertheless initially welcomed the creation of the separate Institute of Chemistry (1877), a professionalizing society for consulting and analytical chemists.

The issue of who should be allowed to become a member of the Chemical Society did not go away when Armstrong stood down from being secretary in 1883 and became president for two years. Women were now taking degrees and practicing chemistry. Should they—active members of the growing chemical profession—not be admitted as members of the Chemical Society, too? Armstrong was now to play a distinctive role in preventing this from happening.

24. Florence Restaurant menu, annotated by Herbert Humphrey, 1 November 1918, papers of H. A. Humphrey, Imperial College Archives, B26. A forty-eight-page notebook records the club's proceedings from November 1872 to January 1875, AP2.183.

✳ 7 ✳
The Admission of Women into the Chemical Society

> Women have sought in recent times to prove that they can compete successfully with men in every field; they claim to have succeeded, but the claim cannot be allowed.
> —HEA 1904b, 14

The question of admitting women as fellows (that is, members) of the Chemical Society first arose in November 1880, when Henry Roscoe was president.[1] After discussion, the council instructed the secretaries, Perkin and Armstrong, to seek the opinion of a solicitor as to whether the wording of the society's charter permitted the admission of women and, if it did, what alteration to the bylaws should be made. Armstrong took the initiative and sought the legal opinion of his friend William Phipson Beale (1839–1922), a Queen's Counsel who had a private long-term interest in mineralogy and crystallography. He had joined the Chemical Society in 1867, was a leading member of the B-Club, and attended Miers's classes on crystallography at the Central College. Armstrong would have undoubtedly known what Beale's personal views on women were.

However, as a dispassionate lawyer, Beale made clear his conclusion that the charter did not prevent women from admission, a view that was welcomed by Oxford chemist George Vernon Harcourt (who had eight daughters), as well as by the professor of chemistry at the Royal Veterinary School in London, Charles William Heaton, who also lectured at the London School of Medicine for Women. When the council met on

---

1. Chemical Society council minutes, 4 November 1880, f. 215. For a fuller account, see Mason 1991a and 1991b; Fara 2018; Ayres 2020; and M. Rayner-Canham and G. Rayner-Canham 2020, 297–313. Also important for this general subject are Dyhouse 1976, and Rowold 2010.

16 December 1880 these two chemists recommended a change to bylaw XVII to read:

> Women may be admitted and become Fellows of the Society, and in the construction of the Bye-Laws words importing the masculine gender shall be deemed to include the feminine gender and the necessary change of language may be made applying the form prescribed by the Byelaws and Appendix thereto to the case of women.[2]

All seemed firmly settled—women would be admitted as fellows. But then, following the Christmas break, at the next council meeting, on 20 January 1881, with Roscoe in the chair, "[i]t was resolved that it is not expedient at the present time to admit women as Fellows of the Society."[3]

Because the minutes only record decisions made and not the details of discussions, we shall never be certain of what happened there. However, since Armstrong has been blamed for what happened as a result of further attempts to admit women chemists, he has inevitably and reasonably been identified as the chemist who blocked women's admission in 1881. Later claims that the reason for the decision was that the number of women claimants was so small as to not make it worth the time and money to alter the practices and routines of male fellows is implausible, since no change in the charter was required and the bylaw amendment suggested by Harcourt and Heaton held no financial implications. Clearly, the amendment was not carried at the January 1881 council meeting. Whether the other councillors worried that women might one day want to put themselves forward for election to council, whether the lack of toilet facilities in the Burlington House premises was a concern, whether some feared that the clubby atmosphere of male camaraderie would be destroyed, or whether some exercised simple prejudice, we cannot be certain.[4]

The issue remained in abeyance for five years until 1888, when William Ramsay, who had readily taught and worked with women chemists in both his Bristol and UCL laboratories, brought the question of female membership before the council again. Women were now visibly present in both provincial college laboratories such as Bristol (where Ramsay had begun his research career) as well as in various London colleges, including

---

2. Chemical Society council minutes, 16 December 1880, f. 219.
3. Chemical Society council minutes, 29 January 1881, f. 221.
4. It should be noted that the in-house servants and other employees (scriveners, later typists, cleaners, and porters) would have all been male at this time. It was not until 1885 that a female employee is recorded as working on correcting proofs.

a few in Armstrong's Central College. Many of them were taking London BSc degrees and, most notably, publishing their research findings in the society's *Journal*. But once again the issue was brushed aside. On 18 March 1888, when the council met with William Crookes in the presidential chair and Ramsay urged the admission of women, the motion was "withdrawn after a lengthy discussion, it being judged inexpedient to recommend any such change at present."[5]

Four years later, on 17 November 1892, the issue came before the council yet again when Walter Hartley recommended the election of one of his female students as an "Associate." Ramsay immediately seized the opportunity to offer the motion "that women be admitted as Fellows,"[6] and this was seconded by Tilden. Both chemists had successfully and gladly taught female undergraduates and research students at UCL and the RCS. However, at the following meeting early the next year the motion was rejected.

> It having been moved by Professors Ramsay and Tilden that women be admitted as Fellows, Mr [Robert] Warington [Jr.] and Dr [W. H.] Perkin proposed the amendment—that it is not desirable at the present time to propose an alteration in the bye-laws with the object of admitting women to the Fellowship of the Society.[7]

On taking a vote, this "poison-pill" amendment was narrowly rejected by seven votes to six; but when Ramsay's and Tilden's original motion was voted on, it was rejected by seven in favor and eight against. Despite being cast out by just one vote, and despite the fact that the general membership had not been consulted, the council resolved that

> although there was no objection in principle to the admission of women as Fellows, the case in their favour was not *clearly* established by any considerable number of applications for the Fellowship and that any change involving so radical an alteration in the policy of the Society should be recommended by a unanimous vote.[8]

It would appear, then, that it was women chemists' own fault for not applying in large numbers for admission!

---

5. Chemical Society council minutes, 18 March 1888, f. 103.
6. Chemical Society council minutes, 17 November 1892, f. 212.
7. Chemical Society council minutes, 19 January 1893, f. 226.
8. Chemical Society council minutes, 19 January 1893, ff. 226–27.

Two interlinked grievances now united both provincial chemists (their rebellion as discussed in the previous chapter) and the growing numbers of practicing female chemists, namely that the society was run by an undemocratic elite. However, there was no further opportunity to seize an initiative until 1904, when William Tilden became president. In 1903, Marie Curie and her husband Pierre were jointly awarded the Nobel Prize in physics for their work on radioactivity and radiochemistry. She was, therefore, an obvious candidate for one of the several vacancies for foreign membership of the Chemical Society. Prompted by Tilden, the council asked once again for legal opinion on the eligibility of women as ordinary and foreign members, and noting, as before, that there seemed to be nothing in the charter to prohibit their election.

The lawyer whose legal opinion was sought this time was Richard I. Parker of Lincoln's Inn, who replied on 17 April 1904:

> In my opinion married women are not eligible as Fellows of the Chemical Society and I think it is extremely doubtful whether the Charter admits of the election of unmarried women as Fellows. (1) It would not in my opinion be wise to elect even unmarried women as fellows without first applying for a Supplementary Charter, but if the Society notwithstanding... desire... to elect an unmarried woman as a Fellow, I think they should... modify their bye-laws [to] authorise the election. (2) With regard to Honorary and Foreign Members and Associates the case is on a somewhat different footing. Neither Honorary nor Foreign members nor Associates are Corporators—their position involves no obligation founded on Contract or otherwise. After a modification of the bye-laws expressly authorising the election of women married or unmarried to be Honorary or Foreign members, or Associates, such election would be legal. (3) It is I think unlikely that if a woman, even without any alteration of the bye-laws, were elected to be an Honorary or Foreign Member or Associate, any great objection would be raised. (4) There is, I think, no reason why an application for a Supplementary Charter so as to admit women should not succeed. (Quoted in Mason 1991a, 234)

Parker was merely expressing the common law of the time that married women were not legally independent persons (that is, independent of their fathers or husbands), and hence they were unable to be corporators. Legally, fellows of the Chemical Society were corporators because, having paid a subscription and made promises to uphold the rules and values of the society, they were able to participate in the society's government.

Because this condition did not apply to honorary or foreign members, the society was, therefore, free to elect Madame Curie, which it duly did on 20 April 1904.

This outcome arguably played into the hands of women: how could there be one rule for foreign women regarding eligibility and another for native Britons? Consequently, on 21 October 1904 the council was presented with a petition signed by nineteen female chemists and organized by Armstrong's former pupils Ida Smedley and Martha Whiteley, together with the estimable Anglo-Austrian chemist Ida Freund, of Newnham College, Cambridge.

> We, the undersigned, representing women engaged in chemical work in this country desire to lay before you an appeal for the admission of women to the fellowship of the Chemical Society. In justification of this plea we venture to draw your attention to the share taken by women in research, to its steady growth, and to the need for greater facilities to continue it.

And then, in what they thought their most powerful argument, the leaders noted that within the past thirty years some 150 women had successfully published 300 papers in the society's journals and that the rates of publication had exploded from just 23 in the decade 1873 to 1882, to 42 between 1883 and 1892, then to 172 between 1893 and 1902. And in the mere twenty months from 1903 to August 1904, when the petition was organized, sixty-six papers had been published. Overall, the women claimed to have shown a trebling of output every decade.

> Seeing that the Chemical Society recognises the value of the contributions made by women to chemical knowledge by accepting their results for publication, we are encouraged to point out that their work would be greatly facilitated by free access to chemical literature and by the right to attend the meeting of the Society. We venture therefore to hope that you will consider favourably the question of extending to women the privileges enjoyed by the Fellows of the Chemical Society. (Quoted in Mason 1991a, 234)

The last sentence was probably unwise since it could be interpreted as asking not for full fellowship but merely for a few privileges such as the right to use the library and to attend meetings at which papers were read and discussed. Accordingly, choosing this softer and less controversial option of women as non-corporators, Tilden proposed that the society alter its bylaws to give qualified women all the privileges of fellowship apart from the right to hold office or vote at annual or extraordinary meetings. This change would come at the cost of £400, the price quoted to the society for altering the charter to remove apparent difficulties over the election of foreign members.[9] One can understand the society's reluctance to spend so much on this issue, a reluctance shared by other learned societies. Given the cost, Tilden also saw that the decision could not be taken by the council alone and that the entire membership needed to be consulted. But when he called an extraordinary meeting for 2 July 1904, only forty-five fellows bothered to turn up, and the motion to change the bylaws was lost.

Instead, it was suggested that women could become associate members, as defined by the bylaws. Tilden was clearly angered, as his rebuke in his final presidential address in 1905 demonstrates:

> The number of women desiring admission is but small, and I fail to see any cogent reason, besides the legal one, for excluding them. Some of them are doing admirable scientific work, and all the memorialists are highly qualified. To deprive them of such advantages as attach to the Fellowship simply on the grounds that they are not men seems to be an unreasoning form of conservatism inconsistent with the principles of a Society which exists for the promotion of science. (W. Tilden 1905, 547)

In the interim, Tilden had taken upon himself to organize another petition to the then-president, Raphael Meldola, and the council of the Chemical Society:

> We, the undersigned Fellows of the Chemical Society, being of the opinion that the time has come when the fellowship of the Society should be rendered accessible to women, request the Council . . . to ascertain the wishes of the Society as a whole in regard to this question.

This second petition then proceeded to quote from the quantitative evidence provided by the women in their 1904 petition, and concluded:

> The chemical societies of Berlin and America, the Society of Chemical Industry and the Faraday Society, admit women on the same terms as men, and our Society has found a place for Madame Curie among the Honorary and Foreign Members: we consider, therefore, that the restrictions should be removed under which the Chemical Society

---

9. Chemical Society council minutes, 21 June 1906, f. 93, and 15 November 1906, f. 97.

denies to women chemists the advantages extended to them by the sister Societies at home and abroad. (W. Tilden 1908a)

Tilden succeeded in gathering the signatures of 312 fellows, including ten past presidents, twelve vice presidents, and twenty-nine past and present council members who between them included thirty-three fellows of the Royal Society, which, ironically, was experiencing its own dilemma over the election of women to the fellowship. Armstrong was not a signatory. A strongly worded editorial by astronomer Sir Richard Gregory in *Nature* derided "the banging, barring and bolting people" who were preventing the admission of women but printed an extensive letter from the two secretaries of the Chemical Society, Martin O. Forster and Arthur W. Crossley, that, while purporting to fairly and squarely present the case for and against admission, ended with the damning conclusion:

> Briefly stated, the position of those unfavourable to the admission of women is that, whilst gladly offering to those women who already have become chemists measures which would give them the benefits derived from attendance at the meetings, they deem it inexpedient publicly to encourage women to adopt chemistry as a professional pursuit, since such a course would tempt them into a career in which they may ultimately not find employment in view of the over-crowded state of the profession. (Gregory 1908a, 227)

Tilden's proposal was put to the vote at council in May 1908 when it came about that, despite the earlier rejection of a postal vote, the question of women chemists' membership should be decided by postal ballot of the total membership. Roscoe seconded the motion and Tilden, Thorpe, Harcourt, and Armstrong were appointed to arrange the terms of the ballot.[10]

This subcommittee reported back that every fellow should be sent a letter putting the argument for and against the admission of women, together with a copy of the petition that had been signed by 312 women. Fellows would be asked to reply "yes" or "no" by postcard. However, William Hodgkinson, professor of chemistry at the Woolwich Arsenal, objected to the petition being included with the 312 names attached (Anon. 1935c). Instead, seconded by Hooper Jowett (an alkaloid research chemist at Burroughs Wellcome Company laboratories at Dartford), it was agreed simply that a prepaid postcard should be sent to fellows along with the arguments

---

10. Chemical Society council minutes, 21 May 1908, ff. 160–61.

for and against (Jowett 1937).[11] It is unclear why this motion was carried, unless it was simply on grounds of cost. Armstrong and Roscoe were appointed to scrutinize the postcards, which had to be returned by 1 October 1908. When Norman Lockyer, the editor of *Nature*, learned of the postal memorandum, he penned an editorial which among other things noted that if the society was only too happy to publish women's research, it was illogical to deprive them of membership when the majority of male members published nothing. Gregory also accused the council of suppressio veri by not circulating the women's petition with the voting forms. "It is a question for the society whether its true interests would not be better served by the transference of its secretarial business to more competent and more judicious hands" (Gregory 1908a, 226).

Much the same was said by Sir Henry Roscoe in a strongly worded letter to the *Times* (Roscoe 1908). Crookes's *Chemical News* also took up the debate, with the analytical chemist Henry Droop Richmond reflecting that an unknown woman would attract more attention than an unknown male chemist and thereby gain employment over a male colleague by virtue of sexual attraction (Richmond 1908). This contention was turned into the grotesque argument by E. G. Bryant that because women would displace men, slowly men would be unemployed and unable to pay their subscriptions, leaving a membership entirely of bachelors and spinsters (E. G. Bryant 1908).

Clearly, the vote hinged on the arguments for and against, presented in the accompanying letter to fellows. The positive argument (probably drafted by Tilden) noted that a small number of women already attended as auditors when papers were read and that no inconvenience was found by their presence. He also thought that the numbers wanting admission was unlikely to be more than about twenty, so that fears that they would come to dominate the council was absurd, especially since the majority of teaching posts in the UK were occupied by males. The coup de grâce was that the women themselves were likely to pay the costs of altering the bylaws.

The negative arguments (most likely drafted by the misogynist Armstrong) were confected to be as strong as able forensics would allow: (1) the cost of seeking an amended charter would not be justified by the small numbers of women likely to seek fellowship; (2) more controversially, given the arduous nature of chemical work it was doubtful that encouraging women to enter the chemical profession would be to their

---

11. Chemical Society council minutes, 18 June 1908. The motion was approved by 12 votes to 4. For the attached letter to members, see Mason 1991a, 236.

long-term advantage; (3) even if the number of publications from women was increasing, there was probably a tendency to overestimate the actual value of their work (it was notable that of the 103 papers cited by the female memorialists, only twenty-three papers were not collaborations with a male chemist); (4) even if these twenty-three women authors had really worked independently of male guidance, "few women have shown marked aptitude for chemical pursuits"; (5) women had already been welcomed as guests at society meetings; (6) if Madame Curie was cited as precedent it had to be remembered that honorary and foreign members had no voting rights and could not stand as officers or council members; and finally (7) the "clinching" statement, the lack of jobs, given the unfortunate employment market.[12]

Perhaps many fellows felt the voting memorandum was biased since, at an extraordinary meeting of the Chemical Society on 22 October 1908, Armstrong, seconded by Meldola, proposed "that the President [Ramsay] inform the meeting that in all proceedings relating to the admission of women as Fellows, the Honorary Secretaries [Forster and Crossley] have throughout acted under the direction of Council." Ramsay then revealed that 1,758 fellows had returned postcards and that 1,094 (62 percent) had voted in women's favor and 642 (36 percent) were against.[13] However, instead of accepting this positive result then and there, a final decision was deferred until another extraordinary meeting on 3 December 1908. At that time, the case presented by the council in 1904 was presented again, and Tilden, seconded by Thorpe, proposed that women should be admitted.

It was only at this point that Armstrong, speaking as a vice president, showed his hand directly by suggesting that since the category of "subscriber" still existed in the society's bylaws, though scarcely used since the 1840s, women should be admitted under that category for an annual subscription of thirty shillings. Under this category, women could attend meetings, use the library, and receive the society's publications, but they would not be eligible for election to council or any of the society's committees, or to vote in such nominations.[14] Armstrong's proposal, which had not been presented as an alternative possibility in the postal ballot, was

---

12. Chemical Society council minutes, 23 June 1908, f. 165. The printed argument was actually issued under the names of the two secretaries, Martin Forster (a pupil of Armstrong) and Arthur Crossley (a pupil of Roscoe and Emil Fischer).

13. Twenty-two registered indifferently; see Chemical Society council minutes, 22 October 1908, f.167.

14. Chemical Society council minutes, 3 December 1908, ff. 174–75.

immediately seconded by his friend Horace Brown, and carried nemine contradicente at the December ordinary meeting.[15] *Nature* printed a satirical poem by Tilden in its Christmas edition:

> Daughters of Eve! So zealous to pursue
> The work of Life by which you seek to live!
> When F.C.S. you claim, as is your rightful due—
> The S alone is what they, grudgingly give!
> Be patient! Time is on your side.
> Reason and justice will your cause defend
> Ignoble spite and arrogance of pride
> Shall meet their retribution in the end. (W. Tilden 1908b)

The "S alone" referred to the "subscriber" status now on offer. Only one chemist saw fit to protest. This was Edward Divers, a pupil of Frankland's who had spent many years teaching chemistry in Japan. At a meeting of the society in the following January, he questioned the legality of obliging women who wanted membership to do so under the terminology of "subscriber" rather than "fellow." Consequently, the council was forced to take legal opinion again. This opinion was forthcoming in March 1909. The advice of Messrs. Bristow, Cook, and Carpmael was that all was above board, because being a subscriber did not entail being a corporator.[16] Unwilling to concede this point, at the next meeting of the council Divers proposed altering bylaw 20 (referring to the common seal and deeds) to read "shall" as "may." This change would have permitted the president to override decisions made by the council, but the suggestion was rejected as "repugnant" to the meaning of the charter.[17]

All the women who had subscribed to the original petition refused to become subscribers, and during the next twelve years only eleven women are recorded as registering for this second-class membership. In another strongly worded *Nature* editorial, Gregory lambasted the Chemical Society for its bungling the issue of "whether in an essentially democratic institution like that of the Chemical Society, the will of the majority is to prevail, or whether it is to be thwarted by the machinations of a self-constituted oligarchy which abuses its trust and makes use of its opportunities

---

15. Chemical Society council minutes, 3 and 17 December 1908, ff. 174, 176.
16. Chemical Society council minutes, 18 March 1909.
17. Chemical Society council minutes, 1 April 1909, f. 195. Divers agreed not to proceed formally with the suggestion.

to gratify its personal prejudices" (Gregory 1909).[18] Armstrong did not reply to this critique, but his attitude toward the status of women, as we will see, had become very clear in his report on American education when he was part of the Mosely Commission in 1903; it would be made explicit in his presidential address to Section B (Chemistry) of the British Association when it met in Winnipeg in August 1909 (HEA 1909d).

The Mosely Commission

The economic depressions of the 1880s demonstrated to many British politicians and scientists, especially chemists, that an industrial war could only be won if British industry and workers' education were radically altered. Fears of the industrial prowess of Germany and the growing prowess of America and Japan, coupled with the inefficient conduct of the Boer War in South Africa that led to revelations of physical malnutrition of British soldiery, invited national comparisons. Armstrong was no exception; indeed, as a German-trained chemist and as a visitor to America in 1897 for the Lawes Agricultural Trust, he succeeded in using international comparisons with authority and without overstatement. He recognized that civilizations had been changed by scientific discovery and the application of science to industry (HEA 1903c). He concluded that the language and methods of practical science had to be taught in schools.

But the panicky comparisons with German institutions instanced by Michael Sadler's *Secondary Education in Germany* (1902), along with the report by the London Technical Education Board's subcommittee on the Application of Science to Industry (1902) and F. A. McKenzie's sensational account of American competition (McKenzie 1901), did not blind Armstrong to the dangers of copying foreign models. One man who did think Britain could learn from overseas models where the grass was, possibly, greener, was the diamond merchant and philanthropist Alfred Mosely (1855–1917). Born in Bristol, Mosely had gone out to South Africa as a young man to seek his fortune in mining diamonds (Anon. 1917). There he had become greatly impressed by the skills of American engineers who operated his and others' mines at Kimberley and who built hospitals during the first Boer War. Back in Britain, by 1900 he had used part of his fortune to investigate the secret of America's growing commercial success and how Americans were educated. Although not an official government in-

---

18. The only comment came from the Kent geologist W. J. Atkinson, who noted that Gregory's critique applied equally to the Geological Society, which, like the Chemical Society, did not admit women until 1919 (Atkinson 1909).

vestigation, the so-called Mosely Educational Commission was supported by Lord Reay, chairman of the London School Board; Michael Sadler at the Board of Education; and Gerald Balfour, the president of the Board of Trade. The plan was for Mosely, together with a group of twenty-six educationists, to spend three months in America to investigate its schools and colleges between October and December 1903.[19] The group was to "to study the educational methods of that country in their bearing on its commercial and industrial life and, if possible, to draw from ... observations suggestions for the improvement of our methods at home."[20] As the printed report makes clear, each visitor made their own investigation, and Armstrong volunteered to edit and assemble the final document.

It would not have been possible for Armstrong and his colleague William Ayrton to obtain permission from the City and Guilds for three months' absence, and thus they spent only four weeks on the tour; however, most of the other delegates who represented medicine, agriculture, mechanical engineering, technical education, and the general world of education stayed the full course of Mosely's hospitality.[21] The delegation arrived in New York on the USS *Philadelphia* on 10 October 1903. The American press was greatly intrigued that British educationists had come to study American institutions of learning. Following ten days of visits in New York, the party moved to Washington, where President Theodore Roosevelt honored them with a reception in the White House. From there the delegates traveled to Baltimore, Philadelphia, New Haven, Boston, Niagara Falls, and finally, Chicago, where the party split up and went their separate ways until reassembling in New York in early December. Armstrong and Ayrton left the delegates in Connecticut at the end of October in order to return to their teaching duties in London by 7 November 1903. Armstrong justified this departure in the final report by stating that he had

19. Armstrong was initially hostile to Mosely's intentions (HEA 1901b, 7).

20. Quote by Revd T. A. Finlay, one of the members, in HEA 1904a. For Armstrong's diary of the visit, see AP2.130a. Mosely had also paid for a party of British businessmen to visit America to investigate the country's economy in 1902. Later he sent a large party of British schoolteachers to America to learn from what they saw. This visit was reciprocated in 1908–1909, when Mosely paid for American high school teachers to come to Britain. Their tours of British schools were arranged by Armstrong, who was subsequently thanked with an engraved testimonial (AP1.504).

21. Letters dated from 28 September to 22 October 1903, concerning Armstrong and Ayrton's negotiations for leave of absence are to be found in the City and Guilds archives (CLC/211) in the London Metropolitan Archives; Armstrong arranged for his son Frank and Moody to cover his teaching during his third week of absence.

already learned a great deal about American education when he crossed the entire American continent twice during his visit in 1897.

As a member of this commission, Armstrong found, like most of the other delegates, much to praise in the standard of buildings, organization, and equipment of American schools, and a praiseworthy "absolute belief in the value of education both to the community at large, and to agriculture, commerce, manufactures, and the service of the state" (Lupton 1964, 42). On the other hand, he expressed nothing but contempt for American teaching methods, except in the areas of medicine and manual training. His biggest criticism was the feminization of American youth through the practice of coeducation, as well as the prevalence of female teachers. He feared that coeducation would sap virility, as the late marriages and low reproductive rate of American graduates seemed to suggest to him. Women teachers, he claimed, were of historically proven unoriginality. "Those who have taught women students are one and all in agreement that, although close workers and most faithful and accurate observers, yet, with the rarest exceptions, they are incapable of doing independent original work" (HEA 1904b, 14). This attitude was clearly prevalent among the other delegates when they mentioned women teachers; only Ayrton stood by women by insisting that he disagreed with his colleagues (HEA 1904b, xxiv).[22]

From his later speech in Winnipeg, Canada, in 1909, it would become clear that Armstrong had thoroughly absorbed the eugenics teachings of Francis Galton and Karl Pearson. It was imperative that Britons retained "a genuine steel, of tough race and tempered stock" (HEA 1909c, 449). Referring directly to the Chemical Society's resistance to women's fellowship, he stated:

> The subject has been brought before the chemical world in England recently by the application of a number of women to be made Fellows of the Chemical Society. Many of us have resisted this application because we were unwilling to give any encouragement to the movement which is invariably leading women to neglect their womanhood, which is in itself proof that they do not understand the relative capacities of the two sexes and the need there is of sharing the duties of life. If there be any

---

22. Irish historian Úna Ni Broiméil has argued that the whole report was a gendered script whose message was that of Anglo-Saxon social and political superiority (Broiméil 2015). This message was certainly the one understood by one delegate, the Reverend Herbert Branston Gray, the warden (headmaster) of Bradfield College, who subsequently wrote an influential book on public schools and the Empire (Gray 1913).

truth in the doctrine of hereditary genius, the very women who have shown ability as chemists should be withdrawn from the temptation to become absorbed in the work, for fear of sacrificing their womanhood; they are those who should be regarded as chosen people, as destined to be the mothers of future chemists of ability. The argument is applicable generally: it is surely desirable in all cases of declared ability that the education of girls should be directed so as to produce not merely minimum disturbance of the woman's attributes and charms but full understanding of the unique position of responsibility she occupies in the scheme of life. (HEA 1909c, 451)

This section of Armstrong's Winnipeg address received wide publicity in British newspapers in September 1909 under headlines such as "Problem of Woman. Her Higher Education a Drawback" (Anon. 1909), and there was considerable correspondence from angry women in the national and provincial press. Typical responses came from two Londoners. Muriel Nelson found Armstrong's views "extraordinary," for if they were true, they would apply equally to the masculine side.

> A child will inherit tendencies from its father equally with its mother. And on these grounds we might just as well say if we want our country to be filled with genius, whenever a Shakespeare, a Lavoisier, a Beethoven appears, withdraw him at once from the divine task. Set him to breed children that his genius may be multiplied by the State! Let us have science by all means, but for heaven's sake, let it be tempered with sanity. (Nelson 1909)[23]

Ada Mitchell of Highgate rounded on Armstrong for placing women on the level of animals by suggesting that their primary purpose in life was to produce children. If that were really the case, then Nature would not have endowed them with brains of such a quality as to entitle them to be made fellows of the Chemical Society. Since women slightly outnumbered men in the population, were not unmarried women entitled to become absorbed in chemistry or any other science? She drove home the economic argument: in an industrial world, women had actually been forced into competition with men by economic conditions.

---

23. Florence Barger, on the same page of the *Yorkshire Post*, thought Armstrong's role in the Chemical Society's vote "not a performance to be proud of."

> Those industries which formerly occupied [women's] time are not now carried out in the home. This is due to the invention of machinery. Seeing, therefore, that women are out in the world, they claim that they shall at least receive fair treatment and share the same advantages as men. . . . It seems to me rather arrogant even for the President of a Section of the British Association to be so sure he knows women's place in the scheme of things. (Mitchell 1909)

Armstrong must have seen many such letters in the press, but he remained convinced of the rightness of his hereditarian position. He had found a fresh arena in which to air his prejudices, namely by tying women chemists in with the unruly nature of the suffrage and suffragette movements. When he became one of the founders of the Biochemistry Club in 1911, it based its membership rules on those of the Physiological Society, which had been founded during the anti-vivisection campaign in 1876. Its bylaws did not prohibit the election of women members, though in practice none were elected until three in 1913.[24] Given that the eminent Frederick Gowland Hopkins encouraged women to engage in biochemical research in his Cambridge laboratory, the question of their election to membership arose in 1912. Armstrong wanted the biochemistry group to remain a social and dining club and not become a formal society. During an angry confrontation at a meeting of its council, Armstrong resigned from its committee and stormed out of the meeting (Goodwin 1987, 15–16).

As his own son, Frank Armstrong, admitted brutally in a memorial lecture in honor of his father, Armstrong remained suspicious of the female teacher all his life: "He had little interest in in the higher education of women, holding the conservative view that every woman should be a cook" (E. F. Armstrong 1941, 84). A woman's place was in the home, he felt: hence her educational training needed to enhance her femininity, not destroy it. Its emphasis had to be on domestic science and household chemistry (Bayliss 1983). He was convinced that the independence of women that had developed during his lifetime was superficial and unreal. Women teachers ought to be married: "the sexless creatures who too often engage in the vain task of training our daughters in the present days of higher education are a real danger to society" (HEA 1915g).

As the work of Marelene and Geoffrey Rayner-Canham has shown, by the 1900s there were hundreds of women chemists who had trained in Oxbridge, in provincial colleges, and in colleges of the University of Lon-

---

24. Namely, Ida Smedley, Harriet Chick, and Muriel Wheldale.

don (M. Rayner-Canham and G. Rayner-Canham 2020). A number of them held doctorates from the University of London. A large proportion had found no difficulty in having their work published in the Chemical Society's *Journal* or in the independent *Chemical News*. They had access to provincial libraries that were the equal of the Chemical Society's. What they lacked was the right to have the letters "FCS" (Fellow of the Chemical Society) after their names and to participate in the government of the Chemical Society.

The women chemists' attempt to obtain full membership, despite being supported by Ramsay and Tilden and many other leading British chemists, was continually thwarted by Armstrong, who, playing Fifth Business as secretary and later as vice president, exploited the society's foundational charter through his careful choice of legal advice. At the last moment, in 1908, when it seemed a vote would prove favorable to women's election, Armstrong and his pupils exploited a little-used loophole in the society's statutes that would save any legal expenses by electing women to the inferior membership category of "subscriber." Armstrong's views on women's role in history were made explicit in his report on American education as a member of the Mosely Commission in 1903 and again in 1909 in Winnipeg when he addressed an overseas meeting of the British Association. There he revealed his adherence to the eugenic teachings of Galton and Pearson that women's role in society was to protect their womanhood for the preservation of human culture.

Although their campaign initially ended in failure, the question of the election of women members to membership never went away, and with the passage of the Sex Disqualification (Removal) Act in 1919, the society was legally obliged to reopen the question. The bylaws would have to be changed to comply with the Act and a general meeting held to consider the exact changes of wording.[25] The revised bylaw came into effect on 1 May 1920, and the first two women, Martha Whiteley and Ida Maclean, were duly admitted as fellows. We have no record of comment from Armstrong, but it is clear from his miscellaneous writings that he never changed his views on women's place in nature (Anon. 1933c). In any case, the Chemical Society elected only two female chemists to be council members during his lifetime. That was in 1926, when Martha Whiteley was elected to council; Ida Maclean, by then Ida Smedley, followed in 1931. It was not until 2012 that Lesley Yellowlees became the first female president of the (renamed in 1980) Royal Society of Chemistry.

---

25. Chemical Society council minutes, 3 April and 15 May 1919, ff. 137, 142.

∗ 8 ∗
The Heuristic Method

> All who seriously study the history of education in our times must agree that, although it may be long ere we can cry *Eureka! Eureka!* of an ideally perfect system, recent experience justifies the assertion that we shall hasten the advent of that desirable time if we seek to minimise the didactic and encourage heuristic teaching.
> —HEA 1898, 389; quoted in Brock 1973, 110

If the Royal Society of Chemistry today were to name its Education Section after a distinguished chemical educationist, paralleling what it has done with its Dalton, Perkin, and Faraday Sections, Henry Armstrong would be a logical choice, despite his regrettable views on gender relations as described in the last chapter. It is arguably his contributions to science education that have had the greatest long-term effect, rather than his chemistry—notwithstanding its importance at the time. While historians of science honor Armstrong for his work on organic chemistry, historians of education remember him for his campaign to improve science teaching in schools by the use of what he called the "heuristic method."

Science Teaching in the Nineteenth Century

"Scientific education," wrote the young assayer and economist William Stanley Jevons in 1856, "is one of the best things possible for any man, and worth any amount of Latin and Greek. It tends to give your opinion and thoughts a sort of certainty, force and clearness, which form an excellent foundation for other sorts of knowledge less precisely determined and established, provided you do not allow your mind to become completely formed to science" (Black 1973). Although the rhetoric of the mid- and late-Victorian advocates of the introduction of science into the British school and university curriculum agreed with Jevons's view that science was mind-strengthening and mind-broadening—indeed, that even better than Latin or Greek, it promised a genuinely "liberal education"—the long-term English tendency (one must exclude the Scots) was toward Jevons's fear of specialization. By the turn of the twentieth century, state-run British secondary education for all had been introduced, reflecting and perpetuating the curricula and values of the elite, private, fee-charging (so-called "public") schools; accordingly, instruction and examinations in the separate disciplines of chemistry and physics and, increasingly, biology—including teaching laboratories—had become the norm. School science teachers were therefore compelled to "form" their pupils in science whether or not their future careers lay in science.

In the 1850s British scientists began to form an effective pressure group that urged a fuller appreciation of, and a more systematic teaching, of their interests. At the center of this propaganda group were the nine men who together formed the "X-Club" in 1854 (Barton 2018). The members, most of whom examined the courses set by the government's Department of Science and Art, included the biologists Thomas Huxley, Joseph Hooker, and John Lubbock, the chemist Edward Frankland, the physicist John Tyndall, the mathematician Thomas Archer Hirst, and the philosopher Herbert Spencer. The epicenter of their campaign was the Royal Institution, where Tyndall (and, among others, Michael Faraday and William Whewell) had delivered the first broadside in a series of public lectures on the value of science in education in 1854. Spencer's remarkable review of essays on education, in which science appeared as the universal panacea of education and civilization, belong to the same period (Spencer 1878) and were republished as a collection in 1862, a few years before the report of the government's Clarendon Commission, primed by the arguments of a group of scientists that included Hooker and Faraday, urged the adoption of science teaching by the English "public" schools.

There followed a decade's debate over the relative merits of classics, mathematics, and science as formative elements in a liberal education. The discussion, together with the scientists' use of faculty psychology and warnings of foreign economic achievement catalyzed by a superior education, interested many Broad Church clergymen, like Charles Kingsley, Frederic Farrar, and James Wilson. Kingsley's lectures to the boys of Wellington College, with their emphasis on observation, were to be frequently quoted by Armstrong as authoritative arguments for practical-science teaching. Farrar, then a teacher at Harrow School, published his important *Essays on a Liberal Education* in 1867; one of the essays included in the book was by the future "Nestor of science masters," James Maurice Wilson (1836–1931), at the time a young mathematics teacher who had deeply impressed the Clarendon Commissioners by his practical-science classes at Rugby School. His curriculum, based on the science of common things, also impressed Farrar. At the British Association for the Advancement of Science meeting at Nottingham in 1855, Farrar launched the first of its educational campaigns, in the form of a committee composed of himself, Wilson, Tyndall, and Huxley, "to consider the best means of promoting scientific education in schools." Its report, issued in 1867 (which was presented to Parliament), distinguished scientific *information* from scientific *training*.

This distinction, which the phrenologist and secular educationalist George Combe had identified as existing between what he called *positive instruction* and *instrumental instruction* (meaning the training of the phrenological faculties), was the antecedent of Armstrong's distinction between *scientific facts* and *scientific method*. The practical conclusion expressed in the report of Farrar's committee was that there existed a common core of factual information that covered the laws and phenomena of everyday life and familiar experience, and that formed an essential part of every citizen's knowledge. Here was the seed of the twentieth century's general science movement, Armstrong's and John Hall Gladstone's (1827–1902) call for teaching the science of everyday life, as well as Huxley's startlingly successful textbook *Physiography: An Introduction to the Study of Nature*. On the other hand, the committee recommended special training in physics, chemistry, or biology, in order to stimulate the dormant areas of the mind that were neglected by mathematics and classics teaching. These subjects were to train human powers of observation, classification, and induction.

Under the Elementary Education Act of 1870, most science teaching took place under the auspices of the government's Department of Science and Art (DSA). Through the rigorous financial control of their examination results, such classes, whether taken in the evening or at day schools, tended to be nonpractical, textbook oriented, and imbued with rote learning. Both Frankland and Huxley tried to stimulate practical-science teaching by holding summer schools for science teachers during the 1870s, but the practical instruction they encouraged, and the practical textbooks that followed in their wake, solely emphasized experimental manipulations and demonstrations.

Farrar's report of 1867 alerted the BAAS and the British government to the needs of scientific instruction. Together with the X-Club's campaign for the "endowment of research," it led directly to the investigations of the important Devonshire Commission on Scientific Instruction and the Advancement of Science, which deliberated between 1872 and 1875. The commission's sixth report (1875), masterminded by astronomer and *Nature* editor Norman Lockyer, dealt with scientific education in schools and recommended that all public and endowed Schools should allot some six hours a week to science teaching. The commission also recommended the transfer of the Royal School of Mines, together with the former Royal College of Chemistry, to a new site in South Kensington. In 1881, following pressure from Huxley and Major John Donnelly, the School of Mines was transformed into a "normal" school for the training of science teachers, but it reverted to a pure science college, the Royal College of Science, after Huxley's retirement in 1885.

With the advent of secular school boards in 1870, science returned to the curriculum of the elementary schools, influenced by the liberal-education controversy of the 1860s. The government's new, complicated Revised Code of Regulations awarded grants to the upper standards of board schools for certain optional or "special" subjects that included mechanics, animal physiology, botany, agriculture, physics, and chemistry. By 1880, the code had been further modified to permit the teaching of elementary science or geography as optional class subjects above the first standard, while specific subjects like mechanics could still be taken from standards IV to VI. In practice, these specific subjects were relatively unpopular and fared badly in elementary schools against the competition of English and geography. Notable exceptions occurred in those schools that were governed by the school boards of London, Liverpool, and Birmingham, where scientists like biologist Huxley and chemist Gladstone offered advice and active encouragement. There, the boards developed coherent, graded object lessons and introduced the economically successful method of peripatetic science demonstrations.

When Armstrong began teaching chemistry in 1870, it seemed to him that his medical students at St. Bartholomew's Hospital and working auditors at the London Institution had been completely dulled by having been subjected to authoritarian, didactic teaching and textbooks. He concluded that his students were unable to think for themselves and were prepared only to memorize facts and formulae for examination purposes. His predecessors had successfully brought science into schools, colleges, and evening classes, but had stifled students in the process because no one had paused to ask how science ought to be taught or how the science curriculum should be sequentially arranged. He himself had been thrown in the deep end by Frankland and told to find ways of improving the analysis of water. And at Leipzig, under the benevolent but scathingly critical eye of Kolbe, Armstrong had been made to see the importance of self-education through laboratory research.

Armstrong's solution for the improvement of science teaching was the heuristic method.

Armstrong's Heurism

In chapter 2, we discussed Armstrong's reading of Richard Trench's *On the Study of Words* and his experience as a witness in a legal case as having been formative. The patent action made him alive to "the need of searching cross-examination and judicial consideration of every item for or against a proposition" (HEA 1920c). The same method of self-cross-examination was also to make him a formidable critic of Svante Arrhenius's theory of ionization in electrolysis. Armstrong was to label this methodological use of knowledge "scientific method," and he decided that it was best learned in a laboratory setting by an autodidactic process he called "heurism." The word came from Archimedes's reputed exclamation, "Eureka!" (εὕρηκα, heúrēka, "I have found [it]"). "Heuristic methods," claimed Armstrong, "are methods which involve our placing students as far as possible in the attitude of the discoverer" (HEA 1920c). What a child or a student can find out for himself, he best remembers because he truly understands. Moreover, because the student-investigator is genuinely interested and involved in finding a solution to a problem, he will learn about it more efficiently than if he is given the solution didactically.

Although similar ideas and practices had been advocated earlier by the Swiss educationist Johann Heinrich Pestalozzi and others, it seems that Armstrong developed these conclusions largely by intuition and from his experience as a research chemist. Similarly, it seemed obvious to him that children would grasp certain ideas better at different ages and that science syllabuses should be age-graded accordingly. The more thorough investigations that Jean Piaget and other psychologists have made on children's cognitive developments in the twentieth century have supported Armstrong's ideas, though the literal application of heurism throughout schooling would be painful to both pupil and teacher (HEA 1903b; Inhelder and Piaget 1958; Shulman and 1884 1966).

It was at the International Conference on Education on 5 August 1884, convening in the still-incomplete Central College, that Armstrong, as the newly appointed professor of chemistry, first began to make public his views on science education. Armstrong's challenging paper on the teaching of natural science, which was delivered to the section on Technical Teaching under the chairmanship of Philip Magnus, argued against specialization and recommended practically oriented instruction (HEA 1884, 69–82; Brock 1973, 74–89). He seemed pleased at the paper's reception along with the Finsbury syllabus he described.

However, the principal discussion following the paper centered on the incompatibility between his syllabus and public examinations. How would one earn a school grant if an instructor followed Armstrong's practical course? Schoolteachers would not dare experiment for fear that the Department of Science and Art would not pay for their science classes. In reply, Armstrong's assistant at Finsbury, John Castell Evans, stated that in his experience it could be done; students would still pass the examinations. He revealed that he and Armstrong had taught a class of boys at Cowper Street Middle Class School since 1880 and obtained good results in the physical science examinations set by the DSA, even though they had adopted a program of individual problem-solving analogous to that of Armstrong's chemistry course (J. Evans 1884). Despite this evidence, the conclusion drawn from the discussion was that the system of payment by results made Armstrong's suggestion unworkable.

Still, Armstrong's appearance at the Education Conference showed that he was no mere idealist. He spoke from experience: he had consulted teachers and headmasters beforehand; he had taught schoolchildren and older students himself; he had studied examination syllabuses and timetables; he had developed new methods that worked both at the London Institution and Finsbury College. He was poised to gain yet more experience at the new Central College and by training teachers himself.

Another forceful speaker at the 1884 meeting was John Miller Dow Meiklejohn (1836–1902), one of the most prolific of Victorian school-textbook writers. An Edinburgh graduate, his younger days were spent in the study of Immanuel Kant and foreign languages. After some years as a private tutor in the Lake District, he settled in London as a journalist and writer. In 1869 he published *The English Method of Teaching to Read* in collaboration with Adolph Sonnenschein. His reports as assistant commissioner for the Endowed Schools Committee for Scotland, to which he was appointed in 1874, brought him to the attention of Dr. Andrew Bell's trustees. They endowed a chair of the "Theory, History, and Practice of Education" at St. Andrew's University in 1876 and appointed Meiklejohn to the position (Meiklejohn 1876).

It was in this professorial capacity that Meiklejohn was invited to deliver a paper at the Education Conference on the same day that Armstrong read his paper. A glance at the conference timetable shows that both men read their papers at the same time to different sections. Consequently, Armstrong (despite later statements to the contrary) did not hear Meiklejohn's humorous and stimulating paper on "Professorships and Lectureships on Education." He probably first read a synopsis of the lecture in the *Educational Times* in September 1884, and a full version when the conference proceedings were published at the end of 1884 (Meiklejohn 1884, 103). Armstrong then found Meiklejohn's essay "eminently suggestive" because it reinforced his own conclusions about "pupil-centred" teaching, or "their *finding out*, instead of being merely told about things" (HEA 1898, 390).

Meiklejohn's concern was with the purpose and function of those professionally concerned with the "science" of education—those academics we should now call educationists. His paper was, in fact, a revised version of the inaugural lecture he had given at St. Andrew's in 1876. Educationists, he argued, had to make comparative studies of the growth of young minds; they had to study and teach the history of education; and, above all, they had to study the methods of education.

> What is the permanent and universal condition of all *method*? It is that it is *heuristic*. Man is by nature a seeking, inquiring, and hunting animal; and the passion for hunting is the strongest passion in him. This view has its historic side; and it will be found that the best way, the truest method, that the individual can follow, is the path of research that has been taken and followed by whole races in past times. (Meiklejohn 1884, 103)[1]

Such a natural heuristic method, which he traced back to Edmund Burke and Pestalozzi, was healthier, more real, than the methodological diseases of encyclopedism, secondhand teaching, the tyranny of books or of teaching over the heads of the young, or the didactic infamy of telling instead of teaching. Of course, Meiklejohn's interest was to argue, ironically, that the humanities teacher should appropriate this scientific method.

According to Charles Browne, Armstrong's former Central College colleague, Meiklejohn first used the term "heuristic" in 1860 (C. Browne 1954, 1). He certainly used it in his inaugural lecture as if it were familiar to his audience (Meiklejohn 1876, 30). Earlier usages can be found, though Armstrong claimed that the term had still not reached the dictionary in 1898. It is unclear when he first began to deploy the word, but it was probably in a lecture to the International Conference on Technical Education at the Society of Arts in June 1897, and in an essay commissioned by Frederic Spencer in the same year (HEA 1897c and 1897d; Brock 1973, 99–108). In the essay for Spencer's book, Armstrong made his first extensive attack on the examination system as detrimental to learning (HEA 1897d, 223). Armstrong's special report on the heuristic method for the Board of Education in 1898, however, soon gave the term wide currency. By the time

---

1. The term "heurism" had been used before in an educational context (W. Ross 1858), namely in the sense of setting exercises for homework in mathematics or English grammar.

of Meiklejohn's death in 1902, "heurism" was recognized as the war cry of those who believed all teaching should be by means of carefully directed inquiry, as illustrated by one of Armstrong's Central pupils, William M. Heller, who had become a school inspector for the Irish National Schools (Heller and Ingold 1905).

Armstrong's heurism embraced four theses: First, what a child or student finds out for themselves is best remembered. The second point relates to motivation and interest: if a student is interested and realizes that something is worth learning, they will do it more efficiently. Third, but less clearly, the learning situation must be graded—children will learn certain things better at different ages. Fourth, a carefully written record of their findings will help children to correlate their mental and verbal understandings and to relate a student's scientific awareness to their study of the English language. The whole emphasis was to be on *doing* in order to *understand*. Precisely what was learned was less important, at least in the first instance, than the method involved in learning, thinking, or finding out about something. Armstrong believed, as the history of science suggested to him, that once a person had learned the methods of experimentation, they could continue to use these techniques to "find out," and so tirelessly acquire by their own efforts the mass of information that other teacher-centered systems of instruction forced into children by rote, demonstrations, and the exigencies of examinations. How could he convince the teaching profession?

In 1885, a year after his talk at the International Conference on Education, Armstrong became president of Section B, Chemistry, at the Aberdeen meeting. Over half of his long presidential address was devoted to education, and it was a signal that he intended to use the BAAS as a propaganda machine for his ideas. His theme was an echo of the accusations by successive chemists—Frankland, Perkin, Roscoe, and others—that British chemistry was poor in quality compared with the growing output from German academic and industrial laboratories. The root of the problem, Armstrong argued, was that science teachers had not been trained in the methods of research themselves. "No amount of instruction, such as is ordinarily given in the mere theory and practice of chemical science, will [ever] confer the habits of mind, the acuteness of vision and the resourcefulness required of an efficient chemist in any works, any more than the mere placing of the best tools in a workman's hands will make him a skilful operator" (HEA 1885, 946).

The solution, Armstrong suggested, was to emulate the German example and create an atmosphere of research within a school of higher education. Even this approach would not be sufficient unless students came to such centers better prepared. At the moment, he complained, their mathematical knowledge was unpractical, they could not draw or read foreign languages, and the scientific parts of their minds were "deadened from want of exercise." Compared with evening class students at Finsbury College or female day students, the bulk of students were drained of motivation, manifested by "their marked inability, often amounting to downright refusal, either to take proper notice of what happens during an experiment or to draw any logical conclusion from an observation." Only a wholesale revolution of the curriculum by the introduction "of a *rational system* of practical science teaching into all schools" would revive motivation and morals. Otherwise, it was better that no science at all be taught in schools. Thus, Armstrong thought it unnecessary to introduce more science subjects into the school curriculum. The aim should not be to produce junior specialists but to develop faculties, to encourage the aspirations of the young "by inculcating broad and liberal views of our science, not an infinite number of petty details." Doing so meant abandoning the obsession with qualitative analysis, for example. Chemistry had recently undergone striking internal changes through the knowledge of the periodic law and of structural organic chemistry: such changes provided another impetus for revolutionizing its teaching (HEA 1885, 948).

In the summer of 1886, no doubt with the discussion of a "rational system" of instruction in mind, Armstrong wrote to both John Hall Gladstone and to his friend at the RCS (and former chemistry schoolteacher) William Tilden to suggest that the subject of science teaching in schools be brought before Section F. Tilden replied that he had already tried to persuade William Crookes, president of the Chemical Society that year, to discuss the subject in Section B, but that Crookes would "not bite."[2] Gladstone welcomed Armstrong's suggestion but pointed out that plenty of opportunity for discussion already existed whenever his elementary schools committee reported to the British Association.[3] However, the subject of Gladstone's committee was limited to elementary schools, whereas Armstrong wanted to debate the fate of science, especially chemistry, teaching over the whole range of school education. In 1887, at the BAAS's Manchester meeting, he succeeded in establishing a new committee to consider and report on "the present methods adopted for teaching

---

2. Tilden to Armstrong, 2 June 1886, AP1.499. Two years earlier, Tilden had congratulated Armstrong on his appointment at the Central and advised him to concentrate on his chemistry and to avoid keeping "so many things going at once" (Tilden to Armstrong, 30 April 1884, AP1.498).

3. Gladstone to Armstrong, June 1886, AP1.189.

chemistry." The committee comprised the chemists Armstrong, Raphael Meldola (Armstrong's successor at Finsbury College), Arthur Smithells (Leeds), Gladstone, W. J. Russell, A. G. Vernon Harcourt, Pattison Muir, W. R. Dunstan (a pharmacist), and the chemistry schoolteachers John Thomas Dunn (Gateshead), Francis Jones (Manchester), and William Shenstone (Clifton College).

In December 1887, this committee circularized five hundred headmasters and training college principals, asking them why they taught chemistry, what difficulties they experienced, and what teaching methods they found to be most efficient (HEA 1888b). Difficulties included expense, the low value put upon the subject compared with other subjects in the curriculum, the examinations syllabuses, and the absence of good textbooks and qualifies teachers; but above all, there was general difficulty over the lack of good laboratories. On methodology, there was a surprising unanimity that the subject should be taught experimentally, and a general feeling that there was too great an emphasis on qualitative analysis and manipulation in existing practical courses. The committee itself concluded that something "should be done in the direction of promoting a more uniform and satisfactory treatment" of chemistry teaching. To this end, it promised to suggest new methods of teaching the subject.

Impatient of further prolonged discussion, and brimming with constructive ideas, Armstrong volunteered to produce some suggestions for a course of elementary instruction in science. These suggestions were presented at the Leeds meeting in 1889. His syllabus covered six stages appropriate from the ages of seven to sixteen. The most important aspect of this heuristic syllabus was the way it abolished the traditional catalogue approach to the chemistry of "elements, oxides, and salts" that dated back to the time of Lavoisier, and replaced it with real problems such as "why does iron rust?" Gone, too, was any emphasis on qualitative analysis (HEA 1889; 1903b, 300).

The basis of his recommendations was the 1884 Finsbury College syllabus, developed into a "more consistently logical form" as a result of practical trials at Finsbury and the Central College, as well as feedback from successful object lessons in London Board schools. In stage 1 (seven to ten-year-olds), object lessons using real objects that were part of a normal child's experience formed the foundation for later studies and allowed the introduction of geography and natural history.[4] This stage was clearly indebted to the "physiography" that Huxley had recommended. Once

---

4. Armstrong made no reference to age groups. I have added ages so that readers can compare these recommendations with twenty-first-century curricula.

children had gained proficiency in elementary arithmetic, they were ready for stage 2 (ages ten to twelve), which taught "lessons in measurement," including the use of the balance, the determination of relative densities of common materials, and the measurement of temperature. This stage then led on to elementary physics and meteorology. Stage 2 was clearly influenced by the teaching experiences of Arthur Worthington at Clifton College in Bristol (Worthington 1881 and 1886).

At stage 3 (ages twelve to thirteen), the physical and chemical effects of heat on common substances (including foodstuffs) were studied, accompanied by the inculcation of note-taking. The open-ended conclusions of this stage were supposed to be that the heat changes involved more than mere destructive affects. The problem of chemical change, which was entered into more detail in stage 4 (ages thirteen to fourteen), swept away the traditional catalogue approach. Armstrong recommended instead a concentric approach that began with the rusting of iron, leading logically (by the deductive method) to the analysis of air, to the phenomena of combustion and oxidation, and to the application of the concept of "analysis" to substances like chalk; to the discovery of carbon dioxide, sulfur dioxide and hydrogen; and, finally, to the concepts of acidity, basicity, and the element (Eyre 1958, 270).

By this point, pupils would possess a fair knowledge of the basic substances of chemistry and their properties, and of the idea of composition. Most students would stop at this point. Those who continued on to stage 5 (ages fourteen to fifteen) could proceed to a more sophisticated quantitative study of substances, partly by lecture demonstrations (a point often ignored by extreme heurists and by heurism's critics) and partly by individual experimentation, until the basic laws of chemical composition and the technique of volumetric analysis were achieved. It is noticeable that Armstrong did not introduce the atomic theory, or chemical symbols and equations, until the sixth and final stage (ages fifteen to sixteen), which made a closer investigation of the physical properties of the three phases of matter. This attitude toward theory was fully in line with nineteenth-century attitudes toward the concept of atomism.

Although Armstrong recommended that most of the experimental investigations should be made by the pupils themselves, he allowed that the course would still be valuable even if given entirely by demonstration. The complete course would be time consuming, he admitted, as well as incompatible with existing methods of assessment. In a further report at the Leeds meeting of the British Association in 1890, he discussed such difficulties (HEA 1890c, 299; HEA 1903b, 345). The criticism of time consumption would disappear if the fallacy of thinking that "sufficient training in a scientific subject [could] be imparted in the course of a term or two" were overcome, and if it were realized that the course was not meant to be entirely practical work. The laboratory did not need to be elaborate: a simple table for the balances; a waxed table or bench fitted with gas taps; a fume hood; a sink; and a muffle furnace. The most expensive items of equipment needed were a set of good balances and weights. Six of these, sufficient for a large class, would cost about £18. He thought that the total cost of apparatus for several classes of twenty-four pupils would not be greater than £50, provided that porcelain vessels and not platinum ones were used.

Armstrong's syllabus had attracted widespread attention by 1891 when he lectured the College of Preceptors on the "teaching of scientific method" (HEA 1891a; 1903b, 367). Following its approval by the Headmasters' Association (founded in 1890), it began to receive favorable attention as far away as Germany and Japan (Terakawa and Brock 1978). By then, he had promoted heurism in two practical ways. In 1888, a Cambridge- and German-trained chemist named Charles Maddock Stuart (1857–1932) was appointed the first headmaster of St. Dunstan's College, a new school at Catford, close to Armstrong's home at Lewisham (HEA 1933c and 1933k). Given the extraordinary appointment of a scientist headmaster at Catford, his interest in local affairs, and the fact that one of the new school governors was an honorary secretary of the City and Guilds Institute, it is not surprising that Armstrong should have quickly made himself known to Stuart and "influenced him from the beginning." Stuart soon became a "staunch, practising advocate of heuristic tenets" (Eyre 1958, 113). Inevitably, Armstrong became one of the school's governors. According to Armstrong's son, Frank, a pupil at the school, Stuart inspired over a dozen boys to take up chemical careers. In due course, Frank succeeded his father as a governor of the school in 1937.[5]

The Catford school offered a way to overcome accusations of idealism, impracticability, or personal unfamiliarity with school teaching. Together Armstrong and Stuart, assisted practically by the former's Central student William M. Heller, put the heuristic system into operation at St. Dunstan's and gained first-hand experience of the difficulties. The method was soon extended to elementary mathematics and geologically focused geography lessons, and eventually to manual work with wood and metal. Among Stuart's staff was Thomas S. Usherwood, a mechanical engineering graduate from the Central, who was placed in charge of manual instruction at the school. Armstrong was so impressed by his instruction in woodwork,

---

5. William Davison (Bursar) to Frank Armstrong, 25 October 1937, AP2.517.

carpentry, and metallurgy that he persuaded Usherwood to move to Christ's Hospital public school when it opened its Horsham site in 1902. There he taught mathematics as well as practical engineering and manual skills, along the lines that Sanderson had established at Oundle. Armstrong later described Usherwood as "a man of remarkably broad outlook and culture" (HEA 1936n, 477; Usherwood and Trimble 1913). Usherwood's daughter, Hilda, became a distinguished organic chemist but was better known for her marriage to Christopher Ingold in the 1920s (Brock 2011b).

The success of this method of teaching in a school environment encouraged Armstrong to enlarge the scope of his campaign in a second way, through the training of teachers. Already in 1887, he had lectured on chemistry to teachers in government-funded elementary schools. In 1896 he was given permission to hold Saturday morning classes for London science teachers at the Central College. Nearly two hundred teachers attended the first two sessions, and the course was repeated over the next three years (Eyre 1958, 135). Among those who attended these sessions were several of Armstrong's Central students who had become teachers: Grace Heath, a science mistress at the North London Collegiate School (M. Rayner-Canham and G. Rayner-Canham 2017, 46, 65); L. Edna Walter, who eventually became a government schools examiner; Martha Whiteley, a chemistry graduate from Royal Holloway College who was then teaching science at Wimbledon High School for Girls, and who subsequently became a lecturer at Imperial College; and an Oxford graduate named Hugh Gordon.

Gordon had begun organic research with Armstrong in 1887 but soon found his teacher's educational activities more interesting. In 1891, after he had held informal classes for teachers in the Surrey Council schools, Gordon was appointed a peripatetic science demonstrator for the London School Board. Encouraged by Armstrong, who longed to carry "the war into the camp of elementary education" (HEA 1894a, 632), and by John Hall Gladstone, Gordon began to hold training sessions in heuristic methods in a deserted rice mill at Whitechapel. These meetings, which were continued by Gordon and by his successor William Heller in 1894, were well publicized. Hence Armstrong's introductory talk to the teachers of Tower Hamlets and Hackney, which was given at the Berners Street Board School in October 1894, reached the pages of *Nature* (HEA 1894a). Unfortunately, when Heller left his peripatetic employment in 1897, three years after Gladstone's retirement from the London School Board, the board failed to continue the system. On the other hand, the methods learned were transplanted to Ireland on the recommendation of the Irish Commission on Manual and Practical Instruction in Primary Schools. The subject "General Elementary Science" became a compulsory part of Irish education, with Heller as its chief organizer.

Characteristically, Armstrong felt that Gordon's greatest achievement as a teacher was not so much his "marvellous success, [as] his introduction of a proper [chemical] balance" into schools, compared with the poor "four-shilling" affair recommended by Worthington (HEA 1894a, 632). In 1893, Gordon published an *Elementary Course of Practical Science* in the excellent Macmillan Science Primer series edited by Henry Roscoe. This volume was the outcome of the syllabus he had introduced and nurtured in fifteen London Board schools between 1891 and 1893, but it had been based originally on Armstrong's British Association syllabus. Gordon's syllabus was accepted by the Education Department from 1893, while the rival DSA agreed that preparation for their examinations should favor experimental work. Gordon's text was the first of many to adopt this heuristic system. Gordon claimed that the method of teaching "should be that of suggestion combined with a minimum of demonstration—that of asking questions, not of answering them" (Gordon 1893, xv). The teacher should give lessons of forty minutes at least twice a week, and the pupils, say six at a time, should be required to perform experiments in front of the other pupils. Gordon's book was designed to be read by the pupils themselves and skillfully interleaved relevant items from newspapers on weight standards or meteorology in order to show the relevance of the experimental work.

In 1897, when Heller became headmaster of the Municipal Technical School in Birmingham (today's Aston University), over forty schools in Tower Hamlets and Hackney had adopted the Armstrong-Gordon-Heller elementary science course (HEA 1898). Armstrong was furious when, using Heller's resignation as an economy, the London School Board abandoned his practical work in its schools (HEA 1897e). For, by then, the cause of heurism had been widened nationally, through the interest of the Incorporated Association of Headmasters, which had appointed a committee on science teaching to consider the formulation of a syllabus that would be suitable for both boys and girls. Needless to say, since Armstrong was a member of this committee, the headmasters' syllabus, which was adopted and published in 1896, was closely modelled on the British Association's course of practical work (HEA 1895a).

The same committee consulted the Oxford and Cambridge Local Examination Boards in 1895, with the effect that the syllabus was approved for the award of scholarships and exhibitions offered by County Councils. (Whether the examination boards had understood the point of heurism was another matter, thought Armstrong.) Of all the opportunities presented to Armstrong for publicizing the heuristic method, probably most important was a long essay he was asked to contribute to the Board of Education's widely read *Special Reports* in 1898 (HEA 1898, 389).

Family Experiments

Armstrong also began to practice what he preached with his own children. He and Louisa had seven children who survived to adulthood. By 1898, Frank, his eldest son, had passed through St. Dunstan's College and the Central College and was about to study chemistry in Germany; Clifford, who was also educated by Stuart at St. Dunstan's, was set to become an engineer; and two daughters, Edith and Annie, had received a conventional education from a governess before being sent to Blackheath High School for Girls, which taught science heuristically. They were destined for marriage and family life. This left the three youngest children: Robin, who had just begun school at St. Dunstan's; Nora; and Harold. They were at this time aged twelve, ten, and eight respectively, and they were conveniently sound materials for a part-time educational experiment on how to develop age-graded elementary-science concepts in young children. Armstrong described his adventures with them with showmanship and pride at a conference of science teachers in 1900.

For Christmas in 1897, he had given the youngest child (Harold) a storybook entitled *The Monkey That Would Not Kill*. In the tale, a mischievous monkey had survived drowning, despite being attached to a heavy stone, because the author claimed the stone weighed less in water. To test the story's credibility Armstrong encouraged the three children to devise simple tests with home apparatus, and to keep independent records of their investigation and results. Much of the charm of the children's notebooks lies in seeing how each child tackled the problem raised in a different way and how, in deducing Archimedes's principle, they also formed ideas about density, averages, and percentages, besides realizing the importance of cleanliness and accuracy in measurement (HEA 1903b, 394; C. Browne 1954, 14; Eyre 1958, 135, 275).[6]

These experiments with his youngest children were actually an extension of observational and recording activities that he had practiced on

---

6. Four of the original six notebooks are in AP2.E.1–2 (Harold), AP2.E.3 (Robin), and AP2.E.4–5 (Nora). Eyre rightly thought (1958, 136) that if they had been published, they would have become reference works "illustrating what can be done willingly [and] enthusiastically by judiciously prompting and guiding the young inquiring mind."

family holidays with his older children, Frank, Clifford, Edith, and Annie. As Frank recalled, their first holidays were spent at Margate:

> We were taught at an early age to walk: to study the changes in the soft chalk, in that district very rapid; to look for fossils; to visit churches and to learn the history of the Roman occupation and of the Cinque Ports . . . A visit of the Geological Association took us to Swanage one Easter (1895) and there and then the cottage on the pier was secured for the summer and for many future holidays. Here the family was trained in nature study—to collect, to observe, to record as part of a normal enjoyable holiday. The transference of affection to the Lake District came later . . . Here [my father] had his happiest days and he returned to Manesty Farm near Grange,[7] with its unique views . . . as often as possible each year for many years. (E. F. Armstrong 1941, 375)

The later weekend sessions of the younger children with their father were occasionally witnessed by Charles E. Browne (1865–1961), a chemistry student and demonstrator of Armstrong's who had left the Central College in 1898 to teach science at the Robert Gordon College in Aberdeen. He returned to London in 1899 at the personal invitation of Armstrong to become science master at Christ's Hospital school. Before the move to Horsham, Browne taught chemistry in a disused dormitory of the school, and it was there that he adapted Armstrong's heuristic ideas to the needs of school classes of up to fifty pupils.[8] Practical work by pupils became the accepted basis of the junior-school science (chemistry and physics). Pupils worked in pairs on a series of related problems; the results were tabulated and discussed by pupils and teacher and recorded in individual notebooks. By 1907, the scheme had worked so well that Armstrong was able to turn his attention to the provision of a manual workshop for metal and woodwork.

In 1908, Christ's Hospital's heuristic methods had to be drastically curtailed when the Board of Education forced the school to adopt the School Certificate examination system as a criterion of its proficiency. Browne compromised by introducing periods of heuristic activity to fertilize the rest of the scientific teaching that was governed by the school certificate's

---

7. Manesty Farm lies at the foot of Cat Bells and Maiden Moor, by Derwentwater; it is now a suite of holiday cottages.

8. Browne left Christ's Hospital in 1926 to become a science tutor at the London Day Training College. There, he influenced several teachers in the 1930s to continue the heuristic method (C. Browne 1962; Rodd 1968).

syllabuses. These methods were continued at the school after Browne's departure by John Bradley and Gordon Van Praagh. American project methods and the Christ's Hospital experience were, together, the indirect antecedents of the Nuffield Foundation's science teaching projects of the 1950s (Van Praagh 1949; Coulson 1970; Waring 1979).

Triumphs and Attainments

During the 1890s, the institutionalization of heurism was aided in an unexpected way. Fears of economic depression and international trade competition were continually expressed by scientists (including Armstrong) and concerned industrialists. Following the Samuelson Commission on Technical Education, which heard evidence between 1882 and 1884, the chemist Henry Roscoe, together with Thomas Huxley, founded the National Association for the Promotion of Technical Education. Its purpose was to lobby industrialists and Parliament to provide local authorities taxing powers to create facilities for technical education. The passage of the Local Government Act in 1888 made this aim possible, and in 1889 the new local councils were empowered to levy a penny rate for technical education. The curricula of local technical colleges were to be controlled by the DSA. Fortuitously, these local powers coincided with the temperance movement's successful appeal to tax spirits. In 1890, a Customs and Excise Tax was allotted to local authorities, who began to use it to build laboratories and lecture rooms for science and technical education in the knowledge that instruction would be supported by grants from the DSA (Sharpe 1971). In this way, Armstrong's heuristic campaign was complemented financially and architecturally by the new buildings and laboratories of the "secondary" schools and technical colleges. Practical-science teaching on a wide scale became possible.[9]

On the other hand, when he investigated the kind of technical training that the London County Council was supporting in 1894, he was sharply critical. In London, the Technical Education Board had set up evening classes in chemistry at Battersea Polytechnic. He had attended one of the classes on the conservation of matter and the laws of chemical combination and found that the lecture lacked any practical basis. It was all

---

9. Armstrong reviewed the Royal Commission on Secondary Education [Bryce Report] in 1895, deploring the fact that the only scientist on the commission was Sir Henry Roscoe and that nobody from the scientific community had been interviewed (HEA 1895c).

cookery-book stuff and did nothing to train the workman (HEA 1894b). The adventurous heuristic teaching he had helped establish was not being extended into other educational areas. It was from this time onwards that Armstrong began to use the *Times* newspaper regularly to broadcast his opinions on educational matters, usually writing from the Athenaeum rather than from the Central College.

Thus, by 1902, when Armstrong paused to review the situation before the newly founded Education Section of the British Association, he could survey with pride the success of the heuristic campaign: a corps of heuristic-minded teachers; St. Dunstan's and Christ's Hospital and their demonstration that conviction and skill solved the difficulty of large classes; a record 1,165 school laboratories throughout Great Britain (Abney 1903, 875); domestic science taught in girls' schools; and a new enthusiasm for practical technical education. The *Journal of Education* hailed him for acquiring "the dignity of a classic" within living memory, while *School World* noted that "the new subject 'Armstrong' had displaced the older fashioned chemistry" (Fry 1902). Not that Armstrong was complacent in his twenty-five-page-long address to the British Association (HEA 1902d). He castigated the English lack of imagination, the dreadful state of military education, the awful examination system, and the dangers posed by textbooks (HEA 1900a; 1900b; 1902c; 1903d; 1904b). Indeed, complacency would have been misplaced; by 1918 it could be said that "in many schools more time was spent in laboratory work than the results can justify" (Thomson 1918).

In 1903, Armstrong collected his papers on heurism together in a book for Macmillan, *The Teaching of Scientific Method and Other Papers* (HEA 1903). As Arthur Smithells pointed out in an otherwise enthusiastic review, Armstrong would have done better to systematize and coordinate his message:

> The arrangement is probably as good as it could be, provided that nothing were feasible but the mere reprinting of twenty-three occasional addresses, but it is impossible not to suppose that the constant reiteration of doctrine, and the continual reappearance of almost the same words, will deter the reader who sits down to read the book solidly through... But Prof. Armstrong has not chosen the persuasive method of Matthew Arnold. He is vigorous almost to violence, red-hot, scathing, scornful, uncompromising and incessant. He is no respecter of persons or institutions, however eminent, however ancient. He is absolutely impartial in his iconoclasm. (Smithells 1904, 289)

The book was reprinted unchanged in 1910 with the addition of three further educational papers. Armstrong issued a second edition of the book in 1925. Reiteration of his beliefs was to remain a consistent fault of his writings during his long retirement.

In that second edition, Armstrong added a preface that reflected on his long campaign, which had begun as a Fifth Business catalyst for the expansion of science teaching (HEA 1925s). It had been a period of strife, flux, and evolution in education; however, as Laurie Magnus (the journalist son of Philip Magnus) said in a book of collected essays on education, Armstrong's arguments (HEA 1901c) had been "lucid, accurate and plausible."

> If British education is to be saved from the consequences of its own errors and the mistakes inherent in its nature and constitution, it must travel on the road of salvation to which Dr. Armstrong directs it. (L. Magnus 1901, 21)

Early Opposition and Criticism

In general terms, criticism of Armstrong's heuristic methods prior to 1900 was based on predictions of its impracticability, often based upon a superficial knowledge of Armstrong's detailed work on a suitable syllabus. Later criticism (discussed in detail in chapter 12), based more on hindsight, was more objective and constructive.

Until 1884, Armstrong was known only as a chemist. It was by no means unusual for a chemist, or other scientists, to take an interest in education—the activities of John Hall Gladstone and Henry Roscoe are ready examples. But since Armstrong was at the peak of his research career between 1880 and 1900, inevitably some of his friends and colleagues, as well as his family, felt that he gave up too much time to the cause of improving science teaching, at the expense of his chemical research. However, his research record shows no evidence that his science was neglected. In fact, he published well over 150 papers in this period, on such subjects as the nature of chemical change, a theory of residual affinity, and naphthalene chemistry, as well as powerful criticisms of the new ionic theory. To be sure, many of these papers were collaborative efforts, but he was no absentee collaborator.

Yet sheer output may mask the real point of this advice, for both friends and family were perhaps thinking more of the quality than quantity of Armstrong's output, and especially of his trenchant criticism of the ionic theory, which he began in the same decade as he first advanced his ideas on heurism (HEA 1927s, xix). His Kolbe-like invective made him many enemies, especially amongst the young school of physical chemists whose intellectual leader in England was William Ramsay.

As we saw in a previous chapter, Armstrong and Ramsay clashed over the existence of the noble gases in the late 1890s, and Armstrong lost this battle. They also disagreed strongly on the question of women in science. Not surprisingly, Ramsay did not advocate Armstrong's heuristic notions. In three lectures to the College of Preceptors in March 1891, he analyzed Armstrong's British Association syllabus in detail. Its lessons on measurement (stage 2) were, he said, devoid of interest unless given by gifted teachers, and the *information* to be garnered from this part of the course was desultory. The quantitative stage (5) was

> unsuited to the capacity of an ordinary class [of students] . . . , the conclusions to be drawn are those which it has taken men of genius 150 years to deduce, and it is not to be expected that an average schoolboy should make out such deductions for himself. Either he must be told, or he will read ahead. . . . Children have small powers of reasoning; and it is inadvisable to set them problems in advance of their powers. Again, the difficulty of securing teachers competent to put such a scheme in practice would be unsurmountable. (Ramsay 1891)

Their different view was aptly summarized by Ramsay: "I believe in facts and I believe that the basis of theory is not grasped until the facts on which they rest have been learned" (Ramsay 1891). In addition, Ramsay quoted the critical opinions of several schoolmasters, including William Shenstone and Tudor Cundall at Clifton College and Francis Jones at Manchester Grammar School.

To Armstrong, Ramsay's viewpoint seemed reactionary and imbued with a perverse inability to understand the importance of training by scientific method. Yet it remains true that many of Armstrong's heuristic disciples did overemphasize method as opposed to content—rather as the contemporary nutritionists failed to discover vitamins because they emphasized the energy value of foodstuffs rather than the qualities of fresh foods themselves.

Entrenched in the arguments of both the older instrumental education of the 1840s and those of the new "practical" education of the 1870s was "faculty psychology." For Armstrong, teaching the physical sciences would "develop a side of the human intellect which . . . is left uncultivated even after the most careful mathematical and literary training: the faculty of observing and reasoning from observation and experiment" (HEA 1884, 69; Brock 1973, 74). Faculty psychology—the idea that the mind could be analyzed in terms of various powers or faculties—originated in Greek philosophy. It seemed to gain powerful empirical support by the clinical observations of Franz Gall and Johann Spurzheim at the beginning of the nineteenth century, observations whose social and educational implications were developed by George Combe in the phrenological movement. Although the complex multiplications of entities by the British phrenologists and phrenology's debasement by mesmerism and occultism led to its discreditation, the phrenologists' belief that full development of all the faculties—of the imagination, judgement, reasoning, perception, memory, and a host of others—depended upon a person's environment and training continued to be a useful for educationists to justify a particular curriculum.

By formally training one faculty, or a part of it, it was claimed that the whole mind was improved for future use in any situation. Although the theoretical justification of instrumental educationists' emphasis on "the 3Rs" was faculty psychology, this rationalization was largely an afterthought. The need to teach reading, writing, and arithmetic was self-evident, without deduction from theoretical faculties (Selleck 1968). Yet so firmly was faculty psychology entrenched amongst mathematicians and classicists in the public schools "that a substantial group of new educationists, concerned to introduce extensive reforms in both the methods and the curriculum, were forced to argue their case on the basis of faculty psychology" (Selleck 1968, 45). The unquestioned assumption made by reformers like Armstrong (on top of the unquestioned acceptance of faculties and their trainability) was that training could be transferred. Heuristic methods trained the faculties "of thoughtfulness and power of seeing; accuracy of thought, of word and of deed" (HEA 1903b, 190). Outside school, these attuned faculties would respond not merely to special scientific stimuli but to the general stimuli of everyday life and work.

Such a theoretical justification for heurism was fated to be overtaken by the development of Herbartianism on the one hand and of experimental psychology on the other. To both, Armstrong first turned a blind eye. Regarding the former, since those who followed the educational philosophy of Johann Friedrich Herbart (1776–1841) emphasized the role of the teacher in molding the child, and since they stressed the significance of books and literary studies, a conflict with practical educationists like Armstrong was inevitable. For the last forty years of his life, Armstrong would thunder against the revival of the literary cult.

Heurism in America

Meanwhile, in the United States, the American progressive educational program associated with the names of Colonel Francis W. Parker (1837–

1902) and John Dewey (1859–1952) were moving in parallel directions. For although neither Parker nor Dewey were chemists (Parker was a secondary school teacher and Dewey an academic philosopher), their educational philosophies were strikingly similar to Armstrong's: in their attack on a morphological curriculum whose primary concern was "classification, in naming parts and describing forms, rather than in developing an understanding of function and functioning" (Rugg 1927, 22); in their attack on stereotyped recitation and grading methods; in their Herbartian emphasis on teaching for interest and relevance to a student's mental stage of development; in their emphasis on a socially relevant education for an industrial society; and in their claims for manual training, for experience, for learning by doing.

Indeed, in Dewey's laboratory school at the University of Chicago between 1894 and 1904 and in his book *The School and Society* (1900), there is much reminiscent of Armstrong's experience in England. Children's work with textiles led on to investigations of a scientific and historical nature, while cookery was "the natural avenue of approach to simple but fundamental chemical facts and principles and to a study of plants as articles of food" (Garforth 1966, 66).

Curiously, neither Dewey nor Armstrong ever cited one another. Armstrong, unlike four other members of the Mosely Commission, did not visit Chicago in 1903, but since he edited its *Report*, he must have been aware of their favorable account of Dewey's laboratory school (HEA 1904b, 123, 203, 279). However, as his diary of the tour reveals, Armstrong was annoyed by the way Mosely had organized the trip, and he was convinced before he left England that there was nothing to be learned from the Americans about education. The "Uesanians" (that is, U-S-A-nians), as he dubbed Americans, had made mistakes, like allowing a preponderance of female teachers, that the British should avoid at all times. He praised the standard of educational buildings, the organization and equipment of the American schools, and the "absolute belief in the value of education both to the community at large, and to agriculture, commerce, manufactures and the service of the State," but he dismissed as worthless all American teaching methods except for those in medicine and manual training (Lupton 1964). There was little evidence, he asserted, that John Perry's "practical methods of teaching mathematics and geometry," which he had pioneered at Finsbury College and used later at the Central College, and which were coming into vogue in England, were appreciated. The old academic methods of teaching science seemed to prevail almost exclusively.

His biggest criticism, as we saw in the last chapter, was of the feminization of American youth through coeducation and the prevalence of female teachers (HEA 1908a). Armstrong regarded the independence of women that had developed in his lifetime as superficial and unreal. Nevertheless, Armstrong did all he could to promote the teaching of domestic science to girls, as a subject which bore directly on their future occupation. Women deserved to be taught methods of accurate thought for the successful management of the household. As we saw, such views were shared by all members of the Mosely Commission with the notable exception of Armstrong's Central colleague, William Ayrton.

Armstrong was wrong both in claiming that America was backward and in denying that Britain had anything to learn from America, though it is fair to say that Dewey had little impact on English education, which went its own independent way until the 1920s (Selleck 1972, 113). With isolated exceptions, like Helen Parkhurst's Dewey-inspired "Dalton Plan" of monthly project work to be completed at the pupil's individual pace—which was enthusiastically adopted in England during the 1920s, arguably because Armstrong's heuristic schemes in chemistry had made teachers more receptive—English and American curricula went their independent ways. They are examples of parallel, not interconnected, curriculum innovations (Macdonald and R. Walker 1976). The *"virus heuristicum"* was operating on both sides of the Atlantic simultaneously, attacking the often administratively convenient formalism of earlier science education (Selleck 1972; Cremin 1962).

Assessing Armstrong's Heurism

For Frankland, Huxley, and Armstrong, what was called a liberal education could only produce a truly cultured person if science were included in the curriculum, for other subjects left several faculties dormant and untrained. However, given the growing specialization of the British educational system (promoted by the obsession with examinations), it was easier to justify science teaching economically in terms of vocations and the industrial importance of science—especially as a result of German and American competition and the First World War. This more utilitarian view of science, particularly of chemistry teaching, was reinforced by the Herbartian lobby, which not only undermined faculty psychology but justified science teaching by the argument that the curriculum ought to be a simplified microcosm of the community's larger interest: economic survival and prosperity. This tension between the need to train a scientific manpower and the aim of making science part of everyone's culture was to be the hallmark of twentieth-century rhetoric concerning science teaching.

In his emphasis on experience, Armstrong was more Pestalozzian than Herbartian. He distrusted the English Herbartians for their emphasis on literary culture, since this emphasis threatened the place of science in the curriculum. In 1904, Armstrong himself was directly challenged. His critic was Frank Hayward, a Herbartian school inspector, who felt that heurism was a poor method because he regarded a child's greatest need "not that his mind should be exercised, but that his mind should be fed with a rich repast of historical and biographical ideas" (Hayward 1904, 76). In 1908, Hayward returned to the downfall of the "formal or faculty training" in the widely read journal *The School World* (Hayward 1908), and although he was really attacking Armstrong's disciples for their "purblind" advocacy of training for accuracy and observation, Armstrong saw this critique as a direct confrontation. In his reply to Hayward, he poured scorn on the pretensions of that attack. The idea that mental faculties could be trained was not yet discredited in his eyes; it was still a working hypothesis (HEA 1908c, 477–78).

At the Dublin meeting of the British Association the same year (1908), Armstrong was positively optimistic that a more practical and less literary outlook was prevailing in British schools, and he was particularly excited by the curricula of the new naval schools at Osborne and Dartmouth (HEA 1908f). Armstrong did not return to the subject again, although for the remainder of his life he reserved the more vitriolic portions of his extensive vocabulary for psychologists. Formal training continued to be used in the practical arguments of educationists and teachers and, in the end, during the last decades of Armstrong's life, psychologists grudgingly admitted that the doctrine of formal training held an element of truth that traditionally had been presented in a psychologically invalid manner (Burt 1938). A person's capacities were improved by formal training, but the transfer of capacities only occurred in a limited fashion between related fields. The admission was to help the English revival of heurism in the 1930s.

Criticism of Armstrong's heuristic method of learning in the late Victorian and Edwardian periods came both from chemists like William Ramsay and educationists like Frank Hayward, who, inspired by the writings of Herbart in Germany, believed that the education of children primarily revolved around learning moral principles best inculcated through literary and philosophical studies. "Faculty psychology" appealed still to both chemists and Herbartians, though for Ramsay scientific method involved learning facts as well as experimentation that confirmed facts and theories, rather than acquiring information by personal discovery.

Despite such contemporary criticism, Armstrong's writings, along with his persuasive powers promoting the creation of schools that embraced the heuristic principle, helped to transform the teaching of chemistry and of science in general, while he was a full-time professor at the Central College. His writings and arguments provoked both thought and action. This was also to be the case with Armstrong's decades-long opposition to the new physical chemistry that was replacing his beloved organic chemistry as the foundation of chemical studies, as we will discuss in the next chapter. By the time he retired from the Central, Armstrong had become equally well known for his opposition to the ionic theory that Arrhenius had introduced in the 1880s as he was for promoting the learning of the scientific method through laboratory and workshop studies.

✸ 9 ✸
Ionomania

> Let this be my considered message at the end
> of seventy years of constant study.
> —HEA 1936k

Compared to his important contributions as an educationist, Armstrong may be more remembered today for his lifelong rejection of the pathbreaking ionic theory introduced by Swedish chemist Svante Arrhenius in his doctoral thesis in 1884, which transformed our understanding of the nature of substances in solution. Until the 1880s, chemists had not seen the concept of solutions as in any way complicated. A solute simply dissolved or dispersed in a solvent. The discovery of electrolysis at the beginning of the century had not materially altered the picture. Michael Faraday had suggested that when a solution was integrated into an electrical circuit whose electromotive force was exerted by a battery, molecules of the solute took up separate electric charges under the influence of the positive ("anode") and the negative ("cathode") poles, to create positive and negative "ions" (either "anions" or "cations") that were slowly attracted toward the respective pole of opposite electrical charge. But there was no thought that the solute ions were already positively or negatively charged *before* the battery was connected.

A Theory of Chemical Change

The fundamental intellectual problem for chemists since time immemorial has been why chemical change occurs. Why do purified substances whose natures are known sometimes react together to produce new products and sometimes remain inert and refuse to change? A central theme of all of Armstrong's scientific work after 1885 was a hypothesis that he hoped would explain chemical reactivity or inactivity; his intent was to develop this idea into a successful general theory of chemical change. In 1881, Armstrong was present at the Royal Institution when German scientific polymath Hermann Helmholtz delivered the Faraday Lecture to the Chemical Society. In this famous lecture, Helmholtz suggested that most chemical compounds were held together by unit charges of electricity, as Faraday himself had speculated when introducing empirical laws of electrolysis in the 1830s.

Armstrong, who had studied Faraday's collected works as a student at the Royal College of Chemistry, gave much thought to Helmholtz's lecture and, in his presidential lecture to the Chemical Section of the British Association meeting in Aberdeen in 1885, he speculated about the nature of chemical change. A few months before the meeting, Armstrong had read a suggestion of Spencer Pickering (1858–1920) that so-called molecular compounds could be explained as being due to partial values of valence that were expressed by certain elements. The expression of such "residual valencies," Pickering claimed, would account for the existence of water of crystallization in salts and for the existence of the polyhydrates in solutions that were the central feature of Pickering's investigations. Pickering had also suggested that the relative inertness of some elements might be caused by their inability to generate such residual affinities. His ideas would explain the relative inactivity of many organic compounds (Pickering 1885). In addition to Faraday, Helmholtz, and Pickering, a fourth influence on Armstrong's Aberdeen address was the work on combustion by Harold B. Dixon (1852–1930), a student of pioneer kineticist Augustus Vernon Harcourt. Dixon had investigated explosive combustion of gases and found that oxidation proved impossible under very dry conditions; reactions occurred only in the presence of traces of water (Dixon 1884). More broadly, combustion only took place in the presence of a third body, either water or, more generally, a catalytic impurity. This catalytic agent was necessarily decomposable into a positive and negative element—in the case of water, into hydrogen and oxygen.

In Armstrong's Aberdeen speech, he suggested that at high temperatures (the equivalent of high electromotive force), water played the role of an electrolyte that took the form of a molecular complex with the other reactants, which then resolved into the final products. This association was a form of "reverse electrolysis," as he explained to the Royal Society a year later.

> The presence of water may be necessary, not because it is essential to have an electrolyte present, but because the occurrence of both molecular interaction and electrolytic conduction depends on identical

molecular and intermolecular conditions. The chemical interaction takes place entirely independently of the water molecules, and these latter serve only to separate and keep apart the fundamental molecules of which the interacting bodies are composed. (HEA 1886b)

Armstrong was suggesting that all chemical reactions may be electrolytic in character and follow the same conditions as that in a battery, namely, by the cooperation of three molecules, at least one of which had to be an electrolyte. In the case of the combustion of hydrogen and oxygen, the two molecules acted as "depolarizing" electrodes, with a trace of impure water acting as the electrolyte (HEA 1893b). The fact that pure water was not an electrolyte made it necessary to postulate that it had to be impure, effectively acting as a catalyst.

He conceded that further studies of a large range of gaseous reactions would be necessary to give credence to his explanation or to refute it. In fact, Armstrong was to repeat elaborated forms of his theory of chemical change over the next forty years (HEA 1893a; 1895d). With the exceptions of silicate chemist Joseph Mellor and Armstrong's former student Martin Lowry (Mellor 1904, 274; Lowry 1915, 534), each of whom described the theory in detail, most chemists ignored it. The physical chemists William Ramsay and James Walker offered a critique (Ramsay and J. Walker 1893) to which Armstrong made no comment. The most important effect of the theory on its author was that Armstrong's belief that chemical change occurred through the association of reactants led him to discredit the dissociation theory that Arrhenius had first proposed in 1884. He held on to this "associationist" view into his eighties.

The work of William A. Bone in the 1920s cast doubt on Herbert Baker's putative proof that reactions needed the presence of a catalytic "seed" like water to make a reaction go. For example, Bone showed that carbon monoxide did not need the presence of water vapor to combust. The trigger, Bone showed, was actually a "minimum spark energy" necessary to bring about explosion. Even gas samples dried in phosphorus pentoxide for six months exploded when a sufficiently high charge was applied (Bone and Townend 1927a). This finding heralded the end of Armstrong's idea that water (or some other solvent) had to be present to catalyze a reaction; it also heralded the study of the kinetics of reactions and the differentiation between homogeneous and heterogeneous reactions.

It emerged that it was not the desiccation of the reactants that was significant but the nature of the surfaces on which the reactions were carried out. In 1923, Ronald Norrish showed that the rate at which ethylene and bromine combined was directly influenced by the particular surface of the reaction vessel. If the glass was coated in paraffin wax, the reactions barely began, but when uncoated glass was used the reaction readily proceeded (Norrish 1923; Berry 1946, 150). Norrish made no reference to Armstrong. Although Armstrong lived to witness the refutation of his hypothesis, he also made no comment to Norrish. But he could take comfort in the fact that Norrish had agreed that a catalyst was required to "activate" the reaction and that it did so by somehow influencing the distribution of polar forces within the molecule—an idea that Lowry (who had suggested the work to Norrish) had developed from Armstrong and under the stimulus of Arrhenius's idea of an activation energy.

The Rise of Physical Chemistry

In 1877, having become a lecturer at the University of Amsterdam, and having made a name for himself in the field of stereochemistry, Jacobus Henricus (Henry) van 't Hoff began to study the affinity forces that held solutions together.[1] One of his Dutch colleagues was the botanist Hugo de Vries, who was investigating the phenomenon of osmosis, whereby the amount of water in a plant cell continuously varied by the ability of the water to flow in and out of a cell's membrane. The pressure that needed to be applied to prevent water moving through a membrane was the "osmotic pressure." The plant's protoplasmic cell walls only allowed water to pass into the cell and were impermeable to any salts outside. De Vries had already found that the semipermeable cell membrane functioned as a balance that allowed a plant to maintain an equilibrium between the concentration of nutrients inside and outside the cell. He referred to the concentrations of the solutions inside and outside the cell as "isotonic" whenever they were at the same osmotic pressure.

From de Vries, van 't Hoff learned of recent quantitative studies by a German botanist, Wilhelm Pfeffer, who had used an artificial membrane of copper ferrocyanide. Pfeffer had studied the osmotic pressures when water and a sugar solution were separated by this membrane and found that the pressure was quite considerable. If the concentration of the sugar was double, the pressure was correspondingly doubled. The same was true if the temperature of the sugar solution was doubled. Van 't Hoff immediately saw a direct analogy between the behavior of solutions and

1. For a simple account of the background to Armstrong's work on solutions, see Kendall 1939a. This charming account of solution chemistry was derived from Kendall's Royal Institution Christmas children's lectures in 1938. For a more nuanced account of Armstrong's vehement opposition to ionism, I have used Dolby 1976 and Rice 2004.

that of gases. Since the behavior of gases with respect to volume, pressure, and temperature had long been mathematized, he now proceeded to treat solutes as if they were gases, the solvent merely acting as a medium that allowed the pseudo-gaseous state to be exhibited. These results appeared in 1884 in the first modern text of mathematized physical chemistry.

Meanwhile, in Sweden, the independently minded University of Uppsala chemistry graduate Svante Arrhenius (1859–1927) moved to the University of Stockholm to take a doctorate involving the study of the conductivity of a variety of electrolytes. The research had not been suggested by his supervisor but was chosen by Arrhenius independently. Arrhenius was intrigued by the fact that the molecular weights of organic materials like sugar could not be determined by vaporization because they decomposed before they could be volatilized. He hoped, instead, that such molecular weights might be determined by adding them to salt solutions and measuring the decline of conductivity. Since the molecular weights of substances like alcohol and glycerin had been determined independently, and because they also decreased the conducting power of salt solutions, it should be possible (he reasoned) to develop an algorithm for the molecular weights of substances like sugar. The hypothesis did not work in practice, but Arrhenius's measurements caused him to ponder what previous chemists and physicists had asked: why, in fact, do solutions conduct electricity? Faraday's long-accepted model was that the radicals of salts (anions and cations) were simply driven in opposite directions by the electric current and deposited, separately, at the anode and cathode terminals. Arrhenius's revolutionary explanation was that, once in solution, the molecular components of salts separated merely as a result of having been dissolved before the solution was exposed to an electric current. Positive ions were immediately attracted to the cathode, and negative ions to the anode, simply because they were already charged.

While this thesis provided a neat explanation of electrolysis, it did so by making the then-radical proposal that molecules decomposed into their charged ions as soon as they were dissolved in a solvent, a process Arrhenius called dissociation. Much of the comedy of errors in the disputes over ionization that followed in the next thirty to forty years arose from the fact that opponents failed to grasp that Arrhenius thought of ions as different chemical species from the molecules from which they were derived. The hypothesis immediately posed other problems: why did some solutions conduct electricity efficiently and others poorly; and why did the conducting power appear to increase if solutions were diluted? Arrhenius's strategy was to label good conductors *strong electrolytes* because more of the molecules disintegrated into ions when dissolved; poor conductors like acetic acid, which he labelled *weak electrolytes*, failed to completely ionize. Experimental determinations also showed that hydrogen ions [H^+] and hydroxyl ions [OH^-] moved between electrodes quickly because they were produced by strong acids and strong bases, which were the best conductors.

Arrhenius was only too aware that his radical ideas would not go down well with his thesis examiners. Consequently, he chose to mask the idea of ionization and padded out the thesis with masses of experimental data. The thesis was accepted in May 1884 with the lowest grade of *non sine laude approbator*, "approved not without praise." In principle, this grade deprived him of the right to teach in a Swedish university.

It is doubtful whether Armstrong ever knew of Arrhenius's relative disgrace. What saved Arrhenius's reputation was the fact that he sent copies of the thesis to a number of European chemists, including van 't Hoff in Amsterdam, Wilhelm Ostwald in Riga, and Oliver Lodge in Liverpool. Ostwald was sufficiently impressed to visit Arrhenius in Sweden in August 1884 and offer him a docentship at Riga. Prompted by this prestigious foreign offer, Uppsala University appointed Arrhenius as a lecturer in physical chemistry in November, and in addition awarded him a fellowship that he used to travel around Europe during the next five years visiting all the leading chemists, including Armstrong. Arrhenius's meeting with van 't Hoff in Amsterdam proved important since it enabled the Dutch chemist to dovetail his ideas on osmosis with those of Arrhenius on solutions. Arrhenius's theory of dissociation helped to explain why dilute solutions did not conform to the osmotic pressure equation the Dutch chemist had developed. Dilute solutions of sodium chloride, for example, gave twice the calculated value for osmotic pressure. The reason was now clear: sodium chloride dissociated into two ions. The two theories of van 't Hoff and Arrhenius proved complementary, since van 't Hoff's osmotic model now seemed to confirm the validity of Arrhenius's model of dissociation, and vice versa.

Armstrong was not immediately aware of these events, since they took place when he was developing his teaching and research at the Central College. He was also busy elaborating his own speculative electrical theory of chemical change. It was Lodge who alerted Armstrong to the radical new way that Arrhenius, Ostwald, and van 't Hoff were developing the idea of solution chemistry into a general theory of physical chemistry. Here was a subject for review and discussion. At the British Association meeting in Montreal in 1884, six papers on solutions were presented to Section B and, as a result, the association commissioned a report on the physical constants of solutions.

At the annual meetings of the British Association, a number of chemists had begun to debate whether solutions were chemical mixtures or real chemical combinations of solute and solvent. For example, William Tilden argued that "a solution contains a mixture of several hydrates, which formed water of crystallisation in the solid state and whose constitution of which depends partly on the temperature of the liquid, and partly on the proportion of water present" (Tilden 1878). Other chemists, such as Dmitrii Mendeleev from 1886 onwards, developed complicated hydrate models, but they were opposed by William Nicol, a lecturer at Mason's College, Birmingham, who argued that the solution was simply held together by intermolecular forces (Nicol 1883). Because of his fame as the discoverer of the periodic law, Mendeleev caught the attention of chemists by plotting the specific gravity of solutions against the percentage of solute in solutions. The various discontinuous straight-line graphs he obtained (similar to Pickering's suggestions mentioned above) were evidence of the existence of large numbers of different hydrates (polyhydrates) in solution.

Armstrong was sufficiently intrigued by this evidence to suggest that his Central student Holland Crompton (1866–1931) should investigate whether the presence of hydrates in solution affected the electrical conductivity of solutions. Crompton's graphical plots appeared to confirm the Russian's argument for the formation of complex hydrates in water. Armstrong added an addendum to Crompton's paper (Crompton 1888, 125), drawing attention to his own previous ideas expressed in his presidential address to the Chemical Section of the British Association in 1885. Electrolytic and chemical action were obviously interdependent, the result of the interaction of complex molecular systems.

In 1886, Armstrong presented a paper to the Royal Society on his own view of electrolysis from the chemist's perspective (HEA 1886b). He pointed out that neither pure hydrochloric acid nor pure water conducts electricity—that is, they were not electrolytes—yet they conduct a current when mixed together. It followed, he argued, that it could not be supposed that HCl was the active agency while the water was merely the agency that separated the HCl into its ions. In Armstrong's opinion, the mixing demonstrated that electrolytes were molecular aggregates of the solute and solvent. It further followed, to his mind, that electrolysis was not caused by a dissociation of the solute, but rather by an aggregation of solute and solvent that then allowed a rearrangement in favor of the appearance of ions. He maintained this opinion for the rest of his life.

Meanwhile, Arrhenius began a correspondence with Lodge that led the latter to publicize Arrhenius's views at the British Association's meeting in Birmingham (Lodge 1886). A year later, at the meeting in Manchester, Lodge presented the electrolysis committee's report containing a direct comparison between the views of Arrhenius and of Armstrong. After the public reading of the report, Armstrong replied that Arrhenius's results equally justified his own model of electrolysis (HEA 1887c). He also drew attention to his more general theory of chemical change in which a third (catalytic) agent had to be present to promote any chemical reaction, including electrolysis. At this stage (1887), Armstrong was fully prepared to compromise between the two explanations of electrolysis. With becoming (and uncharacteristic) modesty, he wrote:

> In conclusion, I would add that I urge these pleas on behalf of my hypothesis with the greatest diffidence, feeling that I am unfortunately unable to fully appreciate the force of the mathematical and physical arguments. (HEA 1887c, 357; Dolby 1976, 318)

In 1896 Armstrong was excited to read a paper on osmotic pressure by Birmingham physicist J. H. Poynting, who suggested that the pressure arose from the association of ions with solvent molecules (Poynting 1896). Perhaps Poynting could accomplish what he (Armstrong) did not have the mathematics to do? He wrote *Nature*:

> If Prof. Poynting would seriously devote himself to putting such an [association] hypothesis into mathematical form, he would be rendering the one great service that is required of physicists in this connection. All that is needed, I imagine, is to show that the equations deduced from the one hypothesis are equally compatible with another diametrically opposed to it—surely a small thing to mathematicians. (HEA 1896a)

There is no evidence that Poynting took up the challenge (nor did any other physicist), and it was another three decades before the ionists developed equations that took the solvent directly into account. In any case, as Poynting's fellow physicists Oliver Lodge (Liverpool) and William Whetham (Cambridge) pointed out, the ionists were not blind to the fact that a chemical association between the free ions and the solvent was more than likely. Their point was that Arrhenius's mathematical model of dissociation worked well for dilute solutions and was leading physical chemists to new insights (Lodge 1896; Whetham 1896).

Clearly, Armstrong was not yet in the uncompromising position of opposing physical chemistry tout court. Indeed, he stated explicitly in his 1896 letter to *Nature*:

> All are agreed that Arrhenius and van 't Hoff and their satellites have rendered inestimable service by their generalisations, and the consequent applications they have made of them; certainly the world has shown its esteem of their work. Moreover, there can be no doubt, as I stated not long ago in my presidential address to the Chemical Society, that in so far as *weak* solutions are concerned a law has been discovered which is broadly true in *mathematical* form; yet I have no hesitation in asserting that that the fundamental premises on which it is based are destitute of common sense in the opinion of those who look at these matters without leaving chemical experience out of account; and I venture to think that his is not only their position, but also that of many physicists. (HEA 1896a, 78)

We should not be surprised, considering these views, to learn that he attended meetings of the Faraday Society in its early years.[2] The electronic theory of the atom and of valency still lay a decade in the future, so there were no good theoretical grounds for adopting a theory of ionization by dissociation. On the other hand, Arrhenius's theory was still a mathematically sound way at this time of saving the experimental phenomena—at least as far as dilute solutions were concerned.

Nevertheless, a plot was laid by Ostwald's Scottish student James Walker (1863–1935) for the British Association meeting at Leeds in 1890. His idea was to invite Arrhenius, Ostwald, and van 't Hoff to the meeting, at which a number of papers on "theories of solution" were to be presented. Walker's plan was that if the papers of "the three Musketeers" of ionism (as Armstrong was to call van 't Hoff, Arrhenius, and Ostwald) came at the end of the program, the opposition of orthodox elders would be overwhelmed. However, as James Kendall, a pupil of Walker's from whom he heard the story, recalled:

> The plot was a dismal failure. It was only the old-timers who remained in the lecture room to doze or drone while antiquated theories were discussed, the young enthusiasts were out in the corridors, clustered around Ostwald and van't Hoff (Arrhenius . . . could not attend himself, but sent a paper that was read by Walker). Before the end of the meeting all the coming chemists of Great Britain had followed the lead of

---

2. The Faraday Society was founded in 1903 to promote the further study of physical chemistry. Armstrong joined immediately, probably revering the name rather than the members' approach. See Sutton and Davies 1996, 14.

William Ramsay and James Walker, and were enlisted under the Ionic banner. The diehards sadly dispersed. (Kendall 1939a, 170)

It is a nice story, and it is certainly true that Ramsay and Walker were early converts, but the more general conversion did not take place overnight, as implied here.[3]

In fact, at the Leeds meeting, probably in private discussions during the long weekend, Armstrong had protested at the typical ionic equation:

$$H^+ + Cl^- + K^+ + OH^- = K^+Cl^- + H_2O$$

for its implication that HCl and H_2O were totally different molecules in their behavior. How could the former undergo more or less complete dissociation, whereas the latter remained unaffected—when all chemists knew that chlorine and oxygen were comparable elements? Ostwald had actually defended the ionists' position by saying that it only ran contrary to chemists' *feelings* and that *feelings* were subject to change over time (HEA 1927s, 150–51). Ostwald was right, as the history of science has shown repeatedly. But Armstrong would have no truck with such a retort: physicists, including physical chemists, were ill-equipped to understand the *facts* of chemistry. He drew on Helmholtz's conservative view:

> Nernst has thrown himself zealously into the newest applications of physical chemistry, as worked out by the Dutchman van 't Hoff and advocated with great vigour by Professor Ostwald of Leipzig in his *Journal* [i.e., *Zeitschrift für physikalische Chemie*, founded in 1887]. These theories have led to a multitude of demonstrably correct conclusions, although they imply some arbitrary assumptions which do not seem to me proven . . . But thermodynamic laws in their abstract form can only be grasped by rigidly trained mathematicians and are accordingly scarcely accessible to the people who want to do experiments on solutions and their vapour tensions, freezing points, heats of solution, etc. (Helmoholtz letter to unnamed correspondent from 1891, quoted in English translation in Koenigsberger 1906, 340; HEA 1927s, 152)

To Armstrong, physical chemists had thrown experimental chemistry to the winds "and were proceeding on hypothetical let-it-be-granted prin-

---

3. Nothing like this anecdote is hinted at in the anonymous account of the meeting published in *Nature*, 25 September 1890, 530–32.

ciples. The physico-chemical school, in fact, has never been a school of chemists."

As Huxley had famously said, "a beautiful theory" was easily "spoilt by a nasty, ugly little fact." The chief ugly little fact in the case of the ionists was their neglect of the solvent and the implicit assumption that it was merely "a screen serving to keep the ions apart." Despite his rejection of dissociation, Armstrong (as vice president of the Royal Society in 1902) fully supported the award of the Davy Medal to Arrhenius that year. Contrariwise, despite his opposition to the ionic theory, Armstrong was awarded the Davy Medal himself in 1911 for his contributions to organic chemistry.

Armstrong and Arrhenius must have met in 1904, when Arrhenius was invited to lecture on his dissociation theory at the Royal Institution. Indeed, it is plausible that the lecture may have been arranged by Armstrong in his capacity as one of the RI's managers. In his lecture, Arrhenius referred to Armstrong's opposition in friendly terms (Arrhenius 1904). The event was a formal occasion, so there was no discussion, though undoubtedly the two men discussed their differences in private. The two men always remained on cordial personal terms.[4]

The sluggish conversion of chemists to the ionic theory is illustrated by the meeting of the British Association at Dublin in September 1908, when there was a furious debate in the Chemistry Section. The context was actually a debate on "the nature of chemical change" organized by Armstrong, but it rapidly developed in Armstrong's hands into an attack on ionic theory and the general tendency of physical chemistry and mathematical analysis to neglect the facts of experimental chemistry. The result, he asserted, was vague theorizing that lacked sufficient grounds in reality. Oliver Lodge argued against Armstrong from the physicist's perspective, while William Ramsay thought Armstrong's viewpoint lacked force because he neglected the fact that ions were probably associated with electrons. On the other hand, several of the other speakers felt that there was some point to Armstrong's criticism of ions.[5] Additional opposition to ionism at this time was exerted by Louis Kahlenberg (1870–1941) at the University of Wisconsin, ironically an Ostwald PhD (1895), who pointed out that the theory offered no account of solute behavior in nonaqueous solutions.

---

4. Arrhenius was invited to Armstrong's home on more than one occasion. In December 1912, Arrhenius sent New Year's greetings and joked: "I hope the coming year will answer all your expectations (except perhaps regarding the theory of electrolytic dissociation)" (postcard inserted in the Armstrongs' visitors' book, RSC archives).

5. This paragraph summarizes a report in the *[Evening] Standard*, 5 September 1908: 5.

Fair Hydrone

By 1909 Armstrong had become unhappy with the way that the ionists seem to be converting young chemists (Kahlenberg, for one, excepted). In that year he published "A Dream of Fair Hydrone (A Chemical Idyll)" in the quarterly *Science Progress* (HEA 1909a). The piece was a simplified version of the paper he had given to the Royal Society in 1908, made into a metaphorical and lyrical essay. He opened the essay with methane, the simplest hydride of carbon, which he stated was "the absolute foundation stone of organic chemistry." After several pages of exposition of basic organic chemistry and the rules of valency, he explained optical activity based on the asymmetric carbon atom. He then compared the properties of the four hydrides ClH, OH_2, NH_3, and CH_4. Hydrogen oxide (water) was clearly anomalous in that it "liquefied without compulsion" because "the molecules of the oxide are gifted in a high degree with attractive forces which lead them to combine *inter se*." The two hydrides ClH and NH_3 were more soluble in water than CH_4 because their residual affinities allowed them to attach more hydrogen atoms of the water solvent. For example, when HCl dissolved in water, the unstable molecule H_2O-(H)(Cl) was probably formed.

He then introduced the reader to his new nomenclature distinguishing between water as a gas (that is, steam), which he called *hydrone*, and liquid water, which (on the analogy of ethylene chains) might be imagined to be polymerized molecules of hydrone:

$$H_2O = OH_2 \quad \underset{\text{Dihydrone}}{H_2O\!\!\overset{OH_2}{\wedge}\!\!OH_2} \quad \underset{\text{Trihydrone}}{\begin{array}{c}H_2O\!\!-\!\!OH_2 \\ | \quad\quad | \\ H_2O\!\!-\!\!OH_2\end{array}} \quad \underset{\text{Tetrhydrone}}{\begin{array}{c}H_2O\overset{OH_2}{\wedge}\!OH_2 \\ | \quad\quad\quad | \\ H_2O\!\!-\!\!OH_2\end{array}} \quad \underset{\text{Penthydrone}}{\begin{array}{c}\overset{OH_2}{\wedge} \\ OH_2 \; H_2O \; OH_2 \\ | \quad\quad\quad | \\ H_2O \quad\quad OH_2\end{array}} \quad \underset{\text{Hexhydrone}}{\begin{array}{c}OH_2 \\ \wedge \\ OH_2 \; H_2O \; OH_2 \\ | \quad\quad\quad | \\ H_2O \quad\quad OH_2 \\ \vee \\ OH_2\end{array}}$$

Thus, the general formula for hydrone in the liquid state became $(H_2O)_x$.

If water consist on the one hand of simple molecules which have a strong tendency to cling together and on the other of relatively heavy inert molecules, such as those of penthydrone $(OH_2)_5$, it is easy to understand why it boils at such an elevated temperature as 100°. (HEA 1927s, 132, 136)

There was no mention of ionization in the "Chemical Idyll." This subject followed in the sequel, "The Thirst of Salted Water, or The Ions Overboard," in the April 1909 issue of *Science Progress* (HEA 1909b). This new piece continued in the form of its precursor with extensive quotations from Ruskin's essay "Of Truth of Water" in his *Modern Painters*, which led Armstrong to reflect that if the painter had difficulty in portraying water, chemists had done little better, especially "in representing it by the thin and unattractive symbol H_2O—still more by *speaking* of it with heartless vulgarity in these latter degenerate days as *Aitch-too-oh*." He claimed that since 1885 he had urged that it was essential to consider the role that water played in aqueous solutions. Only now, in 1909, were there signs that he was being listened to.

> In our own country, the physico-chemical dovecot has evidently been somewhat disturbed of late; seemingly its inmates are beginning to study the gentle art of hedging; unfortunately, most of them lack the courage to release their hold of the Ostwaldian petticoats, to which they have long been accustomed to cling that they are unable to stand alone. As was said of the lambs of little Bo-Peep—if let alone for a while longer, doubtless they will come home and leave their tales (of ions as well as of repentance) discreetly behind them. (HEA 1927s, 145)

Rhetoric aside, he then got down to a serious explanation of what he thought happened when chemicals were mixed with water. For a start, the freezing point of water was lowered while its boiling point was raised. In some cases, the solution conducted electricity; in others, it did not. As Alexander Williamson had argued in his theory of etherification in 1851, molecular exchanges took place continually in solutions. In 1857, Armstrong wrote, Rudolf Clausius had suggested that electrolysis occurred with electrolytes because violent kinetic collisions between molecules caused their partial separation into ions. Arrhenius in 1883 had then come forward "as a Whole-Hogger" and assumed that complete, or nearly complete, dissociation took place; he suggested that this dissociation explained the increase of conductivity with dilution. Arrhenius's theory did not become known until 1885, "and even then, did not catch on." What had caused the change was the villain van 't Hoff:

> In 1885, van 't Hoff, assuming the character of the hatter, invited us to a scientific Mad Tea Party, at which he out-hatted the Hatter by gravely assuring us that "I see what I eat" *was* the same thing as "I eat what I see": introducing us into topsy-turvydom, he insisted that liquid was to

be treated not as a liquid but as a gas; that the attractive forces in play between the molecules of solute and solvent were not really attractions: the solute was to be thought of as banging about and hitting things, as if it were gasified. (HEA 1927s, 149)

Van 't Hoff's reputation as the author of the tetrahedral carbon atom had done the trick; combined with Arrhenius's "juvenile enthusiasm," and the "aid of floods of Ostwaldian ink," the ionization theory (a joke in Armstrong's Carrollian metaphor) was launched into the chemical world. Armstrong obviously blamed van 't Hoff for being "misled by the observation of a parallelism into the assumption of any actual resemblance between gaseous pressure and the peculiar conditions of stress in a solution."

Armstrong then laid down a number of arguments against the ionic theory: (1) When molecular activity increased with dilution, hydrolytic activity actually diminished; how could the same ions determine both behaviors? (2) Enzymes control hydrolysis by catalytic action; this behavior was specific to the enzyme and the specificity was not explicable by the ionic theory. (3) Baker's work on dry reactions had shown that chemical activity was not dependent on ionization but on association with water molecules in a conducting system. (4) The ionic theory limited ways of explaining chemical change to electrolytes and, therefore, had nothing to aid the organic chemists in explaining chemical reactivity. And finally: (5) Irish physicist George FitzGerald had argued that the theory failed to cover the behavior of concentrated solutions. More basically, FitzGerald had written:

> It is a most remarkable thing that osmotic pressure should be even roughly the same as what would be produced by the molecules of the body in solution if in the gaseous state, but to imply that the dynamical theory of the two is at all the same, or that the dynamical theory of a gas is in any sense an *explanation* of the law of osmotic pressures is not at all in accordance with what is generally meant by the word "explanation." This so-called explanation is not a dynamical explanation at all, it is only a very far-fetched dynamical analogy. (FitzGerald 1896)

All these arguments remained pertinent, and even the most ardent ionists were aware that the theory needed extension to cover the full spectrum of solution behavior. But this process remained incomplete until substantively addressed in the 1920s.

Having outlined all these objections, Armstrong turned to his own explanation of the *thirst* of salted water: namely, "the superior activity of composite electrolytes." The answer, he claimed, had come to him at the British Association meeting in Bradford in 1900.

> The assumption made is that the formation of composite electrolytes by the dissolution in water of substances such as the acids, the alkalies and salts generally, is a process involving the *distribution* of both constituents of the composite electrolyte ... and that the occurrence of electrolysis involves the interaction of the two kinds of complex while under the influence of an electromotive force. The extent to which this distribution occurs determines the activity of the composite electrolyte. (HEA 1927s, 160)

In the case of nonelectrolytes, he assumed that only water molecules were distributed to form a complex.

The essay was rounded off with a moral, namely that all the major channels of chemical communication had been taken over by the "high priests of the ionic cult" and their jargon filled the journals and textbooks. Any different point of view, such as Armstrong's, had been written off, and in this way "ionomania" was perpetuated through textbooks and schools of chemistry.

Armstrong was determined not to give up the fight. His attitude toward ionization and its supporters was reflected in his reviewing. For example, he dismissed George Senter's *Textbook of Inorganic Chemistry* in 1911 as an example of that class of books "that ought never to have been born ... it is difficult to understand how a person gifted with the knowledge and intelligence to compel such a treatise can have been led to waste his energies in composing it." Although Armstrong did not say so explicitly, he was no doubt aware that Senter, a trained pharmacist, had taken his doctorate with Ostwald in Leipzig and published *Outlines of Physical Chemistry* in 1909. Senter's treatment of inorganic chemistry, in Armstrong's opinion, was a cram book, a scissors-and-paste job, and mere pabulum to pass examinations,

> with sections irrelevantly introduced such as are found in the more modern treatises on so-called Physical Chemistry. In other words, the dogmas of the ionic dissociation school are interlarded with conventional tit-bits of information which commonly pass as chemistry in academic circles, though nowhere in real life. (HEA 1912a)

Senter wisely left Armstrong's views to the judgment of others, and gauged by the book's longevity in print, that judgment was that it served a useful purpose. Even Armstrong had admitted that the book was well printed, written in a clear way, and contained a fund of information. The historian can only conclude that Armstrong's harsh opinion stemmed from his bias against physical chemistry.

Armstrong's prejudice again became evident after James Walker noted the jubilee of Arrhenius's theory in 1910 (J. Walker 1910). Armstrong seized on Walker's statement that "Whatever be our views of the origin and nature of ions, we must . . . have recourse to the notion of degree of ionisation." How, Armstrong asked, could one use the notion if one did not know what the notion was? If the Royal Society's massive, ongoing catalogue of scientific papers were ever completed, he went on, a dictionary of terms would be needed, one of which would be "ionisation." He, therefore, invited Walker to define the word, which Walker duly did in the same issue of *Nature* (HEA 1910b). Walker's reply is an interesting defense of "saving the phenomena" rather than producing a physical theory.

> Prof. Armstrong and I look at *ionisation* from different points of view. He is chiefly interested in an interpretation of the process and phenomena of ionisation in terms of the kinetic molecular theory. I am chiefly concerned to have a theory, whatever its exact mechanical interpretation, which is capable of being mathematically formulated and of acting as a guide in quantitative investigation. My position, in short, is that of the astronomer who is content to have Newton's law for practical purposes, and to take only a speculative interest in theories of the nature of gravitation. (J. Walker 1910)

For Walker, it was the equations and their resulting heurism that mattered. This stance, of course, confirmed what Armstrong had believed all along: physical chemists were not concerned with real chemistry at all. Again, he demanded Walker explain what "ionization" meant. We must remember that there was still no electronic model of the atom and that spectroscopists still had no theory of the origin of spectra. Ionization, therefore, seem to imply that sometimes, for little apparent reason, molecules split into their respective *atomic* or *elementary* constituents. Walker did not make further reply, probably because he knew Armstrong would only prolong the argument.

Instead, there was a sharp attack on Armstrong's backwardness by Alfred Stewart (1880–1947), the author of *Recent Advances in Physical Chemistry* (1909). Stewart pulled no punches, accusing Armstrong of belonging "to an antiquated school which believed that scientific discoveries are made by forming definite ideas of things, even though these cannot be seen and handled" (Stewart 1910). This comment would again have confirmed Armstrong's impression that for physical chemists, "anything goes." A more temperate suggestion came from the Cambridge physicist and future philosopher of science, Norman Campbell (1880–1949):

> Ions are particles supposed to be present in some media such that, when the medium is placed in an electric field, the particles have a finite average velocity relative to the medium along the direction of the field. (Campbell 1910)

This proposition would cause no offense since it was merely a modern rendering of Faraday's electrolytic theory. However, Campbell went on to say that "ionization" was currently being used in two senses: the first referred to the number of ions present in a unit volume of the solution; and the second referred to the unknown process whereby the ions were produced. Since several processes had been suggested, the term referred to no particular mechanism.

Rather surprisingly, Armstrong seems to have made no comment when the father-and-son team of William and Lawrence Bragg proposed that X-ray analysis of common salt revealed a collision pattern that they interpreted as demonstrating that salt crystals consist of separate ions of sodium and chlorine arranged in a cubic pattern (Bragg and Bragg 1913). It must be remembered that the Braggs solved very few crystal structures before the 1920s, and even though they were awarded the Nobel Prize for Physics in 1915, their work made little progress until after WW1.

The 1920s saw an international effort by chemists to extend the ionic theory so that it would account adequately for the behavior of strong electrolytes. The so-called Debye-Hückel-Onsager theory modelled a sodium chloride solution as completely dissociated, and any deviation from ideality was attributed to interactions between the sodium and chlorine ions and between them and the water solvent—just as Armstrong had always argued. However, Armstrong would have taken little joy from the way the theory used abstruse mathematical symbols, and he would have disapproved equally of what would have been to him unreal "correction factors" such as the activity and osmotic coefficients. The Faraday Society held a discussion on the theory of strong electrolytes at Oxford in April 1927 in which Niels Bjerrum, Johannes Brönsted, Erich Hückel, Lars Onsager, and Martin Lowry all took part. Armstrong did not attend or comment on the meeting, nor was he mentioned in the highly mathematical discussions (Lowry 1927). The attempt to produce an ever-tighter fit between a mathematical theory of strong electrolytes and the actual experimental behavior of solutions has continued to this day.

Armstrong remained attached to his chemical-dream and salted-water essays and actually republished them unaltered in his *Essays on the Art and Principles of Chemistry* (HEA 1927s), long after the basic ionic theory had been adopted in schools and universities across the world. The book attracted few reviews, one of which was a damning dismissal by Manchester University physical chemist Harold Dixon, writing anonymously.

> While we regard parts of this book as beneath the dignity of the science, the art and the principles of which it professes to expound, we cannot repress some feeling of sympathy with the author for the courage he shows in resisting superior forces. We can admire Professor Armstrong's defence of the name of "oxygen," we can appreciate his insistence that water acts chemically when dissolving salts; but is anything gained by crude lampoons on van 't Hoff, Arrhenius, and Ostwald—and indeed on the whole school of modern physical chemistry founded on their work? . . . Is not this language a little reminiscent of a celebrated criticism of the "hallucinations" and "Phàntasie-Spielereien" of a certain Dr J. H. van 't Hoff, and does not our modern Censor rather out-Kolbe Kolbe? (Dixon 1927)[6]

Dixon then proceeded to outline Armstrong's hydrone theory of chemical change and show, convincingly, that "the difficulties of following the hypothesis of electrolytic dissociation seem small in comparison." If Armstrong complained that his voice had been ignored since 1885 and that he felt like "one crying in the wilderness," Dixon concluded brutally, Armstrong had only himself to blame:

> The voice of one crying in the wilderness—as prophesied by Isaiah—surely was listened to. Are we not told that not only Jerusalem went out to hear, but "all Judaea, and all the region round about Jordan"? The voice has been heard, but is not the answer more nearly that pagan rejoinder, *Credat Judaeus* [*Apella non ego*; from Horace: "the Jew Apellus may believe it; I don't"]? (Dixon 1927, 37)

---

6. As mentioned, Dixon's article was published anonymously, but the editor's marked (online) copy identifies the author. Dixon was quoting Kolbe's 1877 cruelly satirical (and outrageously misguided) critique of van 't Hoff's stereochemistry (Rocke 1993, 329).

Solution Chemistry

Despite Armstrong's failure to vanquish the ionists, his opposition to Arrhenius's dissociation theory had beneficial results for his research programs at the Central College. On the one hand, he was led to collaborate with pupils on a series of twenty-five papers on the processes operating in solutions between 1905 and 1913; and on the other hand, he embarked on a study of the origin of osmotic effects that led to four papers in collaboration with his son Frank between 1905 and 1910.

Armstrong began the study of solution chemistry in 1905, with the aim of finding experimental evidence that water and other solvents were chemically involved in the process of solution and, from thence, in the process of electric conduction. The relevant series of papers had a turbulent history. The first eleven papers (1906–1910) were published in the Royal Society's *Proceedings* rather than in the *Journal of the Chemical Society*. However, by 1910 Armstrong decided that the series would become better available to readers of the *JCS* if he offered it a summary of his thoughts on solutions. However, the summary was rejected by the Chemical Society's Publications Committee. The consequence was confusion. While part 12, a paper with his New Zealander student Frederick Worley, appeared in the *JCS*, Armstrong's résumé of the research, part 19 (!), appeared in the *Chemical News* on 3 March 1911, *before* the appearance of parts 13–18. The Worley-Armstrong paper also had to carry a footnote to satisfy the Chemical Society's Publication Committee's dissatisfaction with Armstrong's unique nomenclature (HEA 1911b).

Parts 14 and 15 duly appeared in *JCS*, after which there was, it appears, another complaint over nomenclature. In anger, Armstrong submitted the subsequent parts—16, 17, and 18, along with his résumé of previous work, part 19—not to *JCS* but to *Chemical News*. That résumé contained the following astounding footnote:

> This communication most unfortunately was submitted with others to the Society for the Propagation of Christian Names among Chemists [i.e., the Chemical Society]. I have withdrawn it and others to which my name is attached as a protest against the refusal of the Publications Committee of the Society to permit me the liberty which, in the interest of scientific progress, I contend must be allowed to every author of scientific communication—that of expressing himself in the terms chosen by himself. The previous communications of the series (I-XI) have been published in the *Proceedings of the Royal Society*—the one

Society that unites the interests of all branches of experimental inquiry. I have asked nothing more than that I may not be prevented from using the language I have used in these previous communications—language which I used of set purpose. (HEA 1911d, 97)

When *Chemical News* published part 13 on 31 March 1911, Armstrong added a clarifying note saying:

> The communication of which this is the immediate sequel appears in the March issue of the *Journal of the Chemical Society*, together with Nos. 14 and 15. The withdrawal of the articles now published in this journal—this being the fourth—was rendered necessary solely by the [*JCS* editor's] refusal of my request that I might be allowed to qualify a noun by an adjective—the noun alcohol by adjectives such as methylic, etc. . . . This to a Fellow of over forty years standing who was an active member of the Geneva Conference on Nomenclature! (HEA 1911c, 145)[7]

This was not to be the only time that he would cross swords with *JCS* editor John Cain.

Armstrong's résumé of his students' work on solutions replayed the history of his position and thoughts about chemical change. He began by mentioning his 1886 Royal Society article on electrolysis and chemical change which, amongst other things, had argued against Clausius's suggestion in 1857 that *some* molecules placed in a conducting situation were broken apart (dissociated) by kinetic bombardment by the water (or solvent) molecules. Armstrong had argued instead that forces of residual affinity came into play and brought about the formation of water complexes in solutions. He had continued to express this view in various reports on electrolysis written for the British Association up to 1889. By then he had cast doubt on Arrhenius's views, and he had received support from FitzGerald's Helmholtz Memorial Lecture in 1896.

By this date, as we saw earlier, Armstrong had reached the conclusion that "the hypothesis of electrolytic dissociation is an offence against common sense and in no way a necessary assumption." He was happy to accept (as Claude-Louis Berthollet had argued a century before) that only a certain proportion of a solute was "active" in solution. The real problem

---

7. The minutes of the Chemical Society (ff. 287–88) contain a tipped-in letter from *JCS* editor John Cain dated 4 March 1911, reporting that HEA had withdrawn papers 13, 17, 18, and 19. No discussion was reported.

was how to explain the "activation" in the first place. The problem with Arrhenius's hypothesis, as he had argued in his presidential address to the Chemical Society in 1895, was that it only applied to electrolytes, and even then just to dilute solutions. On the other hand, Armstrong's association model was widely and generally applicable to all chemical reactions, from benzene substitutions to the selective agency of enzymes. The existence of enzymes had already prompted him to study their selectivity, and he had found that the explanation lay in association, not dissociation. Enzymes involved hydrolysis (that is, chemical reactions in the presence of water in which the water molecule was decomposed). In this way he had been led to the series of investigations of solutions, parts 1–30, between 1906 and 1913. The series spilled over into the Royal Society's *Proceedings*, despite the fact that referees like William Tilden deemed some of Armstrong's comments "too confrontational" and felt that the work might be better placed in the *Philosophical Magazine*.[8]

As Armstrong had argued in 1886 (HEA 1886b), the increase of molecular conductivity in solutions was surely due not to dilution per se (as Arrhenius contended) but to the specific influence of the solvent making its presence felt, because by increasing the amount of solvent "it becomes more and more disentangled from its fellows" in the presence of complex molecules formed from the simple water molecule. His complete theory had been presented when writing his essay "Chemistry" in 1900, published in the *Encyclopaedia Britannica* in 1902. He also put it forward again in a paper to the Royal Society on the origin of osmosis in June 1906, as well as in part 6 of the solution studies (HEA 1908b). And for good measure, he had also explained it to a more general scientific readership in *Science Progress* in 1909.

He reiterated that he found it impossible to accept van 't Hoff's kinetic explanation of osmosis; rather, he contended, osmosis was caused by an as-yet-unidentified attraction within the solvent system that disturbed the equilibrium. He contended that osmotic pressure was more akin to a hydraulic pressure than a gaseous one. It followed that physical chemists had been wrongly comparing solutions made up to different volume strengths, whereas it was really essential to compare one and the same proportion of solvent. It was for this reason that all the Central College papers in the solution series compared the masses of the solvent, not the total volumes of the solution. In this way, his student Robert John Caldwell (d.1915) had shown how anomalies between concentrated and dilute solutions mostly

---

8. Tilden's report on part 25, 18 October 1912, is in the RS archives, RR/19/8.

disappeared. The use of solvent mass solutions had also been advocated and used by American chemist Harmon Northrup Morse (1848–1920).

Although chemists subsequently followed the Central College chemists' (and Morse's) experimental procedure, they ignored Armstrong's further drastic suggestion regarding nomenclature. "To secure precision and avoid confusion," Armstrong said that the term "water" should only be applied to the liquid state of the molecule H_2O; its gaseous form, OH_2, was to be named "hydrone." He doubted whether chemists would ever know the exact composition of the molecular compounds in water; but analogy offered some possibilities. (Ironically, although denying the validity of van 't Hoff's analogy between the gas laws and osmotic pressure, he was happy to use speculative analogies when it came to the composition of water.) Because oxygen could act as a tetrad element (that is, it could display fourfold valency like carbon), oxygen could be compared to carbon in its behavior, the only difference being that molecular unions of OH_2 were fewer and less firm than those of CH_2. Armstrong's entire model was speculative: he offered no empirical evidence for its details and no mathematical analysis that generated equations fitting some, if not all, the experimental findings, as the ionists had achieved.

Armstrong went even further, averring that the nature of acidity had to be revised. He recalled that Lavoisier had chosen the term "oxygen" because he thought that what had hitherto been called "dephlogisticated air" was the determining cause of acidity. Chemists soon deemed the word ill-chosen when it was shown that not all acids contained oxygen, but they nevertheless stuck to the nomenclature. Armstrong now suggested that Lavoisier had been half right. He followed Williamson in insisting that sulfuric acid was really the name for sulfur trioxide, and that only the solution of this gas in water was, strictly speaking, hydric sulfate. The ionists were treating the hydrogen ion (H^+) as the cause of acidity; but in Armstrong's view the cause of acidity was the association of a gaseous oxide with water. The distinction could be made by insisting that chlorhydric acid was the correct term for a solution of hydrogen chloride gas in water, hydric nitrate etc. In short: oxygen had not been misnamed! As we have seen, such a change of nomenclature was not approved by editor John Cain and the *Journal of the Chemical Society*.

It followed that the apparently "simple" reaction of an acid with an alkali to form a salt and water was really more complex since it actually involved a reaction between two hydrolated compounds. Armstrong asserted that his students' papers in the series of papers on the processes operative in solutions offered confirmation of his claims. For example, parts 6 and 7 in the series on solutions showed that the amount of "extra" water formed was determined by the proportion of two complex hydrols present in solution. Armstrong summed up the solution series as demonstrating that:

> In all cases of chemical activity displayed by substances in aqueous solutions, the change that takes place is initially one in which hydrolated molecules of the interacting substances are concerned, strength and weakness corresponding respectively to high and low proportions of hydrolated molecules. (HEA 1911d, 110)

In an addendum, he treated the case of nonaqueous electrolytes in an identical manner but qualified it by suggesting that whereas in water the solvent and solute became hydrolyzed and distributed, with different solvents (and in cases of fusion) the molecules only distributed amongst themselves. The overall conclusion that "the properties of water seem to be singular in all respects" had, he asserted, been confirmed time and again by all subsequent chemical and physical research.

By 1910, in fact, the ionists had already conceded that positive and negative ions in aqueous solutions became hydrated. Armstrong differed only in saying that it was only the negative ions that were hydrated. In a Faraday Society discussion on "the constitution of water," arranged by James Walker in April 1910, Ostwald's pupil Paul Walden discussed whether water itself was an electrolyte (Walden 1910). Walden differed from Armstrong in suggesting that conduction of a current occurred because mixtures of acids and alkalies formed oxonium salts, like

$$OH_2 + HCl = OH_3Cl$$

on the analogy of the way acids reacted with ammonia gas to form ammonium salts:

$$NH_3 + HCl = NH_4Cl$$

Under the influence of an electric current, these oxonium salts then resolved into ions that were surrounded by atmospheres of water:

$$OH_2 + HCl \leftrightarrow OH_3Cl \leftrightarrow OH_3^+aq + Cl^-aq$$

Walden concluded that "water as an amphoteric body will only become an electrolyte and a better conductor when it is dissolved in a liquid of a *distinctly acid or basic* character, that is, when it will form with this solvent a *kind of salt*" (Walden 1910, 76; Bousfield and Lowry 1910, 88).

With this quasi-reconciliation of the ionists and non-ionists by 1910, why did Armstrong continue to combat the physical chemists who specialized in ionism for the rest of his life? One answer is that until the 1920s, he did not believe the evidence coming from crystallographers and from the electron theory of the physicists that ions could exist in a free state. Indeed, Armstrong still toyed with the idea that discharge tube experiments did not necessarily imply subatomic particles like electrons. If impurities were present in the discharge tube, they might catalyze reactions according to his theory of chemical change, and the phenomenon interpreted by J. J. Thomson as the electron might be chemical and not physical. The other explanation for his antagonism toward ionism was his continuing distaste for what he saw as the random, wild extravagances of the mathematically trained physical chemists.

The Uesanians

The ionization theory continued to be Armstrong's bête noire throughout the 1920s. With the rising ascendency of American chemistry over German and British chemistry, his principal targets now were mainly American physical chemists (U-S-A-nians—as he called them). Chief among them was Wilder Dwight Bancroft (1867–1953), the founding editor of the *Journal of Physical Chemistry*, the first English-language journal in the new field. A pupil of Ostwald and van 't Hoff, Bancroft taught at Cornell University from 1895, and he quickly became one of the most eminent American physical chemists (Servos 1990 and 1994). Like Ostwald, Bancroft saw physical chemistry as playing the crucial role in chemistry in providing core linkages between all the sister sciences.

By contrast, for Armstrong organic chemistry was the core of chemistry. Armstrong was amused when Bancroft published a sixty-eight-page monograph on "the electrolytic theory of corrosion" as a supplement to the *Journal of Physical Chemistry* (Bancroft 1924 and 1925). For Armstrong, Bancroft's historical treatment of the subject was decidedly wanting. He asked: what was the point of publishing when little or no use was made of what had already been published? (It will be recalled that he himself had published on the subject.) Why had Faraday's immortal work on electrolysis been forgotten? (HEA 1925j). He wondered at Bancroft's patience to set out "the myth" of corrosion at such length.

> Why cannot our friend Bancroft set the example of thrusting all such cumbersome matter aside and give us his own clear reading of the corrosion story, in the common tongue, in a crisp form? Half a dozen pages

should suffice. We should then have something definite to consider and criticize—if there be anyone left to criticize anything with reference to the facts!

In a final barb, he expressed pleasure that after he (Armstrong) had argued the point for forty years, Bancroft had finally agreed that "corrosion reactions are at the heart electrochemical."

> If the great lawgiver will now go a stage further and issue an edict that chemists shall no longer copy the girls and "bob" all acids, reducing them to the miserable, anaemic, characterless state of the mythical hydrogen ion, we shall be able to deal honestly with molecules again and substitute a rational cult for the [Søren] Sörensen [who introduced pH] phallic [!] worship, which so obscures all acid issues. In fact, we shall be thinking, working chemists once more and not mere dreamers, blind worshippers of a baseless superstition. (HEA 1925j, 762)

Armstrong and Bancroft eventually met at a conference in the Hague in July 1928, but neither man left an account of their conversation.

Early in 1925, Armstrong was invited by the newly founded American *Journal of Chemical Education* to contribute an article on the present state of the education of chemists. The result was a bizarre essay entitled "The First Epistle of Henry the Chemist to the Uesanians."

> Henry, unto the Church of Chemists on the great continent. Grace be made unto you and the peace of understanding. For the gospel is come unto you not in word only and also in practice but in much assistance, so that ye are examples to all. So being affectionately desirous of you, we are willing to impart unto you not the gospel only but also our own soul, because you are dear unto us. . . . Now we exhort you, brethren, warn them that are unruly, comfort the feeble minded, support the weak, be patient towards all men. Quench not the Spirit. Despise not prophesyings. Prove all things. Hold fast that which is good. We pray your whole spirit and body be preserved blameless. We charge you that this epistle be read by all the holy brethren. (HEA 1925q)

There followed several pages on the ethics of belief, education, literature, chemistry, and culture, embellished with hyperbole, paradoxes, aphorisms, and parodies of proverbs that attacked the press, the current teaching of science, the poor literary quality of science writing, and the complete absence of any cultural values in the science of the day.

His principal vitriol was reserved for the ionists. He regarded them as dogmatic and intolerant, just like the early Church fathers—along with those religious fanatics who opposed the theory of evolution. He was currently following the Scopes trial in Tennessee as it was reported in *Nature*, and this was probably why he chose a parody of New Testament language. An unsigned editorial in *Nature* publicized Armstrong's Uesanian polemic in December 1925 by noting how its striking language shouted for attention in the style of Bernard Shaw and G. K. Chesterton. Richard Gregory, the probable author, noted, however, that the essay carried a serious message: that dogmatism in science was the very negation of true science (Gregory 1925, 827).

The opportunity to review the special semicentennial issue of the American Chemical Society's *Journal* in 1926 provided a further opportunity to lambast the Americans' obsession with physical chemistry. He dismissed Bancroft's essay on fifty years of physical chemistry in America as "rambling," and he expressed surprise to learn that he, Armstrong, together with Kahlenberg in America and Wilhelm Traube (1866–1942) in Germany, formed a triptych of "three irreconcilables who do not believe at all in the electrolytic dissociation." Yet, there was hope for Bancroft's sins. "It is easy enough," he quoted Bancroft as having written,

> to point to one factor which has been neglected practically completely and which may be the one which has caused most—and perhaps all—of our [the ionists'] difficulties. For years H. E. Armstrong in England has chided the physical chemists for considering water only as water, whereas it is a complex and variable mixture. This criticism seems well founded; but unfortunately, Armstrong has never succeeded in showing what could be done with his idea and consequently the idea has been valueless hitherto. (HEA 1926r, 807)

Bancroft's point was well made, but Armstrong turned it aside by saying that it confirmed what he had always thought: that publishing papers in the *Proceedings of the Royal Society* was a form of "decent burial" in that chemists never read the journal. Whether or not this was true, it did not meet Bancroft's real point that neither Armstrong's association theory nor his more general theory of reverse electrolysis as the cause of chemical change could be developed into a respectable mathematical theory with quantitative predictions.

In 1923, Armstrong's son-in-law Stephen Miall published an Izaak Walton–Robert Boyle parody entitled "The Compleat Chymist" (a pun on Walton's celebrated 1653 tract on angling and Boyle's 1661 *Sceptical*

Chymist). Using the pen name B. Lagueur (French *blagueur* = kidder, joker), Miall featured Armstrong in the guise of the "compleat chymist" who remained sceptical of the theory of ions (Miall 1922, 308; "B. Lagueur" 1923; HEA 1923c). Armstrong himself penned a witty introduction to the pamphlet in the style of the seventeenth-century writer Joseph Glanville. In Miall's invented conversation between a seventeenth-century chymist, a farmer, a brewer, and another friend, written in the style of Boyle's *Sceptical Chemist*, the text reveals Miall's extensive knowledge of English poetic literature—a talent he later displayed in his anthology *Poets at Play* (Miall 1932). The brewer first mistakes the chymist for a pharmacist and is given short shrift in a topsy-turvy world that abruptly swings from the seventeenth to the twentieth century. But the brewer is delighted because his friend, who supplies him with water for his brewery, has met with a problem in understanding the words "the three-course layer of aleurone cells super-posed upon the starchy endosperm" during a lecture on Zymic Change. The friend, pleased to converse with a chemist, has found too much salt in the water, or that "there is an undue concentration of sodium and chlorine ions therein." The chemist asks whether he's ever seen any of these "ions." Nothing is resolved, but it seems clear that Miall had some sympathy toward his father-in-law's scepticism. Toward the end of 1924 the lampoon led Armstrong to write two Shavian essays for *Chemistry and Industry*, "Why Was I Born So Soon?" and "The Compleat Chymist" (HEA 1924d and 1924e). He had been reading Shaw's preface to *Saint Joan* and had seen the play's premier in London in March.

Armstrong made no further efforts to dissuade chemists from dissociation theory, but he continued to express sardonic references to the ionists in his invited addresses and in his journalism right up until his last months. Although Armstrong's theory of chemical change received some later support from Lowry in his studies of prototropic tautomerism (Lowry 1925), it was fated to be overtaken and extinguished by the electronic theory of reaction mechanisms developed by Christopher Ingold at Leeds in the 1920s and at University College London during the 1930s (Nye 1993).

Osmosis

Because of van 't Hoff's direct application of his work on osmotic pressure to Arrhenius's dissociation theory, it was inevitable that Armstrong would also investigate the phenomenon of osmotic pressure in connection with his work on solution chemistry. He and his son Frank published four papers with the Royal Society on the subject between 1906 and 1911 (HEA 1906b; 1909e; 1910d; 1911a). Although the first introductory papers appeared in the physical series of the *Proceedings* (A), the other three papers appeared in the separate biological *Proceedings* (B) because the research had moved back into the realm of plant biochemistry. Their publication in the biological series was also appropriate because the investigations had a bearing on the work being done at the Rothamsted Agricultural Research Station, where Armstrong was an active trustee of the Lawes Agricultural Trust. The osmotic papers also complemented another line of research that was spawned by the investigations of solutions: an inquiry into the nature of enzymes as factors in plant growth and nutrition. The latter investigation produced a series of twenty-four papers, begun with his son Frank in 1903 and only reaching a conclusion in 1925. Frank had returned from working with Emil Fischer on enzymes in Berlin. This elaborate series was initially reported to the Chemical Society, but as the investigations became more biochemical than chemical, Armstrong again found outlet with the Royal Society's biological *Proceedings* (HEA 1903e; 1904d; 1905a; 1905b).[9]

The initial paper on osmotic effects of 1906 made it clear that, in Armstrong's view, only an associationist, not dissociationist, viewpoint could explain the effects of enzymes and acids as hydrolyzing agents.

> In the case of osmotic phenomena generally, whatever may be the degree of fortuitous coincidence between calculated and observed values when they are treated as gaseous pressure effects [i.e., the van 't Hoff analogy], the effects cannot in reality be due to pressure in any ordinary sense. In liquids generally attractive forces must come into play between the molecules; in solutions attractive forces must also come into play as between solute and solvent. (HEA 1903e, 265)

Armstrong investigated several plant walls or septa in his investigation of osmosis, including barley seeds, and he extended these studies to the hormones that penetrated the leaves of various plants (HEA 1910d; Eyre 1916). In his research into hydrolysis of sugars and glucosides, he found that the rates of hydrolysis were constant and dependent on the amount of enzyme present. The eventual conclusion was that hydrolysis took place on the surface of the enzyme, which itself attracted water and molecules such as lactase, maltase, invertase, emulsin, and urease. The work was sub-

---

9. It is notable that Armstrong headed the papers with the category of subjects appropriate for the International Catalogue of Scientific Literature for which he was the British chairman. This kind of category heading was equivalent to the sort of "keywords" demanded today by many periodicals.

sequently extended to a large number of plants and to the effect of hormones on the osmotic pressure of plant cells, only to be abandoned when Armstrong's laboratory was closed down between 1911 and 1914. Unable to carry out further work himself, Armstrong nevertheless took a keen interest in the way that biochemistry developed during the next twenty years and how it might affect agriculture and nutrition and the human condition generally. His final word on osmotic effects was published by the Royal Society in 1923, even though the referee, James Walker, found it "difficult to accept many of the conclusions."[10]

The Last Word

Armstrong's final discussion of dissociation came seven months before his death. In September 1936, James Kendall, with whom he was on friendly terms, had sent Armstrong an advance copy of his introductory talk to a meeting of the Faraday Society in Edinburgh that discussed "pure liquids and liquid mixtures" (Kendall 1937). Armstrong was unwell and found Kendall's remarks "most amusing reading [since] gleams of sanity are becoming apparent in his philosophy" (HEA 1936k, 916). Kendall's address, Armstrong noted, had begun with a remarkable confession:

> It is even more astonishing to find, by a perusal of current articles, how many research workers who are familiar with all the refinements which the Debye-Hückel theory has imposed upon the classical ideas of Arrhenius, are still totally unaware of the fact that the van 't Hoff analogy between dilute solutions and gases has long ago been discarded.

As Kendall might well have expected, these words of his set the old man off in a rant about how he, Armstrong, had been characterizing Arrhenius's ideas as ridiculous ever since 1885. "Wait and see" had always been his "opinion of the school of ignorance fathered by Arrhenius and Ostwald."

> I have had to wait over fifty years. They put facts aside and fitted their curves to suit their assumptions. Neither took chemistry into account. [But has] van 't Hoff *spurlos versunken* [sunk without trace] as a dilutionist? What is the alternative? Debye and Hückel? This is a plunge from the Apostles into the Athanasian Creed? Sheer dogma. They have been taken far too seriously . . . The fact is, there has been a split of

---

10. James Walker, RS archives, RR/28/55, referring to HEA 1923h. This was HEA's last paper extolling the hydrone theory.

chemistry into two schools, since the intrusion of the Arrhenic faith, rather should it be said, the addition of a new class of worker into our profession—people without knowledge of the laboratory arts and with sufficient mathematics at their command to be led astray by curvilinear agreements; without the ability to criticize, still less of giving any chemical interpretation.

He had never been taken in, just as he had never been taken in by religious dogma when he heard nonsense spouted from the pulpit. "Most so-called physical chemistry is just religious dogma—faith with no works," he declared. He ended the tirade by lamenting that nobody read his work, every line of which spelled association, not dissociation.

Armstrong had not understood that there had been a rapprochement between dissociationism and associationism in solution chemistry. As early as 1900, the American pupil of Ostwald Harry Clary Jones (1865–1916), who became professor of chemistry at Johns Hopkins University in Baltimore, accepted that ions were solvated (Rice 2011, 17). By 1936, the majority of chemists understood that the solvent (be it water or some other liquid) played a role in solution, either via the formation of hydrates or through the hydration of ionic species. Whatever the complexity of the Debye-Hückel model, it took chemical account of the presence of the solvent, as Armstrong had always asserted was necessary.

The publication of Armstrong's work on a nonionic theory of solutions done in collaboration with the final cohort of his Central students proved controversial. Not only was it directly opposed to the views of most physical chemists, it also caused strife with the Chemical Society due to Armstrong's insistence on using nomenclature that the society had proscribed. The research was also difficult to follow by even the most dedicated reader, because its many parts appeared in seemingly random order in the *Proceedings of the Royal Society*, the *Journal of the Chemical Society*, and the weekly *Chemical News*. In retrospect, the result was almost comical. Overall, however, the work, which included significant new data on the phenomenon of osmosis, demonstrated Armstrong's determination to explain chemical change by molecular association rather than by dissociation. This endeavor was by no means unreasonable, though the theory that he espoused has been superseded.

Armstrong in Leipzig, ca. 1868.
Courtesy Imperial College Archives

Armstrong, ca. 1880. Courtesy Imperial College Archives

Armstrong, ca. 1910. From a photograph by
Frederick Hollyer. Courtesy Imperial College Archives

Armstrong above Seathwaite, Cumbria, 1925.
Courtesy Imperial College Archives

Henry and Louisa Armstrong on their golden wedding anniversary, 1927.
Courtesy Imperial College Archives

Armstrong in 1930. Courtesy Imperial College Archives

PART II

10

Semi-Retirement

> Hypotheses, like professors, when they are seen not to work any
> longer in the laboratory, should disappear.
> —Hartley 1971, 199

The century's second decade saw the end of Armstrong's research career and, disappointingly from his viewpoint, a failure by government and other organizations to utilize his talents in fighting the First World War. However, he still played a strong Fifth Business role in the activities of Chemical Society, the Royal Society, the Royal Institution, and the British Association for the Advancement of Science. He also found a new platform to voice his views in the Royal Society of Arts. But his only official engagements were as a governor of Christ's Hospital (a public school in West Sussex) and his membership of the Lawes Agricultural Trust, though he was to make himself useful at India House in connection with the supply of natural indigo from the Indian empire. With time on his hands, and sitting on the sidelines in the Athenaeum (he had been elected to its membership in 1897) as a world war was fought, he found further outlet in expressing his sharp views and opinions in newspapers and the scientific press of *Nature*, and also in *Chemistry and Industry*, where his son-in-law, Stephen Miall, was editor. It was also a decade in which secondary and higher education were much debated, but here too Armstrong could only look from the sidelines, not having been invited to participate directly in various educational investigations and reports. One might say that his talents were wasted and unappreciated. Nevertheless, his views were heard and were sufficiently loud to catalyze and sometimes galvanize both scientific and public opinion.

Although Armstrong formally retired from the South Kensington institutions in 1911, he had several research projects that were uncompleted, as well as a handful of research students who had begun but not completed their doctoral research. Helpfully, T. Edward Thorpe, the UK's chief government chemist, allowed Armstrong to transfer himself and his remaining research students to a small attic laboratory at the Royal College of Science. This arrangement continued when Thorpe retired in 1912 and was replaced by inorganic chemist Herbert Brereton Baker as head of chemistry. The college then decided it should have three chemistry professors, one each for inorganic (Baker), physical (James Charles Philip), and organic (Jocelyn Field Thorpe, unrelated to Sir T. Edward Thorpe) (H. Gay and Griffith 2017, 102). It was only in January 1913 that Armstrong's City and Guilds laboratory was finally closed, an act of vandalism in Armstrong's eyes for which he never forgave Imperial College.

> My laboratory was one of the best and most convenient in the country; the last thing I witnessed before going to Australia, in June 1914, was its being hacked to pieces—several thousands of pounds worth of original fittings was destroyed: had it not been for the long sea voyage and the excitement of travel at the Antipodes and in the East, I doubt I should have recovered from the shock. (HEA 1920c, 18)[1]

Chemical Society Controversies

Armstrong became a regular contributor to the short-lived illustrated monthly *Chemical World* following its launch by textile chemist William Porter Dreaper (1868–1938) in 1912. Unfortunately, the publisher discontinued it as soon as war broke out in 1914. A characteristic Armstrong contribution was an essay on how the optical properties of materials offer clues to their structure. This piece turned out to be an attack on the way physicists like William and Lawrence Bragg were taking on subjects that were really the province of chemists.

> The time must soon come . . . when it will be recognised that it is the office of the chemist to be primarily a chemist rather than a pseudo-physicist or pseudo-mathematician and when it will be admitted that chemistry is in large measure an art, not to be appreciated to the full by those whose soul hankers most after figures and who are hampered

---

1. This account is from a lecture given on 14 September 1920, with Sir Richard Gregory as chairman. The section on the University of London was also published as HEA 1920d. Armstrong had pleaded to be allowed access to a laboratory at the City and Guilds College in exchange for delivering a course of lectures. See HEA to John Wolfe-Barry and Alfred Keogh, 1 and 7 May 1913, AP1.114–15.

by the narrowness of outlook and lack of precedent which seems to be in some measure engendered by physico-mathematical study. (HEA 1914a, 3)

He invoked the spirit of his German mentor Kolbe, "whose worthiness as a chemist and Gediegenheit [solidity] has yet to be recognised by his countrymen," and who was a relentless campaigner against the introduction of "a metaphysical spirit into our science." Instead of remaining content with weighing and measuring in determining constitution, there were now self-described chemists using optical properties without bothering about chemical properties per se.

The closure and destruction of the City and Guilds chemical laboratories were an occasion for an editorial commentary in *Chemical World*, which noted that Armstrong had always been "a hard hitter."

> In this he may not always have been right. What man could claim such a distinction? But we are certain that no chemist in this country has anything but praise for his work which has been accomplished, or a word to say against his honesty of purpose, which has been obvious to all. (Anon. 1913, 2)

The journal then provided a generous illustrated outline of Armstrong's activities from the pen of William A. Davis (1913), citing the theory of chemical change outlined in Armstrong's 1885 presidential address to the Chemical Society as the highlight of his achievement. The panegyric concluded by noting how Armstrong had always followed the advice given him by Edward Frankland and Hermann Kolbe. It was of infinite importance to hear doubt cast upon current belief and authority called into question; "the display of solid grounds for such disbelief was eminently inspiring." Armstrong's own version of Frankland's maxim was: "When in doubt, disbelieve until convinced of truth; and even then, remain on guard."

In March 1914, Armstrong published a curious article in *Chemical World* entitled "A Move towards Scientific Socialism." In this analysis of the current state of chemical journals and the future of abstracting chemical literature, he noted how the privately edited *American Journal of Chemistry* was being amalgamated with the *Journal of the American Chemical Society*. He thought something similar should happen in the UK. At the same time the awful language employed in both British and American abstracts should be improved. When he was young, he wrote, if uncertain how to say something in English, he would ask himself how to say it in German.

That was no longer possible because German chemists were now just as bad as the British. He laid the blame for the breakdown of English style on the Publications Committee of the Chemical Society and the editor of its *Journal*. Unsurprisingly, an impetus for his remarks was that he disliked the way some of own prose was being edited. The article proceeded to attack the pricing of journals generally, and it asserted that papers in German journals had become little more than transcripts of laboratory notebooks. "Most of us who maintained a library of periodicals in our early days have ceased to do so," he continued, because of expense, lack of shelf room, and deterioration of the quality of reports (HEA 1914c).

Armstrong was soon answered by his former pupil Martin O. Forster, who feared that the diatribe was unhelpful to international peace and goodwill, for Armstrong was being offensive to both American and German chemists; like George Bernard Shaw, Armstrong was acting as an irritant for the sake of it. On comparing Armstrong's own chemical prose of forty years back, Forster thought it had declined and would now be blue-penciled even by Armstrong's model editor, Henry Watts. As to the salient point about abstracts, a common abstracting service would be only an ideal because chemists preferred journals to address their own special interests. The kind of "socialist" cooperation that Armstrong suggested would be unworkable because of geography, publication schedules, and finance. In fact, the new International Association of Chemical Societies would be discussing the very problem of international abstracts at its next meeting (a meeting that never happened due to the outbreak of war) (Forster 1914).

Armstrong's diatribe understandably offended the editor of the *JCS*, John Cannell Cain. At the next meeting of the society's council after the issue of *Chemical World* appeared, the council suggested a vote of confidence in Cain and proposed that Armstrong be asked to apologize.[2] Dreaper initially refused to insert an apology in his journal until after Armstrong had written a personal letter to Cain.[3] As for Forster, he remained on excellent terms with his former teacher, and, as we shall see, joined him in one of the Chemical Society's notorious acts, as WW1 entered its grimmer stages.

---

2. Chemical Society council minutes, 26 March and 2 April 1914, ff. 168, 170.

3. Armstrong to Cain, 27 March 1914: "Dear Dr Cain, I much regret that words used by me have been understood to reflect upon your professional ability." Armstrong had meant nothing personal and had written about general issues. He was, he said. profoundly sorry for any annoyance; *Chemical World* 3 (April 1914): 104. The "slander" was also reported in *Chemical News*, 24 April 1914.

The British Association Meeting in Australia

On 27 June 1914, Armstrong sailed from Liverpool for the port of Freemantle in Western Australia. He visited Perth and then made his way to Adelaide, Melbourne, and Sydney for meetings of the British Association that were being held in Australia for the first time. The outbreak of war did not disturb his plans for a leisurely return to London via southeast Asia and the Indian subcontinent. He was not back in London until 21 December, by which time many important decisions regarding the role of science and scientific manpower in war had already been taken. Without a laboratory to work in, and not having been recruited for the war effort despite his talents, Armstrong was destined to be an outsider and, inevitably therefore, a critic of how the British were fighting the war.

Plans for a meeting in Australia had begun to be laid when the BAAS met in Canada in 1909; the meeting finally took place under the presidency of geneticist William Bateson, culminating five years later. It was the Australian spring, and the weather proved fine. As a member of the association's council, Armstrong was awarded a grant to support the expenses of the voyage to and from Liverpool. In each Australian state, elaborate arrangements had been made by local committees for overseas members' accommodation and hospitality in private houses, clubs, and hotels, and in addition to sectional meetings and public lectures and receptions, members went on carefully planned excursions.[4] On 4 August the advance party joined the main party to sail from Perth to Adelaide, where news of the outbreak of hostilities was confirmed and the ship was commandeered by the Australian army and navy. In Adelaide, the official business began with sectional meetings, public lectures, and receptions as well as excursions. But before the official business began, Armstrong had the opportunity to stay with his former London Institution pupil Edward Rennie (1852–1927), who had returned to his native Australia in 1882 to become professor of chemistry at the University of Adelaide. Rennie shared Armstrong's interests in natural history, and the two men spent two days examining Australian flora and fauna together (HEA 1927e). The British Association party moved on by sleeper trains to Melbourne. There Armstrong was awarded an honorary degree by the University on 14 August.

On the same day, Armstrong gave a provocative address to the Education Section on the role of science in government and in the education of its

---

4. For detailed report, with maps of excursions, see Anon. 1914. This account appears to be based on a series of reports by different authors that appeared in *Nature* during the autumn of 1914.

peoples (HEA 1914d). He aptly recalled how Huxley had visited Melbourne in 1847 as a ship's surgeon and how he had subsequently devoted his life to forwarding education. Quoting Huxley's great speech on science education of 1861, Armstrong said the message had still not been heeded fifty years later. "In England," he declared, "what I will venture to term the Oxford spirit still reigns supreme—the spirit of the literary class—the medieval spirit of obscurantism, which favours a backward rather than a forward outlook" (HEA 1914d, 214). How could such an attitude still prevail considering the advances and improvements to people's lives brought about by science and technology? He made it clear that the greatest intellectual advance had been that of Darwinian evolution: "through it we have at last acquired full intellectual freedom, and the belief that it rests with ourselves alone rightly to order our lives, that by it all dogmas have been undermined."

What, then, Armstrong asked, was the future role of science, and, in particular, what bearing did it have on the future "of the white people"? If the British were to avoid the fate of previous civilizations, science had to be promoted as a central feature of education. Unfortunately, British schools were still in literary hands. Arthur Balfour's then-recent argument in *Science and Belief*—that science was founded on unprovable premises and metaphysics—deserved but short shrift. The method of science was that of the law court, based upon the collection and sifting of evidence (HEA 1914d, 210). Taking Francis Galton's eugenical message to heart, the idea of equality of intellectual endowments was moonshine. This point applied, particularly, to the case of women. Here he was explicit: the case put forward in recent years goes back to John Stuart Mill:

> That their disabilities are in no small measure due to the fact that we have neglected their education: give them time to educate themselves and they will be as men in all things. Years ago, at our Stockport meeting [*sic*, Southport, 1903] I ventured to express the difference by saying that woman is not merely female man but, in many respects, a different animal: the two sexes have necessarily been evolved to fulfil different purposes. Nothing is more instructive in the history of modern educational progress than the fact that women have asked merely for what men have: at the universities they have attended the men's courses; not one single course have they demanded on their own account. Higher teaching in relation to domestic science so-called has only been thought of very recently and mainly because men have urged its importance. Most serious and, I believe irreparable injury is being done to women, in London especially, by forcing them to undertake the same studies and to pass the same university examinations as the men; and the damage is done to the race,

not merely to individuals, as the effect of education, whether direct or indirect, is clearly to diminish the fertility of the intellect. (HEA 1914d)[5]

Armstrong then proceeded to suggest that different types of education were needed for different talents. He was far from alone at this time in holding such views (Dyhouse 1976; Rowold 2010).

Later that year, and when Armstrong was still absent from England, the lectures given by John Perry, Gerald Moody, and Armstrong were the subject of independent, highly critical comment by the two professors of education at the University of Manchester, Henry Bompas Smith and the Herbartian educationist Joseph John Findlay. The former had not been present in Australia but criticized the initial reports printed in *Nature*; the latter had been present in Australia but had left as soon as war had been declared (Findlay 1914; Brooks 2004). Findlay had been appalled by the sledgehammer approach of the president of the section (Perry) and the two vice presidents (Moody and Armstrong) in denigrating British education, and by their patronizing attitude toward Australian teachers. What the Australian teachers had wanted to hear about was genuine educational research being conducted in the home country on child psychology, school administration, the teaching of reading, and so on, rather than the egocentric rantings and "frothy libels" of a trio of South Kensington professors who had been saying the same things for the past thirty years. Bompas Smith, who was more critical of Perry than of Armstrong, attacked their inaccuracies and wild generalizations.

> My quarrel with these two distinguished writers [Perry and Armstrong] is that their attitude towards the wider questions of educational theory and practice is essentially unscientific. They seem to me to suffer from a weakness which besets many very able men who have become recognised authorities in their special field of work. They appear to think that they have the right to dogmatize upon questions in which they take an interest, but of which their knowledge is derived chiefly from their personal experience. (H. B. Smith 1914, 466; Duff 1953)

Findlay's and Bompas Smith's sallies hit the mark. None of the three men replied.

Armstrong had harmed his reputation earlier in the year with a gross calumny against the explorer Robert Scott, who had died returning from

---

5. The paper read to the Education Section at Southport in 1903 was HEA 1903f, but it includes no such remark on women.

the South Pole at the end of March 1913 and had become a national hero. At the annual meeting of the Association of Public School Science Masters at Imperial College in January 1914, Armstrong had said that Amundsen's recent account of the Antarctic expedition (Amundsen 1913)

> showed that every moment during the winter was directed to preparing himself for the final attempt to reach the Pole. Every action was carefully thought out. There was not a word of evidence that Scott's party did anything of the kind. The science was evident except at the desk in the English expedition.[6]

This statement rightly drew the fire of Huxley's son, Leonard, a fellow member of the Athenaeum, who gave page-by-page references to the "very careful forethought and planning of the journeys" (L. Huxley 1914).

In Melbourne, Armstrong also attended the joint session of the Physics and Chemistry Sections at which the structure of atoms and molecules was discussed, led by Ernest Rutherford. In the discussion, Armstrong welcomed physicists' attempts to unravel the structure of elementary atoms but said that, for the time being, the ideas were too novel for chemists to either appreciate or criticize. He had long believed, he said, that the so-called atoms of elements had substructure, on the grounds of analogies drawn from organic chemistry. But the chemists' ideas were quite different from those of the physicists. The latter unfortunately "in the past have held aloof from the chemists; they have paid too little attention to their methods and to their results; the movement now in progress is therefore to be welcomed, as it must have the effect of leading the two parties in future to work together to a common end." Then came the riposte. He could not accept the existence of isotopes:

> It is doubtful if it be permissible at present to conclude that elements of different atomic weight may and do exist which are indistinguishable chemically; the observations on which reliance is placed have been made with quantities of materials far too small to permit of such an inference; in the case of the rare earth elements, although very large quantities of material have been at the disposal of chemists, they have only slowly discovered differences by which they are enabled to distinguish and separate them. Though the special methods made use of by physicists are very powerful, they suffice only in certain cases and have

---

6. Armstrong, in a vote of thanks to the association's president, H. B. Baker, in the *Times*, 14 January 1914: 4.

little chemical significance; when physicists resort to chemical methods the work becomes subject to ordinary criteria. (HEA 1914e)

The same caution was necessary in interpreting the X-ray spectra of elements which the young Oxford researcher, Henry Moseley, had discussed at the meeting. There was no justification for using Moseley's work to conclude that "all but very few of the elements are discovered." Armstrong evidently preferred to imagine that most of the known elements belonged to a certain "type" that formed a nearly complete series, the similarity of whose spectra was due to a common radical, like homologous paraffins. Readers will recall Armstrong's vastly expansive periodic table (1902b, 712) discussed at the end of chapter 5.

He then returned to his long-held view that any theory of atomic structure had to be crystallographic; any theory of structure acceptable by chemists had to invoke valency relationships and atomic volume and shape. Because hydrogen had a valency of one it was unable to link with a second element, so that J. J. Thomson's recent hypothesis that hydrogen could exist as a triatomic molecule, H_3, was preposterous and showed why chemists and physicists need better accommodation of views. He then urged the audience to reflect on the causes of apparent variation of valency as well as spatially directed valencies; a proper theory of atomic structure would have to account for these phenomena. At the moment, the Pope-Barlow atom seemed to him the best model from the chemists' point of view. He ended his reflections by noting that the production of helium by radioactive disintegration still required the chemist's interpretation and not just the physicist's.

The group traveled to Sydney on 19 August, and the serious business and junketing were repeated. Finally, trains conveyed members to Brisbane on 27 August, and Armstrong gave a public lecture in the Centennial Hall on "the materials of life." The meeting officially ended on 1 September. Arrangements had been made for the remaining British party to join *HMS Montoro* on 3 September, and Armstrong was thus conveyed to northern Queensland and on to Java, Singapore, and the Indian subcontinent, where he spent time in Ceylon, Madras, Calcutta, Darjeeling, Pusa, and other cities.

During his tour of India, he had acquired considerable knowledge about Indian education, which he summed up in his journal:

> The problem of education in India is a vastly difficult one. At present, it has all the faults of our English System magnified, and is not merely worthless but wicked beyond measure: we glorify "cram" but Indians worship it. The degree is an end in itself; each examination passed adds

to the social value of the student, raises his value in the marriage market. . . . A complete reconstruction of the entire system is required: before all things, it needs to be put into honest hands. . . . But the natives have no desire to make any change, having no wish for real education and caring merely for the status conferred by the degree; and they resist the introduction of any better system, fearing that those brought up under it would be at a disadvantage. (quoted in Eyre 1958, 189)

It was time to leave. Armstrong sailed back to Colombo, Ceylon, where the *HMS Nore* departed for England on 20 November, calling at Aden before entering the Gulf of Suez on 8 December. As the ship sailed by way of Port Said and Malta, Armstrong saw ever more signs of the European war: harbors full of British, French, and Russian warships. They reached Algiers on 15 December and Gibraltar two days later. The ship finally anchored at Portsmouth, without meeting any German naval hostility, at noon on Monday, 21 December. He caught the train to Paddington, where his daughter Nora was waiting for him to conduct him home to Lewisham.

Only Armstrong's wife Louisa and daughter Nora were now living at home. Frank, the eldest son with whom Armstrong had collaborated on enzyme research, had just been appointed technical adviser to Crosfield and Sons, soap and chemical manufacturers in Warrington, where he was soon drawn into vital war work on the catalytic production of acetic acid and acetone from ethyl alcohol. He had married in 1907 and had three young children. Armstrong was to visit Warrington to see his grandchildren regularly during the war. His three other sons, Clifford (who had married in 1907), Robin, and Harold, had each enlisted into the armed forces and survived the war. Robin had served as a registrar at the Great Ormond Street Children's Hospital after studying medicine at Cambridge. There he had made a particular study of children affected by tuberculosis and published observations on the disease in *Science Progress* during his father's editorship in 1913 (R. R. Armstrong 1913). Two of the daughters, Edith and Annie, were already married and living independently with their husbands, leaving only Nora at home to help look after her parents.

Armstrong had not expected to retire until 1918, upon reaching the age of seventy. Had that been the case he would undoubtedly have been involved in the impressive work that Imperial College accomplished during the First World War—notably, perhaps, the work of his former student Martha Whiteley (H. Gay and Griffith 2017). Armstrong was destined to stand on the sidelines of the war effort as a critic, a role he played with passion.

11

The Great War

> Though I have had 50 years' experience as a chemist... I have never once been consulted [in the present wartime]; the only request for my assistance that I have received, since the beginning of the war, came from a German gentleman long naturalized as a British subject.
> —HEA 1915b

Armstrong kept a detailed diary of events and observations throughout his travels in Australia, India, and Ceylon, and he planned to make a book out of it when he returned. Unfortunately, the time was not right for such a book, and publishers were not interested. His planned biography of Ludwig Mond, the German-born British chemist and industrialist, was also abandoned. What was he to do? He no longer had a laboratory or research students. When Armstrong left for Australia in June 1914, his books and apparatus had remained at the Royal College of Science. But when he returned, Imperial College was on war footing; Herbert Baker and Jocelyn Thorpe made it clear that the space was needed and that Armstrong must remove all his possessions. His research career was now over, and the Great War was changing everything. What role could he play? The answer was, of course, as an independent adviser and critic, using the platforms of the Chemical Society, the Royal Society, the Royal Institution, the British Association for the Advancement of Science (which continued to meet until interrupted by war in 1916), and the newspaper press.

The Boycott of German Chemists

At the start of the war, the government passed the Aliens Restriction Act, which required Germans residing in the country, now seen as potential enemies and spies, to register with the police, who were given powers to intern or deport them. Many scientists were affected, notably German-born

Arthur Schuster who, although a naturalized Briton, had just been elected the physical sciences secretary of the Royal Society. The Cambridge mathematician Alfred Basset (1854–1930) urged the Council of the Royal Society to remove Germans from Fellowship status and ensure that elections to the Fellowship were of native Britons only (Basset 1914; Badash 1979; MacLeod 1971). A reply came from Joseph Larmor, a former physical sciences secretary, that Basset was being ungracious; Oliver Lodge went so far as to call for Basset's resignation from the society. This small episode suggests that while patriotic, the majority of British scientists were still respectful of German science and scientists.

At the Chemical Society's first council meeting after war was declared, there was also a discussion concerning whether anything should be done about the German chemists who had been elected honorary members. As with the Royal Society, it was decided to let matters remain as they were. This decision immediately changed when it became known that ninety-three German professors and intellectuals had signed a manifesto addressed "To the Civilized World," which denied Germany's responsibility for starting the war and defended the German invasion of Belgium and the claimed atrocities surrounding the invasion. The fact that the manifesto had been signed by such famous chemists as Emil Fischer changed the neutral views of many British chemists toward Germans. Even so, 120 British scholars, including thirty scientists, among them William Crookes (president of the Royal Society), W. H. Perkin Jr., and William Ramsay, took a moderate stance in a statement published by the *Times* in October 1914.

> We ourselves have a real and deep admiration for German scholarship and science. We have many ties with Germany, ties of comradeship, of respect, and of affection. We grieve profoundly that, under the baleful influence of a military system and its lawless dreams of conquest, she whom we once honoured now stands revealed as the common enemy of Europe, and of all peoples which respect the Law of Nations. (Ramsay 1914a)

Hence, when Armstrong eventually returned home from the Australian meeting of the British Association in December 1914, he arrived at a nation in turmoil over the war and its cultural implications. Aware that his close friend Hugo Müller, a former president of the Chemical Society, had voluntarily resigned from the society as soon as war was declared, he was determined that Schuster should also do the honorable thing and resign his secretaryship of the Royal Society. In a letter to the *Morning Post* (HEA 1915a) and a series of three letters on the organization of science for the war effort, he berated the Royal Society for its putative lack of leadership (HEA 1915b, 1915c, and 1915e). His newspaper complaint was laid before the Society's council on 21 January 1915, but the council decided against calling a meeting of fellows to discuss the issues Armstrong had raised.[1] In private correspondence with Larmor, Armstrong continually urged him to get Schuster to resign.[2]

> If he desires to be one of us, he should be the first to see that the position he has taken up is one which no Englishman, under parallel circumstances, would feel justified in retaining.... [Being] of German birth and extraction, [Schuster] cannot have and exercise our English point of view. Our interests are therefore prejudiced by his attempt to represent us.

It says much for Schuster's strength of character that he made no mention of the war or of Armstrong in his later reminiscences (Schuster 1932).

Armstrong had more success at the Chemical Society. In May 1915, he and Forster proposed

> that in view of the misuse by our enemy of chemical science and of chemical appliances and products, in direct contravention of international agreement, this Council recommends the withdrawal from the roll of Fellowship of the Society of the names of Adolph von Baeyer, Theodor Curtius, Emil Fischer, Carl Graebe, Paul Hendrich von Groth, Walther Nernst, Wilhelm Ostwald, Otto Wallach, and Richard Willstätter.[3]

All these honorary foreign members had been signatories to the German manifesto to the civilized world in 1914. But when put to the meeting, the motion was defeated. That was, however, not the end of the matter, for at the following meeting, the president, Alexander Scott, was forced to make the following statement:

---

1. Harrison to Armstrong, 22 January 1915, RS archives, NLB/51/238. See also letters of support to Armstrong from William D. Halliburton (biochemist), 16 February 1915, AP1.199; William A. Herdman (zoologist), 15 February 1915, AP1.212; and Edwin Ray Lankester (biologist), 17 February 1915, AP1.249. They agreed that the Royal Society needed reforming but thought it unwise to drag the society into the general question of organizing science for the war effort.

2. Armstrong to Larmor, 1 and 8 July 1917 and 29 July 1918, Larmor collection, RS archives, 28–30.

3. Chemical Society council minutes, 20 May 1915, item 23, f. 221.

At more than one of their meetings the Council has carefully considered the question of removing from our lists the names of Honorary and Foreign Members and of Ordinary Fellows who are at the present time alien enemies of His Majesty King George. Whilst regarding with the deepest detestation and abhorrence the ruthless and barbaric methods adopted by the Germanic allies and more especially the debasement of Chemical Science by them as revealed in their latest operations [gas warfare], the Council have to-day by a large majority decided that no steps can consistently with the dignity of the Society and with due regard to British ideas of justice, be taken in the above direction until after the cessation of hostilities.[4]

There the matter rested for another year. In April 1916, Armstrong and Forster presented a stronger resolution to the council:

In view of the provocative misuse by our enemy, in direct contravention of international agreement, of chlorine gas and of the continued calculated application of chemical agents to murderous acts, whereby large numbers of innocent non-combatants, including many women and children, have lost their lives, the Council resolves to make the one practical protest within its power, by removing from the roll of Honorary and Foreign Members of the Chemical Society the names of A. von Baeyer, T. Curtius, E. Fischer, C. Graebe, P. H. R. von Groth, W. Nernst, W. Ostwald, O. Wallach and R. Willstätter, the same being hereby resolved.

The motion was passed, subject to an amendment that the names be struck off for the duration of the war, following which "their position shall be reconsidered." The amendment was suggested because of the question of individual responsibility for German atrocities. How could individuals' names "be removed without personal injustice unless they had been given an opportunity to defend themselves"? Council asked the membership's views by postal ballot.[5] The result was overwhelmingly in favor, and the nine German chemists' names were removed from the membership.[6] These were the honorary members; forty-six foreign fellows of German

---

4. Chemical Society council minutes, 17 June 1915, f. 227.
5. Chemical Society council minutes, 11 May, 18 May, and 16 June 1916, ff. 280, 282, 287.
6. Eugen Bamberger had already resigned early in the war, so he was not included in this list. Willstätter was reelected in 1927, and Nernst, Ostwald, and Wallach were reelected in 1929. Of the other five, Baeyer and Fischer had died in 1917 and 1919 respectively; Curtius, Graebe, and Groth were not reelected.

nationality, including Fritz Haber, were also removed from the membership list in November 1916.[7]

Hugo Müller, the German chemist who had emigrated to Great Britain in 1854 and taken English nationality at the time of his marriage to an Englishwoman in 1878, died in 1915. As a former president of the society he deserved an obituary, and Crookes asked Armstrong, one of his closest friends, to write it. Armstrong took his time. By 1 March 1917, the council resolved that unless Armstrong produced the obituary by the end of the month, the writer would be reassigned. That did the trick, and the obituary was in the editor's hands by 29 March (HEA 1917d). The piece turned out as a wonderful example of Armstrong's ability to tell the story of a life and the lessons it taught.

Armstrong's anti-German campaign sometimes went too far. In November 1917 he sent a letter to the *Evening Standard* that was also reproduced in the *Manchester Evening Chronicle*. In it he accused the dyestuffs manufacturers Clayton Aniline Co. of being run by Germans. Armstrong had got hold of the wrong end of the stick, because many of the company's shares were held by the Swiss Society of Chemical Industry in Basel, and Switzerland was, of course, neutral. Both companies sued for libel, and the matter came to the King's Bench in January 1918. Armstrong claimed he had acted in good faith, naming the (unknown) source of his information. The plaintiffs withdrew the complaint when Armstrong apologized, but he was forced to pay the costs of the plaintiff's action.[8]

On the other hand, in partial apparent contradiction, Armstrong also wrote the following cautionary remark:

> I hold no brief for the German, though I owe much to him; but I am sure the greatest mistake we can make is to misunderstand and underrate the ability of an enemy: least of all must we allow insular prejudice and narrowness to mislead us into the blind belief that we are superior to the rest of the world and cannot learn from it. (HEA 1916a)

The Organization of Science

Despite being a leading older fellow of the Royal Society, Armstrong was no longer a member of its council and thus not privy to the society's war

---

7. Chemical Society council minutes, 16 November 1916, item 31, f. 301.

8. Reported in the *Derby Daily Telegraph*, 24 January 1918, under the headline "A Newspaper Article That Was Untrue." I have not traced the *Evening Standard* article in question.

efforts (Varcoe 2000). A War Committee had been set up by the president, Crookes, in November 1914. Initially, this cohort oversaw subcommittees devoted to chemistry, physics, and engineering, but it was reorganized in April 1915 to form a relatively secret group composed of Crookes, the three secretaries, the treasurer, and a select number of fellows representative of the main disciplines. Three of its earliest members, Arthur Crossley, Percy Frankland, and Martin Forster, had immediately lobbied the Board of Trade to prevent the voluntary call-up of trained chemists so that they could be retained or posted to positions in the chemical industries. This suggestion had already been made to the Royal Society's council by Raphael Meldola in April 1915. Meldola had pointed out, as would Armstrong later, that research was vital in modern industry and that the Germans were well ahead of the British in this respect. Germany had quickly got off the mark to exploit the synthesis of hitherto imported chemicals and, as a result, now dominated the world's markets for organic chemicals. William Henry Perkin Jr. had agreed with Meldola, as had the consultant agricultural chemist Bernard Dyer.[9]

In January 1915, not long after his return from India, Armstrong had publicly urged the Royal Society to group fellows into specialized disciplinary committees and set them to work on ideas for winning the war, but it was only in June 1915 that he became aware of such action by Crookes and the Royal Society's council (HEA 1915b). Nevertheless, he complained that many fellows (like himself) who were well versed in the chemistry and properties of explosive materials had not been consulted. The Royal Society (the "House of Scientific Peers," as he termed it) ought to be advising the War Office. He complained bitterly in a letter to the *Times*:

> Though I have 50 years' experience as a chemist, particularly in connexion with the materials now being used in the manufacture of explosives and of natural and artificial organic products, I have never once been consulted... No doubt, I am properly regarded as merely a retired professor, but I know highly competent younger men among those trained by me who are equally unutilised.

He concluded that the Royal Society must be organized to serve the state (HEA 1915b). This was what William Ramsay had also urged at the beginning of the war (Ramsay 1914b, 221). The Chemical Society, under the presidency of Alexander Scott, had also organized a council that would

9. War Committee, RS archives, CMB/36.

advise the government, but Armstrong was not invited to be one of its members (Scott 1915).

Armstrong repeated his attack on the leaders of the Royal Society and their secrecy in the 15 July 1915 issue of the *Times* (HEA 1915c; A. Fleming 1915a and 1915b). It had taken the Royal Society six months, he complained, to send a questionnaire to its fellows about advice and expertise that could be sent to the War Office. There was further support from the ailing Henry Roscoe in Manchester that the Royal Society should constitute itself the Parliament of Science rather than the British Association. The latter was meeting in Manchester in August 1915, when Schuster was to be elected president; but Armstrong thought that little would come of the meeting to put science on a war footing. In a final diatribe at the end of the year, he declared that Britain was destined for perdition unless politicians like Prime Minister Herbert Asquith took advice from practical men rather than from lawyer-politicians (HEA 1915f).

Armstrong's pessimism was unwarranted. As a result of the pressure from the Royal Society and other scientific societies, in May 1915 Sir William McCormick, the president of the Board of Education, recommended that Parliament should set up a new and permanent organization for the promotion of scientific and industrial research (McCormick 1930). Initially set up under the Privy Council, this new organization became in December 1916 an independent department of state known as the Department of Scientific and Industrial Research (DSIR). It was to well serve the needs of higher education and industry until its abolition in 1965.

But Armstrong never ceased criticizing the Royal Society's wartime activities. When he heard that it was forming a Conjoint Board of scientific societies in March 1916, he castigated it as a waste of time, for "the Royal Society itself is such a board already, the highest that can be secured." The Royal Society, along with the government, was creating a scientific bureaucracy and "through its own weak action," Armstrong averred, it was "shattering the fabric and the public reputation of science" (HEA 1916b). The Conjoint Board was dissolved only in 1923 (Jackson and Watts 1923).

Earlier in the summer of 1915, the Society of Chemical Industry had met in Manchester and heard its president, G. G. Henderson, urge the need for closer cooperation between manufacturers and academic chemists and for a greater appreciation of the application of research in industry. Armstrong took up this theme in the lecture he gave at the meeting, where he urged manufacturers to avoid graduates from the older universities and instead employ men of the right stamp: men of "generous mind, good presence and real ability."

The success of German chemical industry is due to two causes; first, the fact that their universities are practical institutions properly supported by the State, and in touch with the educated community; and, secondly, the factories are in the hands of experts. The development has been from within, though it has received great assistance from public sources. The academic party has worked under conditions of freedom, of *Lehrfreiheit* and *Lernfreiheit*, whereas in England the tradition that it is necessary to be well-read instead of well-practised has prevailed. (HEA 1915d)

Another suggestion he made (in 1916) was that the interests of pure and applied chemists would be better served in fighting the war if these chemists formed a federation of academic and professional chemical societies—an idea that he had previously floated in 1894 (HEA 1916e). His model would have included all the chemical societies of the British Empire; thus it was a good deal more ambitious than the amalgamation achieved with the formation of the Royal Society of Chemistry in 1980 (a merger of the Chemical Society, the Faraday Society, the Society for Analytical Chemistry, and the Royal Institute of Chemistry).

A constant theme of all of Armstrong's speeches and articles during the war was of the uselessness of an Oxford education. He was, therefore, elated when W. H. Perkin Jr., professor of organic chemistry at Oxford, succeeded in introducing a fourth year in the chemistry degree that would be entirely devoted to pure research. In addition, Perkin had persuaded Oxford to institute a DSc degree for a research dissertation (HEA 1917a). Another Armstrong target was the composite University of London, which, he felt, strangled the freedom and independence of the three principal colleges that had joined together in 1908: Imperial, King's, and University College. By admitting so many other different grades of higher education, it had made the University a "pool of academic jealousies." And, by "compelling the worship of the iniquitous examination fetish, the London system has both deprived the teacher of his freedom and imbued the student with false ideals; until this idol has been cast down, there can be no progress" (HEA 1919j). In his view the three principal colleges should be independent universities, with Imperial a "great centre of Physical Science and Engineering; University College, a great centre of Biological Science; and King's College, a great centre for Arts."

Armstrong's vision was partially realized in 2006, when Imperial College withdrew from the composite University of London to become a fully independent institution, thenceforth named Imperial College London.

Gas, Coal, and Oil

Deprived of a chance to play a positive role in aiding the war effort by way of research, as his former colleagues at Imperial College were able to do (H. Gay 2007, 114–46), Armstrong found time to expand on some of his previous interests in the efficient use of fuels, as well as smoke abatement in towns. He also developed a newfound interest in the economics of Indian indigo plantations, now that synthetic indigo was steadily replacing the natural plant dyestuff.

At the Belfast meeting of the British Association for the Advancement of Science in 1902, Armstrong was lampooned for asserting in his sectional address that "a scuttle full of coal excites no emotion in the literary mind" (HEA 1902, 389). Throughout his career he claimed that the future of civilization depended upon coal and that nonscientific politicians and businessmen should treat it with respect and wonder, as chemists did. In 1885, addressing the Iron and Steel Institute, Armstrong had urged investigations on how better to recover useful products from coal. Progress in this area was made during the following decade. In his Royal Society of Arts lectures on "Fuel and Its Future" (March 1908), Vivian Byam Lewes, professor of chemistry at the Greenwich Royal Naval College, drew attention to investigations of coal tar carried out by Lewis Wright in 1886 in which Wright had sought to improve ways of manufacturing coal gas. At the RSA meeting two months later, Armstrong asked: "Would it not be better to deal with [coal] so as to produce a large proportion of residuals, and a different class of coke?" He raised the same question in his second paper to the Iron and Steel Institute in the same year (HEA 1908d).

The provident use of coal became an issue again at the British Association meeting in Sheffield in 1910 when Armstrong advocated the development of smokeless fuel, and again at Birmingham in September 1913 when he arranged that Section B would hold a discussion on how chemists might improve efficiency of using coal and the fuels (like domestic gas supplies) derived from it. Among the speakers were George Thomas Beilby (1850–1924), a Scottish chemical engineer, who spoke on fuel economy and the potential of the low-temperature carbonization of coal; and William Bone (1871–1938), who had just joined the Fuel Technology Department at Imperial College. Bone gave a general account of the uses of coal gas, often called "town gas" (Bone 1913). Armstrong had long hoped that chemists and engineers would find a solution to the production of fully carbonized coal, a smokeless fuel that would virtually abolish air pollution of towns and save the valuable volatile products for chemical industry in a more efficient manner. He was convinced, like the chemist-economist Stanley Jevons before him, that supplies of coal on which the British economy was built would run out.

The development of a smokeless fuel had been considerably advanced by two inventive individuals known to Armstrong and Ayrton initially through their development of instrumentation for the rapidly developing electrical industry. They were F. W. Salisbury-Jones and the engineer sometimes called the "English Edison," Thomas Parker. The two men developed the smokeless fuel they named "Coalite" between 1904 and 1906 at a plant in Wednesbury, near Birmingham. From Armstrong's historical remarks in a later lecture, it would seem that he was a formal or informal consultant on the project (HEA 1921j, 393).

Coalite is a variety of coke produced at lower temperatures. It was given negative publicity in 1907 by the coal gas industry, which claimed that its own coke from gas works was just as smokeless, and cheaper. The bad publicity struck home because the initial production of Coalite varied in quality since its makers were initially unaware that coals differed in the amounts of ash they produced when submitted to low-temperature carbonization in the absence of air. The combustion retorts erected at the works at Wednesbury and Barking also proved to be badly designed.

The company employed Sir William Preece, the former post office electrician, as its chairman, and it was Preece (who knew Armstrong from their joint membership in the Athenaeum) who brought Armstrong in to act as a consultant. Armstrong became particularly interested in the tar produced by low-temperature combustion of coals, and his analyses formed part of his lecture at the British Association meeting in Belfast in 1910. He was called back as a consultant again in 1916 in consequence of a critical talk he had given to the Society of Chemical Industry that year. By then a second Coalite works, modelled upon a prewar German plan, had been erected at Barnsley in Yorkshire. Armstrong found fault with the ovens, which easily became jammed, and pointed to other design errors. In his report in 1919, which he later quoted in extenso to the Royal Society of Arts in 1921, Armstrong was certain that the tars produced, if not the Coalite itself, would pass muster as a fuel for the British fleet (HEA 1921j, 397). Acting on Armstrong's advice, the Coalite company not only improved its smokeless fuel, but perfected a second product, "coal oil," that could be sold to the Admiralty and also used as a motorcar fuel.

But there was also an educational problem regarding fuels, he decided, because engineers were entirely practically minded and lacked analytical skills. They were content to shape materials into useful forms rather than analyze the constitution, formation, and transformation of the forms themselves. Out of all his Finsbury and Central students who specialized in engineering, only one, Herbert Arthur Humphrey (1868–1951), the inventor of the gas-powered water pump, had taken a deep interest in chemistry. Modern industry needed the cooperative skills of the chemist, engineer, and capitalist, as had happened in Germany.

Armstrong's remarks on smokeless fuel at Birmingham in 1913 were attacked ad hominem by the "vulgar" editor of the *Journal of Gas Lighting*, who denied Armstrong's claim that chemists were excluded from the gas industry. In turn, Armstrong retorted at length in the November issue of *Chemical World* that the editor's sweeping tone, "full of bad grammar," was utterly mistaken. As evidence, he cited the case of a friend named Coleman who had been employed as a consultant by the Birmingham Gas Company but not allowed inside the gas works! He apologized if he had upset the industry, but it really was inefficient. He cited the example of the manufacture from coal of "water gas" (a mixture of carbon monoxide and hydrogen), which employed twice as much fuel as needed. The industry still failed to remove sulfur from town gas, which consequently ruined people's furnishings. The editor and "those who live in g(l)as(s) houses should not lightly cast stones." They should look to develop "smokeless" coke as a fuel to replace smoky coal, extract the sulfur, and make a soft fuel for the fireplace. In the future, he predicted, town gas would be diluted with water gas and used for *heating* houses, not for lighting them. Electricity was also in future to be generated at the gas works as a byproduct of gas making (HEA 1913b).

Throughout his career as a chemist and as someone who taught chemistry to would-be engineers, Armstrong had been repeatedly concerned with Britain's reliance on coal as a fuel and chemical feedstock. His anxiety increased as wartime conditions reduced coal supplies, with the implication that both coal and gas supplies might have to be rationed (HEA 1916d; 1918e, 9). Armstrong's proposed solution was control of the coal industry by a National Fuel Board. He returned to the subject, relating it to fuel economy, at the Society of Chemical Industry meeting in Edinburgh in July 1916 (HEA 1916c). Armstrong was one of the several voices whose concern about Britain's use of its coal resources led to the formation of the DSIR's Fuel Research Board in 1917 under the direction of Sir George Beilby.

During the First World War, Armstrong repeatedly urged the prohibition of the use of pure coals for domestic use and the taxing of coal to raise income for research on the efficient production of coke. "There was great hullabaloo," he recalled, when he jokingly (and crudely) called for "a little lynching" of members of the gas industry at a lecture he gave to the Society of Chemical Industry at Nottingham in 1917, and the Society refused to publish the address.[10] Later, at the British Scientific Products Exhibition at Westminster arranged by the British Science Guild in the summer of 1919 (Anon. 1919), he pictured a future with towns fitted with central fuel and power stations that would distribute "all the forms of fuel and power required by the public within their areas" (HEA 1919e, 7).

In his rumbustiously dictated autobiography in the *Times* in September 1918, Admiral Lord John Fisher prophesied that future wars would be fought in the air rather on land or sea. Armstrong agreed but suggested that the army, navy, and air force would need more brainpower in future, and he warned that Germany would rapidly recover its position as a world power despite the vengeance the Allies had placed upon the country's economy. Fisher had argued that the armed forces would need more oil, but Armstrong wondered where this oil would come from, having received (faulty) intelligence from an American friend that American oilfields were already nearly exhausted. Given this evidence, he thought it would be unwise to build ships fuelled only by oil, as Fisher was suggesting (HEA 1919g; Fisher 1919a; HEA 1919h; Fisher 1919b).

Both men were mavericks, and they liked each other's style and panache. Armstrong returned to the subject in the summer of 1920, a month before Fisher's death, arguing again that it was a mistake to use oil as a maritime fuel (HEA 1920b). The trigger was the news that the furnaces on the ocean liner *Aquitania* were to be converted from coal to oil and would need to transport seven thousand tons of oil to complete the return journey from Liverpool to New York. This quantity was enormous, he asserted, and it constituted profiteering by the Cunard Company at the cost of future generations. He reiterated the (false) claim that Americans had already run out of oil and were desperately searching for new supplies across the world, "so much do they fear the coming exhaustion." A further newspaper letter reiterated the desperate need to find an efficient way of carbonizing coal to retain the volatile products so that coal was no longer wasted (HEA 1920f).

Natural Indigo

During the nineteenth century, indigo from Bengali plantations became the world's most utilized blue dye and an important commodity in British colonial trade. It was also the subject of great chemical interest. When Adolf Baeyer devised a synthetic pathway to indigo (specifically, to the dye substance indigotin) in the 1880s, there was initially no cause for alarm

---

10. Mentioned by Armstrong in HEA 1921j, 392–93.

from the owners of Indian plantations. However, by the early 1900s, synthetic indigo production was overtaking production of the natural product. Armstrong had already foreseen the lessons of what he called "the downfall of natural indigo" in 1901. He noted in a long letter to the *Times* that he had heard from India that planters there were holding a meeting to unite in their cause, which Armstrong thought was shutting the stable door when "the steed was stolen," especially since some of them wanted to exchange their indigo for sugar production (HEA 1899b; 1901a; 1903b, 144). It was one of the intentions of the founding of the Imperial Agricultural Institute at Pusa in 1904 to find ways of improving the growth and extraction of the natural dyestuff so that it could compete on equal terms with the synthetic product. However, by 1914 production of the indigo plant had virtually ceased in India apart from the province of Bihar. Whereas in 1898 some 1,700,000 acres had been devoted to indigo cultivation, that area had now shrunk to only 150,000 acres, and the price had fallen from seven shillings per pound to only three shillings.

All seemed lost until the outbreak of war in 1914 meant that supplies of synthetic indigo were impossible to obtain from Germany. The British government, in league with the Indian government, was forced to seek ways of increasing production of natural indigo in India. During his final days in Calcutta in 1914, Armstrong had published a letter in the Indian English-language daily newspaper the *Statesman* in which he argued that it would still be possible to make a living from natural indigo in competition with the synthetic material.

> The recovery of the position of natural indigo may be a costly business, but if one-tenth of the zeal put by the Germans into the production of the artificial pigment be brought to bear on the problem I believe it will be solved satisfactorily. (HEA 1918d)

The letter led to considerable discussion within India, with the result that a conference was organized by both the British and Indian governments in Delhi in February 1915. The meeting was attended by planters, government officials, and trade representatives, and the auditors heard evidence from both Armstrong (by letter) and Bernard Coventry, who by then had become the director of the Imperial Agricultural Institute at Pusa. Both men argued that planters, with the help of scientific experiments, needed to turn the natural product after its extraction into a paste similar to the paste that the synthetic producers sold commercially and at a similar standardized strength and concentration (Kumar 2012, 282). Clearly, the help of a chemist was needed, preferably at the Pusa research institute.

Accordingly, in June 1915 a letter was sent by the Indian viceroy to the India Office in London that, instead of answering the appeal immediately, proceeded to organize another conference in London in September 1915 (Kumar 2012, 285). Among the discussants were Armstrong and Arthur George Perkin (son of W. H. Perkin Sr.), the professor of color and dyeing chemistry at the University of Leeds. Both men supported the idea that had been floated at the Delhi conference, and Armstrong suggested that indigo planters should cooperate and send their extracted products to a central depot where the products could be made into a standardized paste of 20 percent indigotin strength for export in paste form. Armstrong made it clear to planters that they had no option but to cooperate and organize if their businesses were to survive.

Perkin suggested that a supervisory committee should be formed for the future management of research in India, and Armstrong agreed. Inevitably, the two chemists found themselves the backbone of the committee. However, it was not until the beginning of 1916 that India House asked it to recommend a chemist suitable for overseeing the transformation of the indigo industry in India. The choice fell on Armstrong's former research student William Alfred Davis (1875–1939), who, after being taught by Charles Stuart at St. Dunstan's College, joined Armstrong's course at the Central College before beginning his career at Courtaulds to solve problems connected with the chemistry of crepe (E. F. Armstrong 1939). He had returned to the Central in 1897 as Armstrong's demonstrator to the engineers taking heuristic chemistry classes. In 1908, Armstrong recommended Davis as works manager at the Lawes factory for making chemical fertilizers at Deptford, only for him to transfer to the Rothamsted Agricultural Research Station in 1911, where he worked on the formation of carbohydrates in plant leaves. To Armstrong's mind (and Perkin agreed), Davis was the ideal person for the Indian appointment.

Davis leapt at the opportunity and arrived at Pusa in May 1916 (Davis 1918). His first task was to prepare a paste that was comparable to the quality of the synthetic paste used by dye-makers in Britain and Germany. That task was quickly achieved, but at the cost of heavy freight charges that made the Indian product very expensive. "Obviously," Armstrong wrote, "it is undesirable to send water about the world if dry materials are procurable" (HEA 1918d). The India House advisory committee to which Armstrong belonged agreed to investigate the problem of producing a natural-indigo dry paste. Armstrong was joined in the investigation by the chemist Reginald B. Brown (1870–1945), who had worked for BASF on synthetic-indigo production before the war.[11] Production of a dry cake proved a difficult problem and one that was still unsolved by 1918, by which time the reorganization of the British dyestuffs manufacturers meant that Britain now had the ability to produce synthetic indigo. By then, however, Davis had identified a much more serious problem with Indian agriculture itself, namely a shortage of phosphate fertilizer for the indigo lands, as it was both costly and difficult to procure (Davis 1918, 388). The lesson— for inevitably there was a lesson to be learned from the debacle—was that the British and their Empire should have made adjustments to the Indian market when the production of synthetic indigotin had been first announced by German chemical manufacturers. Instead, the agricultural industry had been allowed to go under.

The problem of Indian soil exhaustion was also taken up by Armstrong in a lecture to the Royal Society of Arts in May 1919, wearing his cap as vice chairman of the Lawes Agricultural Trust (HEA 1919e). He recognized that the recently deceased Sir William Crookes had been correct in warning that chemists had to solve the problem of nitrogen fixation if the world was not to starve. This issue had been potentially solved by Fritz Haber's ammonia synthesis from atmospheric nitrogen. However, the shortage of phosphate fertilizer was now another problem. Davis had alerted him to the irregular fluctuation in the annual yields of the indigo crop. Like Walter Leather before him, he put this phenomenon down to soil exhaustion—something Liebig had warned about in the mid-nineteenth century. The lack of Indian soil fertility was, Armstrong suggested, also the cause of the poor nutritional quality of Indian food and the poor milk yields for Indian cattle. Perhaps the cause was, as Leather and others had suspected before Davis's investigations, the predominant exploitation of European agricultural husbandry and the abandonment of traditional native methods that used animal and human dung as fertilizer?

In February 1922 Armstrong received a telegram from Davis announcing: "Coming home. Indigo discontinued" (HEA 1922c, 6; 1922f). Armstrong was appalled, pointing out that coal supplies (the other subject of great wartime interest to him) were going to run out. At that point there would be little coal tar produced and "the raw material from which dyestuffs are made [would] be no longer available to the required extent." Further research on natural indigo was abolished, as was Davis's post at Pusa as an economy measure. All resources were to be put into university

---

11. I have found no details for Brown's career other than that he became a consultant chemist for the dyestuffs industry according to the 1939 census.

education instead. Armstrong protested to the India Office, but to no avail. Saving the natural-indigo industry had proved a hopeless task economically, and by 1922 it seemed more important for India to catch up with Europe's industrial development in chemical industry. On his return to the United Kingdom, Davis worked for some time with Armstrong's pupil John Vargas Eyre at the Linen Research Institute in Belfast before becoming a research chemist with Lever Brothers in Port Sunlight, Mersyside. It was from there that he gave a full account of the indigo story (Davis 1924). He returned to London in 1927 to become chief analyst at the Distillers Company in Epsom, run by John Eyre, where he had plenty of opportunities to socialize with Armstrong.

Nutrition and the Chemistry of Food

The first known public lecture that Armstrong gave was at the invitation of Henry Roscoe in Manchester in 1875. His subject was "Food," and he included demonstrations to illustrate photosynthesis and digestion. Food chemistry remained a background interest throughout Armstrong's life. It was further stimulated when he joined the Lawes Trust Agricultural Committee and became involved in enzyme research. Armstrong went on record that all of his children had been "hand-fed," meaning that they had always been given unpasteurized raw milk (HEA 1912c). He believed that heating milk was likely to modify the constituents of natural milk by destroying enzymes that played important roles in digestion. Bacteriologists, he thought, grossly exaggerated the possibility that bovine tuberculosis could be transmitted to humans through milk. He claimed to know that TB was transmitted through the nasal passages only. His letter to the *Times* on the transmission of TB and the milk supply prompted a reply from the Medical Officer of Health for Newcastle-upon-Tyne, whose name was also Henry E. Armstrong and who had previous experience of having been confused with the London scientist. Health Officer Henry E. Armstrong sharply disagreed with chemist Henry E. Armstrong's views, asserting that it had now been shown that pasteurizing milk had no effect on its nutritive value and that TB could indeed be transmitted from cattle to humans via milk (H. E. Armstrong 1912). Subsequently, chemist Armstrong took care to always sign himself as "Professor" and to give his address to avoid confusion with his northern namesake.

Naval blockades—and the shortage of manpower after men (and horses) were diverted to military duties—produced serious problems for British agriculture and food distribution as war took hold in 1915. Although the government introduced price controls on staple foodstuffs like flour, sugar, and milk, rationing was not introduced until February 1918, following devastating U-boat attacks on British cargo vessels. Armstrong, who was much more concerned with how science could best be organized to fight the war, made no comment about wartime food until February 1918, when he warned about the need for fat in the human diet and how the current methods for the breeding of pigs was exacerbating deficiencies (HEA 1918a). Referring to a report by the Royal Society's Committee on Food (he had not been invited to be a member), he deplored its emphasis on calorific values and lack of consideration of quality. "It were far better to give us less, but more palatable bread and to feed some of the 'offal' now forced upon us to stock," he wrote. Fat was essential to diet. And because we grew no crops that contained quantities of fat, we had to rely on animals. The pig was the key. "The pig has a wonderful power of turning starchy foods into fat, and a truly wonderful power of working up both vegetable and animal waste of all kinds." Fats were also essential to the conduct of war; manufacturing smokeless gunpowder required glycerin, which was made from soap, soap having been made from pig fat.

> The introduction of pig-feeding throughout the country, not only as a war-time precaution, but a measure of thrift, as a means of preventing waste, is a policy to be encouraged by every means in our power. In the future the pig should adorn every set of garden allotment, and be especially catered for. (HEA 1918b)

A month later he turned his attention to variety in diet because of the need for accessory factors that were mostly absent after food was cooked. Such nutrients were essential since milk supplies had dwindled, and especially since he believed sterilization destroyed such factors. He attacked those "amateur" advisers who worried about the quantity of food available on plates as opposed to its nutritional quality. It was absurd that pulses were claimed to be superior to meat because they contained more protein. The issue was that the types of protein were different. While economy via rationing was a vital strategy during wartime, "our diet should be varied in every possible way, particularly in the case of children" (HEA 1918c).

By the end of 1918 there was a severe shortage of milk supplies. Armstrong was worried about the health of children. "We are alive to the fact that the children of the masses are not properly nourished, and milk, as is well known, is the food of children; in future, to improve the food of our children, we must provide a better and far more ample supply of milk" (HEA 1919a). The problem of wheat supplies had been solved, but whereas wheat could be stored, milk could not be brought from a distance and stored (HEA 1919b). Nevertheless, it was not until 1921, long after the war, that free milk was provided in schools.

A related subject for Armstrong's advice was beverage alcohol. In 1909, the Welsh socialist journalist Ernest Edwin Williams (1866–1935) became one of the founders of the True Temperance Association, which propagandized for drinking beers and spirits in moderation (Williams 1896a; 1896b; 1917). When war broke out, the government was subjected to a great deal of pressure by the stricter temperance societies to prohibit the sale of alcohol on the grounds of morale, and the government did restrict the opening hours of pubs in industrial towns. Williams, who saw this restriction as an infringement of personal liberty and freedom, accordingly solicited a suite of essays by a group of anti-prohibitionists to answer an anonymous pamphlet entitled *Defeat*. The pamphlet had argued that the war was being prolonged because of the overconsumption of alcohol by British troops (Williams 1917). The six essayists were the Royal Institution's doughty treasurer, James Crichton-Brown (who argued that from a medical standpoint, alcohol was a good food); Major General Stephen S. Lang (who said that denying the army alcohol would harm morale); the Labour MP and trade unionist Will Thorne (who viewed attacks on alcohol as class warfare); the controversial dean of Durham, Herbert Hensley Henson (who argued that prohibition caused conflict with human equity and thereby had bad consequences); and the Catholic writer G. K. Chesterton (who suggested that spokesmen for prohibition like the eugenicist Caleb Saleeby, who was an adviser to the Ministry of Food, were fanatics).

Williams asked Armstrong, as a leading scientist, to sum up in a sixth and final essay how the chemist saw the issue.[12] There was no question, wrote Armstrong, that temperance supporters would be brighter, more companionable, and more imaginative if they had a drink! Women who supported prohibition laws would do better to improve their husband's home life so as to make drunkenness unattractive. In the past, drinkers had been men with "neither a hobby nor interests outside their business." Drunkenness was bound to recede once social conditions had been improved—he was thinking of better food, housing, education, and amusements. Speaking personally, Armstrong admitted:

> I have never taken spirits of any kind consciously, though accustomed to drink wine and beer from my boyhood onwards, not in any regular way, but only at intervals, when either opportunity has served or the

---

12. Armstrong drew heavily on a talk he had given to the Institute of Brewing at the Criterion Restaurant, Piccadilly, on 26 May 1913. See HEA 1913c, with discussion.

inclination, if not the desire, has come upon me—as it distinctly does at times. Obviously, I inherit a marked power of toleration. I have what appears to be an unusually acute sense of taste and smell. Perhaps I owe the accentuation and retention of these faculties to the fact that I have never smoked. (Williams 1917, 46)

At home, he explained, he usually just drank water, but away from home, "being an epicure as regards water" and usually finding it was nasty, he drank beer. Abroad, too, he commonly drank the local beer "as being the one safe drink the world over." He found the coffee and tea commonly served with meals in the US, Canada, and Australia quite nauseating and felt that it reflected a low standard of nutrition in those countries. But their use at mealtimes surely indicated a universal human desire for stimulants that water alone could not satisfy. Then, clearly though implicitly referring to his alcoholic father:

> In my early days men in my circle were hard drinkers of spirits, and I had much reason to lament the evil effects produced. The circumstance had something to do with my avoidance of spirits, though the main reason, I am persuaded, has always been that they not only do not appeal to my sense of smell but even exact feelings of nausea. (Williams 1917, 47; HEA 1917b)

The effects of alcohol had puzzled him as a chemist since its physiological effects remained unaccountable in chemical and medical terms. There was no proof that alcohol was inimical to life, and he ridiculed insurance companies who argued that teetotalism prolonged human life. His own conclusion as an observer of human behavior was that drinkers were more imaginative, active, and adventurous than teetotallers; and it was only because they were prepared to take greater risks that they were subject to more accidents (HEA 1913c). He concluded the essay for Williams's book by considering the latest scientific work concerning the effect of alcohol on solutions. Since alcohol was so like water itself in chemical behavior, chemists had to look at the constitution and behavior of barley and seeds that were used in brewing and viniculture and compare them to the behavior of human and animal cells. The essay concluded, therefore, with a short lesson in enzyme and hormone chemistry. Because alcohol promoted digestion, he concluded, it was best taken with meals.

All of Armstrong's views and worries about food supplies were summed up in the prestigious Cantor Lectures that he was invited to present to the Royal Society of Arts in 1919 (HEA 1919f). These lectures told the story of food constituents and their assimilation by enzymes and hormones; they also covered the role of what he termed "functional adjuvants," that is, vitamins. These adjuvants had been shown to play vital roles in the prevention of scurvy and beriberi. The lectures included some vintage attacks on the attitude of the Liquor Control Board in wanting restrictions on the sale of alcohol (voluntary moderation was better, Armstrong averred, as he had stated in Williams's booklet) as well as on the anti-vivisectionists' attacks on vital physiological experimentation. The final lecture went into greater detail on the function of fats in nutrition and on ways to improve milk supply than he had been able to give in recent letters to the *Times*.

Although Armstrong favored beer rather than wines or spirits, he attacked the chancellor of the exchequer's proposal to raise taxes on French wines in 1920. Prohibition had been necessary in America, he thought, because of the drunkenness caused by the availability of cheap spirits. Drunkenness was rarer in European countries, where wine was common, so the growing popularity of wine drinking in the UK since the war was to be welcomed. The exchequer's proposal not only damaged French interest at a time when we should be encouraging the entente cordiale rather than any intercourse with Germany, he wrote, it was actually a perverse action. If the exchequer actually decreased the tax on French wines, the increased sales from imports would produce the necessary revenues and would raise the British in the social scale. The sale of wines was to be encouraged "not only on aesthetic and dietetic but also on moral and racial grounds" (HEA 1920a).

Wartime Consultancies and Addresses

The Herbartian secularist Frank H. Hayward (1872–1954) and the socialist and Rudolf Steiner anthroposophist Arnold Freeman (1886–1972) developed a scheme to replace the teaching of religious knowledge in elementary schools by special days devoted to quasi-liturgical ceremonial and celebratory treatments of great personalities from history such as Saint Paul, Alfred the Great, and Joan of Arc, as well as of leading ideas such as the new League of Nations, agriculture, and science. The plans were an expansion of the already-established school programs that celebrated May Day or Empire Day, and they were reminiscent of ideas that Auguste Comte had introduced into his Positivist Church in the nineteenth century. The two men invited comments from Armstrong and several other educationists, which were then published as prefatory matter to their book *The Spiritual Foundations of Reconstruction: A Plea for New Educational Methods* (Hayward and Freeman 1919).

Armstrong was not greatly impressed, seeing the suggestions as "the work of enthusiasts and idealists." Although the scheme might work in the authors' hands (Freeman did employ it in his Sheffield Educational Settlement school), where were the teachers to come from who would be competent to lead such celebrations?

> I know from sad experience, the difficulty of introducing a novel method. My advocacy of heuristic teaching has led to very little—and has been a good deal talked about, but outside my own circle of clever pupils, there are very few who have even a conception of its meaning, let alone the ability to carry it into execution. (Hayward and Freeman 1919, x)

He concluded his brief contribution with a typical misogynist statement: "Very few teachers will be competent to carry such a scheme as yours into practice, especially where women predominate."

This instance is a good example of the way Armstrong tended always to temper praise with criticism—a trait he deployed constantly when delivering votes of thanks to speakers. When his oldest friend, Horace Brown, retired from Worthington's brewery at Burton in 1894, he moved to London to act as a consultant chemist and carry out independent research. In 1916, in honor of his jubilee as a brewery chemist, he was invited to address the Institute of Brewing on his experience of applying science to brewing (H. T. Brown 1916). Armstrong chaired the session and lauded Burton-on-Trent as "the most stimulating scientific centre in the country in the 1870s" (HEA 1916e, 349). In his vote of thanks, he was full of praise for Brown's engaging and valuable reminiscences but noted that, in contrasting Pasteur's methods with those of Lister, Brown had failed to see that they represented "two entirely different and complementary habits of mind—the physical and the naturalistic." However, Armstrong continued, this failure was interesting because Brown himself (a polymath) had both the attributes of a physical scientist and those of a naturalist. Pasteur's work had transformed the brewing industry; therefore, as Brown had indicated, it was regrettable that the Lister Institute (founded in Lister's memory) had no official connection with the industry.

The war decade ended with an invitation to address the Old Students' Association of the Royal College of Science in the Imperial College Students' Union Building. Armstrong delivered a splendid autobiographical sketch of his experiences at South Kensington from the time of his arrival as a student at the Royal College of Chemistry in the summer of 1865 (HEA 1920c). Many of his former students, like Forster and Moody, took part in the occasion, and there was a vigorous debate over Armstrong's condemnations of London University's system of examinations. Here he repeated his proposal, described earlier in this chapter, to dismember the University of London into three independent institutions devoted, respectively, to math, physics, chemistry, and engineering (Imperial); biological sciences (UCL); and the Arts (King's). He was also conscious that Oxford and Cambridge carried social advantages denied to students in London. His solution was to rebuild King's at Kenwood House and endow it with playing fields, gymnasia, debating halls, and student accommodations, where students from all three colleges could board overnight and socialize. Under this federal scheme, students would also be free to attend courses in each other's colleges, so that arts students could acquire some science, and science students some further literary knowledge beyond what they had naturally acquired during their schooling. The important point was that each of the colleges should award its own degrees; this proposition was in opposition to that of Sir Philip Magnus, who wanted the colleges to award their own diplomas but under the umbrella of a common University of London degree (P. Magnus 1920). Armstrong felt that the diplomas already issued by Imperial College and its constituent colleges were equivalent to a degree.

In 1920, Armstrong was also invited to address the Ceramics Society in Stoke-on-Trent—probably by the pottery chemist Joseph W. Mellor. In this rambling lecture, he tried in layman's terms to explain what the art of inquiry was, taking the potter's limestone as his material and the work of Joseph Black as his historical exemplar of research (HEA 1921d). He mentioned meeting the potter Bernard Moore who, Armstrong declared, was "like a musician playing a violin and constantly varying his use of the bow; he varies the fumes of his fire with the conscious object of obtaining mixed effects but instead of playing set pieces he extemporizes, and the variation is therefore always a surprise and he has the joy of the artist in creating novel effects." The lecture ended with reflections on whether state aid to the relatively new research stations (funded by the DSIR) would be effective. These organizations, he felt, had been "bribed" into existence rather than formed by independent action.

> I have no hesitation is asserting that, unless the scientific spirit be carried in the works and until it be made part and parcel of their practice, the movement will be a failure: the mere occasional improvement of a process through information brought in from outside, from a common research laboratory, will not suffice, the information must be used properly, i.e., scientifically, with understanding, if it is to be of real value. This our enemies in the war have long known. (HEA 1921d, 69)

And he noted drily that when chemical factories on the Rhine had been visited by a representative deputation of British Chemical Manufactures in 1919, it had observed that "German chemical industry has been one stupendous organisation for effecting and promoting the application of science to industry." He took heart, however, from the pottery industry, which had developed a hard form of porcelain during the war through the research efforts of Mellor and Moore.

Although Armstrong had always proudly claimed to have been "made in Germany," the publication of the manifesto signed by a group of German intellectuals defending Germany's invasion of Belgium upended the pride he had had in his Leipzig education. Consequently, he took a leading role in the decision to cancel German chemists' ordinary or honorary memberships in the Chemical Society. He was clearly hurt by the fact that he had not been called upon by either government or the Royal Society to play any vital role in fighting the war. Age was presumably not the reason for his exclusion, since there were others in their sixties who were called upon—though the fact that he had no laboratory or students obviously meant that he could not help with chemically relevant studies such as improved armaments, medical supplies, or antidotes to poisonous gases, as many of his former pupils were doing. The probable reason was that many saw him as a maverick with whom cooperation might be difficult. This ostracizing, whether overt or implied, forced him to play Fifth Business once more and become a critic of the organization of the war effort. Others, too, were critical of that effort—particularly concerning the fact that Britain's education system had failed to educate sufficient manpower for science and technology. It was therefore a disappointment to him, as the next chapter shows, that he was once again left on the sideline as a critic when it came to advising the government on the improvement of science education, and when he found his heuristic campaign denigrated.

12

Heurism Denigrated

> We did not see why every child should walk around Ireland
> before he could believe it was an island.
> —C. L. Bryant 1950, 140

Many teachers' organizations and pressure groups emerged between 1870 and 1910. The Mathematical Association, founded in 1897 to improve mathematics teaching in secondary schools, is a good example. A Modern Languages Association and a Geographical Association had similarly been founded in 1892 and 1893 to promote the teaching of these school subjects. In 1900, in order to counteract their feelings of benign neglect in the public schools, where classics were snobbishly thought to be superior to the sciences, the Association of Public School Science Masters (APSSM) was founded. Women science teachers followed in 1912 with their own association. The Classical Association (1903), Historical Association (1906), and English Association (1907) completed the sextet of prewar secondary school subject associations (Layton 1981). In 1919 the APSSM opened its ranks to all secondary school science teachers under the title Science Masters Association (Layton 1981; Price 1994) and began the important journal *School Science Review* on the model of the Mathematical Association's *Mathematical Gazette*, which had been founded in 1894. Since their foundations, these professional organizations have played an effective role in advocating for improved syllabuses and curricula, better and fairer examinations, and laboratory facilities.

Prewar Skirmishes

The APSSM was conceived originally by four Eton College science masters, R. C. Porter, W. D. Eggar, M. D. Hill, and H. de Havilland. They held their first meeting in London in 1901 under the chairmanship of Oliver

H. Latter, a master at Charterhouse, with Sir Henry Roscoe as their first president. Although Armstrong attended this inaugural session and most of the APSSM's meetings during its first decade of existence, it appears that his interventions were not particularly welcomed. It was not until 1910 that he was elected by the members to be the association's president, six years after the presidency of William Tilden. The choice of Tilden rather than Armstrong was surely justified by the fact that he had taught chemistry at Clifton College, Bristol, for several years before becoming an academic chemist.

One of Latter's pupils was Arthur Vassall, who taught at Harrow School. Hill and Vassall became the leaders of the APSSM and were instrumental in getting biology taught in public schools, the momentum of which led to a "General Science" movement in secondary schools (Anon. 1948–49). It was Vassall, in a paper described by *Nature* as of "remarkable lucidity," who made the APSSM aware of the defects of faculty psychology. Citing recent psychological studies and opinion, he denied that powers of, say, observation, were increased through scientific studies, or that training in one scientific field "overflowed" into others. More broadly, he concluded that schools would have to cultivate wider areas of knowledge and interest and pay more attention to content and subject matter. Armstrong refused to accept the experimental finding that was the basis of Vassall's inference (Daniell 1912, 394).

Although the tenets of Armstrong's discovery method were adopted positively in a few schools and, at a more advanced level, in his own Central College, his ideals were often debased into nothing beyond a heavy emphasis upon measurements in physical sciences. The Ministerial Code of 1904 had directed that secondary school instruction was to be given practically in two unspecified sciences. In practice, before Latter's intervention, these came to mean physics and chemistry; astronomy was thought too esoteric, anatomy and physiology were considered too indelicate or bound up with ethical problems of vivisection, and geology was still identified with religious controversy (Cane 1959). Armstrong had always urged syllabuses based upon a science of everyday life, which would include elements of astronomy, geology, and biology, besides physics and chemistry; but in many teachers' hands the enthusiastic probing for the inculcation of "scientific method" left pupils with very little factual knowledge and a depressed scientific vocabulary. Dissatisfaction with this state of affairs found an outlet through the APSSM and the British Association's Education Section L, which Armstrong had founded.

At the British Association in 1902, Charles William Kimmins (1856–1948), who was to become the chief inspector of the London County

Council's Technical Education Board in 1904, praised an earlier BAAS report on chemistry teaching and suggested that similar reports on other sciences were needed. "The biological side should receive as much attention as that already given to the physical side" in schools, he asserted; too much time and money were wasted teaching sciences in separate compartments (Kimmins 1902). In a paper on "a rational curriculum" delivered in 1903, Armstrong tacitly agreed. There was no reason why biology should not be introduced, Armstrong said, and he hoped in future to find schools giving their pupils a broad view of nature, creating "an interest in all that goes on around them." However, "physics and chemistry are the foundations... of scientific belief," he argued. "They underlie all phenomena, all vital changes" (HEA 1903f, 884).

Non-specialization, which had always been a central theme in Armstrong's campaign but remained overlooked in many circles, received further support in 1904 from J. H. Leonard, who once again emphasized the absence of biology in most syllabuses (Leonard 1904; Meadows 1973), and from Armstrong's friend and supporter Arthur Smithells (1860–1939), at the University of Leeds. Smithells presented a school syllabus in which "scientific discipline and scientific method can be inculcated by simple experimental work, based entirely on matters of the household and of daily life, where the information acquired is truly useful knowledge" (Smithells 1906). The syllabus owed much to Smithells's experience with and interest in "domestic science" for girls' schools. But it was T. E. Page of Charterhouse, the secretary of the Incorporated Association of Assistant Masters, who successfully launched a committee of the British Association to discuss the curriculum of secondary schools (Page 1906).

Chaired by Birmingham University physicist Oliver Lodge, with headmaster of St. Dunstan's School Charles Maddock Stuart as its secretary, the BAAS committee included Page, Michael Sadler, Philip Magnus, and Armstrong. In 1907, it was joined by Richard Gregory, a former laboratory assistant of Arthur Worthington's, a textbook writer, a staff writer for *Nature*, and the editor of *School World* (Armytage 1957). Stuart was one of Armstrong's most enthusiastic disciples, and it was to his school that Armstrong sent three of his sons. Gregory chaired a subcommittee that included Latter (but not Armstrong) on the "sequence of studies in the science section of the curriculum" (Gregory 1908b, 526). It found that although nature study was included in the earliest part of most secondary school curricula, this inclusion did not lead to a more detailed study of biology—except for intending medical students as a post-matriculation subject. The subcommittee also concluded that the curricula had a lopsided emphasis on "elementary physical measurements" and "elementary heat," but they showed little evident desire to tamper with this situation. Laboratory work was now universal, though many complained that examinations hindered or prevented heuristic methods from being used. The subcommittee found that opinion of the value of heuristic "problem" methods varied considerably, though their use was especially favored in the early stages of teaching, before a "subject" approach was adopted.

Gregory's subcommittee concluded that laboratory methods had been a great success but felt that there had to be both a greater freedom from examinations—so that teachers could experiment with the curriculum—and a greater emphasis on the interdependence of the physical sciences. Moreover, schoolwork had to be brought into closer touch with everyday experience. This sentiment was echoed again at the same meeting of the British Association by Armstrong and his former pupil William M. Heller in their report on the curricula of elementary schools, in which they tried to produce a better correlation between instruction in physical and in biological science. So-called nature study, they argued, was usually an artificial subject and, in its treatment of valueless objects, "a laborious elucidation of the obvious." It was no substitute by itself for formal and systematic instruction in practical physical science (HEA 1908g, 509). The physiology and biology of Heller's and Armstrong's syllabus appeared in the third and fourth years of their course and rested on a foundation of physical science.

Both Gregory's subcommittee on secondary education and Armstrong's report on elementary schools effectively agreed that the physical sciences taught in schools must be correlated with mathematics and biology. Through essays and the APSSM, Oliver Latter continually stressed the value of biology, and when London University tightened up its medical regulations and made biology a compulsory subject for its first MB examinations (and so effectively made biology compulsory for matriculation), this move was acknowledged to be a triumph of APSSM policy—but it was seen by Armstrong as a threat to non-specialization (Mansell 1976a).

Science for All

On his return from Australia in December 1914, Armstrong engaged immediately in offering unsolicited advice on how chemical products, especially dyestuffs that had hitherto been imported from Germany, could be obtained. As we saw in the last chapter, he also played his part in the war mania by eliminating fellows of German extraction from membership of learned scientific societies. Education was temporally forgotten. Meanwhile, the APSSM had compiled a humanizing syllabus entitled *Science for*

All, which effectively advocated what was soon afterward called "general science" (APSSM 1917).[1]

This syllabus, together with the resolutions on the science curriculum from the wartime "Neglect of Science Committee" (an activist group organized by biologist Ray Lankester), was presented to the "Committee Appointed by the Prime Minister to Enquire into the Position of Natural Science in the Educational System of Great Britain," appointed in August 1916 under the chairmanship of the prominent physicist J. J. Thomson (A.-K. Mayer 2005). Although free of commitments in retirement, Armstrong was notably not chosen as a member of Thomson's team. Simultaneously, Gregory (a member of the Neglect of Science Committee) persuaded the British Association to set up another committee under his chairmanship to examine the place of science in secondary schools and its role in general education. Armstrong, Sanderson (headmaster of Oundle), Vassall, and the seriously ill Worthington served on this committee, which reported in 1917 in time to be noticed by the Thomson committee (Gregory 1917, 123).

The APSSM syllabus *Science for All*, which received wide distribution after the war through publication in G. H. J. Adlam's periodical *School Science Review*, was largely the work of the science masters W. D. Eggar (Eton), F. M. Oldham (Dulwich), and Arthur Vassall (Harrow) (Eggar 1920). Later it was thought to have been composed deliberately in opposition to heurism, but C. L. Bryant, another contributor to the *Science for All* syllabus, denied this.

> Indeed, we welcomed [Armstrong's] campaign in so far as it supported our claims for more time and equipment for practical work in the schools, but we thought that the dear old man was too insistent on one method of teaching . . . [and] it fell to my lot to tell him so. He didn't like it, nor did I; but we remained friends all the same. *Science for All* was neither for nor against Armstrong . . . It just happened that the pamphlet was compiled at a time when relations between us were a bit strained. (C. L. Bryant 1950, 144)

Such strain was partially due to the APSSM's acknowledgment that faculty psychology was discredited. Additional reasons militating in favor of a middle-ground position in the APSSM syllabus were the British Associa-

---

[1]. The syllabus occupies pp. 32–43. The phrase "science for all" had been popularized by geographer and journalist Robert Brown in a popular encyclopedic miscellany for young people (R. Brown 1877–1881).

tion's belief that heurism needed to be tempered by a more informative approach and, above all, its view that that biology should share equally with the physical sciences and not form a lame coda to a science syllabus. Specialized physics and chemistry lessons were not suited to youths whose forte lay in classics or the humanities or who were simply average pupils. Hence, if one believed that science was liberalizing, humanizing, and useful knowledge in an increasingly science- and technology-based society (as most APSSM members did), it was clear that to make science available to everyone, a rather different, more general, form of science teaching would be necessary. Specialization was bad; science was to be taught as part of the general education of everyone, and not as if it were to be the vocation of everyone. (Of course, "everyone" here referred only to the mainly upper-class boys in the major public schools.) Science had to appeal to the *imagination* as well as to reason; this was something that heurism could not or would not do.

Science for All proved to be a successful strategy to give public school science masters status within their schools and science a firm place within the curriculum. Ironically, it also gave teachers a broader base upon which to build a sixth form of boys who would opt for scientific degrees at universities, thus aiding the very specialization the syllabus had been designed to combat. But in achieving status, and despite a modicum of utilitarian information in the *Science for All* syllabus, its stress upon science as a form of high culture inevitably devalued technology. The president of the Science Masters Association in 1920, Arthur Vassall, went so far as to deride students who "desert science for industry" as interested only in commercial gains, not in the search for truth. As Mansell has commented, in this respect "a rank order had been established, sciences for the gentlemen, technology for the players" (Mansell 1976b, 584). This attitude would have mattered little but for the fact that after 1902 the public schools tended to be models for the state-aided municipal grammar schools, which tended, therefore, to devalue the utilitarian aspects of science. *Science for All* was, in fact, science only for the upper and middle classes, and for the few working-class children who climbed the scholarship ladder.

Science for All included six tentative syllabus suggestions in which biology figured prominently, and which moved from cosmology to the gramophone. Suggestions for less able boys in the lower school included an experimental investigation of chalk. Complete freedom was left to the teacher to decide on the precise content and method, but insofar as the syllabus was more extensive than any proposed before, proponents of General Science inevitably tended to favor teacher demonstrations rather than individual pupil assignments.

The Gregory and Thomson Reports

Gregory's British Association committee (on which Armstrong sat) reached similar conclusions. The most interesting part of its unanimous report concerned the method of science teaching; this account revealed that Armstrong now accepted the fallibility of faculty psychology as well as the criticism of form and content in many heuristic schemes of instruction. The unfortunate view had gained ground, according to the report, that "all subjects, in different ways and to different degrees, can be made to give a training in scientific method." However, if it were admitted that the sole task of a science teacher was to train pupils in scientific method and that this method was common to all well-conducted intellectual inquiries, there would be no need for special science teachers!

> The paradoxical conclusion depends upon the assumption that the method of scientific investigation can be regarded as separate from the matter, which is not correct.... No-one ought to expect a training in scientific method acquired in one field of inquiry to be transferable to—that is, to guarantee competence in—a field substantially different from the former. (Gregory 1917, 134)

To Armstrong's disgust, Thomson was able to take such an argument or admission as evidence of the fallibility of heurism.

The Gregory committee stressed the unique value of laboratory work (Gregory 1917, 137). It brought a pupil "into direct contact with reality through his own senses and his own manipulation" and tested his reasoning powers in ways that the best manual training could not. But should laboratory work be performed by the subject method (demonstration exercise) or problem method (heurism)? The latter method, the committee stressed, was not intended "as sometimes supposed, to make pupils discover for themselves laws and principles previously unknown to them, though to some extent this can be done. But rather [it is] to provide a continuous thread of reasoning for the practical work and a definite purpose for whatever is undertaken" (Gregory 1917, 139).

The heuristic method was acknowledged to be difficult: freedom to use both methods independently of external examinations was needed. Unfortunately, however, there had always been a tendency in practical work to limit the scope of science courses. "Increased attention to laboratory exercises has ... often been associated with a very restricted acquaintance with the world of science. The tendency has been to make all teaching a matter of measurement, to the neglect of the human aspects of the pursuit of natural knowledge." This ironic effect of Armstrong's success—making content "narrow and special rather than broad and catholic"—was not lost on J. J. Thomson or his committee.

Finally, Gregory's committee was concerned with the human aspects of science. No doubt urged by Gregory himself and Sanderson, it recommended lessons in the history of science and the discussion of modern scientific issues in schools. At all times, scientific instruction was to be related to the affairs of everyday life. Thomson also appreciated this point. Syllabuses used and recommended by Vassall (at Harrow), Sanderson (at Oundle), T. Percy Nunn (at the London Day Training College), and Armstrong (in a newly developed course on practical food studies) were also reproduced. All these courses included elements from biology as the science that closely concerned human life.

Thomson's government-appointed committee, upon which, astonishingly, no schoolteachers were represented, reported in 1918. All the evidence presented to it had suggested that school science courses had become far too narrow, both because of undue emphasis on physics and chemistry and because of the undue restriction of subject matter. The committee reported: "Principles are often taught without reference to the phenomena of nature which they explain: the course does not satisfy the natural curiosity of the pupils; it may give them some knowledge of laboratory methods, but little idea of wider generalizations, such as the principle of conservation of energy, which are quite within their powers of comprehension" (Thomson 1918, 41). Too many schools planned their courses for the specialized study of science by senior boys, to the disadvantage of those who left school earlier. The growth of laboratory-based teaching had been of the greatest value, but "many teachers have become so dominated by the idea of the supreme value of experimental work that they have left on one side and neglected those sciences which do not lend themselves to experiments in school: the tendency has been to restrict the work to parts of physics and chemistry in which the boys can do experiments for themselves." To be sure, laboratory work was essential:

> But sometimes the performance of laboratory exercises has been considered too much as an end in itself—such an exercise loses the value of a real experiment when it becomes a piece of drill; often exercises succeed each other without forming part of a continuous or considered scheme for building up a boy's knowledge of his subject.... Insistence on the view that experiments by the class must always be preferred to demonstration experiments leads to great waste of time and provides an inferior substitute. (Thomson 1918, 42)

So far, the report offered Armstrong little to quarrel with: he had never denied the value of demonstration experiments, and he was aware of the limitations and enthusiasms of teachers.

However, Thomson directed these limitations toward an explicit attack on the heuristic method itself.

> Much of the waste of time is due to a conscientious desire of the teachers to encourage the spirit of inquiry by following the so-called heuristic method: the pupils are supposed to discover by their own experiments with little or no suggestion from the teacher, the solutions of problems set to them or of problems which they themselves suggest. The spirit of inquiry should run through the whole of the scientific work, and everything should be done to encourage it, but it seems clear that the heuristic method can never be the main method by which the pupil acquires scientific training and knowledge. He cannot expect to rediscover in his school hours all that he may fairly be expected to know; to insist that he should try is to waste his time, and his opportunities. (Thomson 1918, 42)

This representation of the heuristic method was a caricature of Armstrong's intentions. For him, heurism meant "directed inquiry," with the teacher playing a discreet but positive directing role. The method was not meant to be the means by which stores of information were acquired, nor was it to replace entirely other methods of instruction and learning. Even though Thomson did not include these castigations in his important official recommendations in the conclusions of the report, the damage was done.

Not surprisingly, Armstrong was furious. When the report was discussed at the British Association meeting at Bournemouth in 1919, he categorically dismissed it as worthless (which it was not) and unlikely to influence educational opinion (which it did) (Anon. 1920; Freund 1920, 6). He was particularly upset by one section of Thomson's report, which appeared to assert, incorrectly, that heurism "involves the rediscovery by the pupil in his school hours of all that he may fairly be expected to know," whereas it was really a method of learning the art of inquiry (HEA 1919c). Thomson's report would be of little value to teachers, Armstrong thought, "a lost opportunity for examining and utilising experiments already tried."

Gregory, who is sometimes portrayed as an enemy of heurism, in fact strongly supported it, provided there was not an overconcentration of method, and that laboratory work was complemented and supplemented by a broad general course of lessons. Nevertheless, it proved to be Gregory's, and not Armstrong's hour (Gregory 1919, 354). Training in experimental scientific method, Gregory recommended, was to be separate from "a general course of science" based upon descriptive lessons and reading. Supported by Thomson's major recommendation that "natural science should be included in the general course of education of *all* up to the age of about sixteen," Gregory used the British Association, *Nature* (which he now edited), and the *School Science Review* as platforms to urge the teaching of "general science" to all children who were unlikely to attend university (Thompson 1958; Jenkins 2019). Laboratory work was to be "different in nature and intention from work that can be regarded as manual training, or from measurements which exemplify purely mathematical principles. It must be used to illustrate subjects dealt with in the classroom—the 'Science for All'" (Gregory 1920, 93).

Meanwhile, Vassall, the other arch-apostle of general science, attacked those who would "limit the function of science in schools to one end ... a training in inductive reasoning from experimental data obtained by the pupils themselves" (Vassall 1920). His three aims of science teaching were the inculcation of method and a critical faculty, the increase of the knowledge of science, and the stimulation of imagination and the aesthetic sense. On another front, through the influence of practicing scientists like Harold Hartley, physical chemistry (including the ionic theory that Armstrong despised) penetrated the schools and proved an elegant way of using theory to order and unify practical chemistry courses.

In 1924, the Science Masters Association, which the APSSM had become in 1919, published a pamphlet on "General Science" that contained a revised syllabus of *Science for All*. The syllabus was accepted for the Lower School Certificate examination by most of the examination boards. A revised edition of the syllabus was published in 1932. Yet when the British Association investigated a year later, it found that, despite support for the syllabus by the Hadow Committee on the education of adolescents in 1926, general science was only taught in 37 percent of boys' secondary schools, 40 percent of girls' schools, and 42 percent of coeducational schools (Heller 1932 and 1933). The perennial problem with general sciences syllabuses proved to be one of correlation; few teachers felt competent to teach the whole of the syllabus, and it was difficult to teach without compartmentalizing the curriculum into discrete blocks (or terms) of chemistry, physics, and biology. Textbooks failed to solve the problem. Part of the trouble was that between 1920 and 1940, unemployed scientists went into secondary school teaching in large numbers. Trained as specialists, they rebelled against heurism and the general science movement and campaigned for their subject specialisms. They saw themselves as scientists manqué, not as science teachers. For these reasons, as examination statistics show, the subject of general science was never popular in schools, though it undoubtedly stimulated the development of biology teaching (Brock and Jenkins 2014).

In 1929, Frederic W. Westaway published his outstanding and influential book on *Science Teaching*. One of the aims of this school inspector and former teacher was to increase the stock of biology teaching. Whilst he was appreciative of Armstrong's educational achievements, he was explicit in rejecting heurism. He asserted that it was far too slow and unworkable to be a recommended strategy for science teachers. He suspected that Armstrong had never revealed the true secret of his method or even been personally aware of the true nature of his own success with the method—sheer personality, exemplified by "his rather tart impatience toward his students, his refusal to help them one iota more than is absolutely necessary, his amazingly clever and ever-ready questions to meet the needs of the moment, his resourcefulness under all experimental difficulties, his rather 'grumpy manner', and his rare words of praise" (Westaway 1929, 27). The implication was, then, that only teachers with extraordinary personal magnetism could use heurism effectively and that demonstrations were probably better for class discipline anyway.

Such curt dismissal hurt Armstrong. To him the heuristic method was a practical system that worked. It was pedagogically sound and tested by his experience. His own syllabuses did not have all the faults found by Westaway or the followers of general science. He covered syllabuses broadly and systematically and at a level appropriate to the age of the child. Good teachers could be found and trained, as his earlier experience with the Board schools in the 1890s showed. Complaints about incompatibility with examinations were invalid; he would have changed the whole system of examinations anyway. Objections as to costs were exaggerated and compensated for by the need to purchase fewer textbooks.

A seminar on heurism published in the same year as Westaway's book suggested that its spirit was not entirely crushed (Anon. 1929). Yet, not surprisingly, by the time of Armstrong's death in 1937, the heuristic method as originally conceived had vanished, killed by the examination system and the collective criticisms of the new psychology, the general science movement, and brilliant writers like Westaway; while another war and the need for new stringent economies closed the 1930s, some teachers again questioned the British emphasis, or obsession with, laboratory teaching (Lowndes 1940). But small cells of heuristic teaching continued, notably at Christ's Hospital school with Gordon Van Praagh and John Bradley. Their example was to be influential in reviving the heuristic method as part of the Nuffield Foundation's efforts to modernize science teaching in the 1950s.

Armstrong first collected together his educational writings for publication by Macmillan in 1903. He saw the book, *Teaching of Scientific Method*, as a critical contribution to the movement to modernize education following the 1902 Education Act (HEA 1903b). Arranged in non-chronological order, with constructive essays on heurism and its syllabus following didactic essays that castigated the current educational system, the book's impact was weakened by its lack of coordination. Smithells rightly thought it a pity that Armstrong had not distilled his thoughts into a monograph (Smithells 1904). Commenting on its language, he said it was "vigorous almost to violence, red hot, scathing, scornful, uncompromising and incessant." Nevertheless, there was a call for a second edition in 1910 and reprint of this in 1925, in which the author opened the volume with a new prefatory essay that outlined the changes in education that had taken place a quarter of a century after "our winter of discontent." It had been a period of searching criticism and reappraisal, but the rooted individualism of the British character had ensured that nothing was organized and that little use had been made of past experiences. Consequently, he continued, although Britain spent more on education than any other country, there was little to show for it. "But worse than this—by allowing the academic party to control the work, we have suffered the foundations of our old pre-eminence to be sapped and undermined in every direction by constant dogmatic teaching and a lavish system of enforced lesson learning supplemented by inelastic and inappropriate examinations" (HEA 1910a, xviii).

The outlook for the future was not bright, Armstrong concluded. The chances of the British changing their ways were slim. They were living in a technological society, but it was difficult for people to recognize the value of the language of science "when those who are called on to introduce it and supervise the teaching have no true knowledge of the subject in which they are to give instruction and little understanding of its beauty and value" (HEA 1910a, xix). This state of affairs would continue until a far broader training program was introduced into schools, until the phrase *experientia docet* (experience teaches) was adopted in all schools, until the faculty of criticism, rather than acquisitiveness, was cultivated, and until, paradoxically, Britain learned to rate theoretical knowledge much more highly, as other nations did.

Clearly, as many critics said, Armstrong was living in the past and had failed to recognize that, while there was always room for improvement, education had changed considerably since the 1880s when Armstrong first took up his critical stance.

Armstrong was never a lone scientist asking for educational reform. But he was exceptional in making the reform of science teaching a main focus of his life's work. Until his death in 1937, Armstrong wholeheartedly committed himself to promulgating the notion that understanding came through doing. Precisely what was learned (information) was less important, at least in the early stages, than the method (process) involved in learning, thinking, or finding out about something. He believed, as the history of science seemed to him to confirm, that once a person had learned the techniques of experimentation, he could tirelessly acquire by his own efforts the mass of information that other teacher-oriented systems of education forced into children by rote, demonstrations, and the intimidation of examinations.

Hindsight suggests that Armstrong did not sufficiently distinguish between the kind of "discovery of understanding" which a tyro makes in a school or university laboratory ("I get it") and the original discovery of innovation that a research scientist may make. That this point has caused confusion and led to objections to discovery methods is understandable. Nevertheless, until the First World War, heurism was probably the chief catalyst in transforming school science teaching in Great Britain.

✳ 13 ✳
The 1920s

> In seventy or eighty years, a man may have a deep gust of the world; know what it is, what it can afford and what 'tis to have been a man.
> —T. Browne 1716; HEA 1936a, 752

Although Armstrong had long given up laboratory chemistry, during the 1920s he remained interested and attached to the affairs of the Chemical Society; less so for the Royal Society. He continued to attend annual meetings of the British Association but did so as a commentator and reporter rather than as a contributor. He also found the Royal Society of Arts a congenial outlet for expressing his opinions following his Cantor Lectures on foodstuffs of 1919. He became even more widely known for his witticisms when writing for *Nature* and in his letters to the *Times*. While the dissociation theory remained a constant target of his contempt (notably for the way American chemists had accepted physical chemistry), much of his concern was now turned toward environmental issues, including the protection of the landscape and mankind's sources of food and its purity. As friends died, he also became a skilled writer of obituaries, most notably one for his friend James Dewar, to be described in chapter 15. There were occasional potboilers, such as his contribution to the Mathematical Association's book commemorating the bicentennial of Newton's death in 1927, but usually he had something interesting to say even when he repeated what he had said previously (HEA 1927t, 5). Consequently, Armstrong was kept busy with invitations to speak at local divisions of societies, schools, and commemorative events.

At the Chemical Society

In one of his occasional attendances at council meetings of the Chemical Society, in October 1921 Armstrong drew the society's attention to the

"Safeguarding of Industries Act" that had just been passed by Parliament and the potential burden it might pose for research workers. After discussion, the council decided to take the issues up with university MPs. In a letter to *Nature*, Armstrong ridiculed the Act, which placed large taxes on the sale of many basic commodities used by the chemical industries, for example the import of guano. Was quinine exempted because it was used as a contraceptive and the government was trying to reduce the population of our crowded islands?[1] In fact, the National Union of Workers, as well as the British Section of the International Association of Chemists, had already begun to lobby Parliament successfully to ameliorate the most damaging aspects of the taxations specified by the Act.

Throughout the 1920s, Armstrong, together with William Pope (now Sir William, knighted for his war work in 1919), continued to represent the society on the Federal Council of Pure and Applied Chemistry. Pope was now Armstrong's most famous pupil.[2] The Federal Council had been set up in 1919 to deal with the problems of reconstruction that the Allies (British, French, Italian, and American) faced after the war. It held its first meeting in Paris with Pope and the analytical chemist A. Chaston Chapman representing the Chemical Society, and Armstrong's son-in-law Stephen Miall representing the Society of Chemical Industry. The declared objects were:

> To consolidate, as between the different Allied countries, the bonds of esteem and friendship which have already been strengthened during the War; to organise a permanent co-operation between the Chemical Associations of the allied nations; to co-ordinate their scientific and technical resources; and to contribute to the advancement of chemistry in all its different departments. (Chapman 1919, 222)

The International Union of Pure and Applied Chemistry (IUPAC), as it soon came to be called, was one of four unions (the others were for radio science, geodesy, and astronomy) that were the outgrowths of the decision to transform the prewar International Association of Academies

---

1. Chemical Society council minutes, 20 October 1921, f. 15, and 17 November 1921, minute 6. See also HEA 1921h, 241, 271.

2. When Pope gave the presidential address to the annual dinner of the Society of Chemical Industry in October 1921, Armstrong led the vote of thanks with the remark that Pope exemplified the worth of his method of education. "The result was that Sir William never swore, never referred to colloids, or even to the concentration of hydrogen ions; he was free from all the bad language of the day, and spoke in honest English, which everyone understood [laughter]" (HEA 1921g, 452).

into a more restricted International Research Council (IRC). It was to be geographically limited to just sixteen countries, excluding Germany and Austria and their wartime allies (Fox 2016b). Although Armstrong had attended the meeting in Brussels in October 1919 at which these organizational decisions were made, and although he attended further meetings of the IUPAC during the 1920s, there is no record that he played any decisive role in its deliberations. The reconciliation of German academies with the IRC was to remain a problem until after the Second World War; on the other hand, the IUPAC began to welcome German delegates again starting in 1930—a decision which received no comment from Armstrong.

Armstrong still remained very much concerned with the economic viability and success of natural indigo production in India (HEA 1922c; Travis et al. 1992). By the 1920s, natural indigo had become a niche agriculture whose sales depended upon customers who preferred the tone of the natural dyestuff. At a council meeting on 16 February 1922, he moved that the Chemical Society should send a letter of protest to the secretary of state for India against the cessation of research on the cultivation of the indigo plants.[3] A letter was sent and reached the attention of the Indian government, with approval.[4] When an English farmer recommended the use of lucerne (alfalfa) as forage rich in nitrogen and organic matter, Armstrong agreed and said this technique had been tried on his recommendation for indigo crops in India (HEA 1924c). Armstrong summarized all of his and Davis's efforts to save the Indian indigo industry in a long paper to the Royal Society of Arts in April 1922. He made two significant points. First, if (as seemed likely to him) coal supplies diminished or failed, it would become impossible to extract the necessary basic organic chemicals needed for the synthesis of indigo; and second, German industrialists were synthesizing and marketing pure indigotin, not indigo itself. Since the two were not chemically identical, it was important to maintain the Indian trade (HEA 1922e).

Smokeless fuel also continued to concern him. When the London County Council announced postwar plans for the electrification of London, he pointed out that Londoners also needed gaseous and smokeless solid fuel—the latter especially since the navy had moved over to fuel using petroleum products. The use of coal must be rationalized, he reiterated. He represented the Chemical Society when a deputation of the Coal

---

3. Chemical Society council minutes, 16 February 1922, f. 48; 30 March 1922, f. 61. For background, see Kumar 2012.

4. Chemical Society council minutes, 30 March 1922, f. 61.

Smoke Abatement Society lobbied the Minister of Health in April 1922.[5] When the first World Power Conference was held during the Wembley Exhibition in 1924, Armstrong was but one of the over four hundred participants. The huge meeting, which greatly impressed H. G. Wells, had been organized by W. D. Dunlop and several electrical manufacturers to consider the future of the world's energy resources. Armstrong spoke about smokeless fuels and the need to develop them in powdered form (Wright, Shin, and Trentmann 2013).

> London would be best supplied with smokeless fuel from the collieries near at hand in Kent. It may well be that the supply of electricity might with advantage also be relegated to these collieries. I suggest that gas should be piped to us, electricity wired to us, from Kent. [The Kent-London railways] could well convey to us all the smokeless fuel we need. (HEA 1921e)

In 1924, Cambridge physicist and philosopher Norman Campbell started a fascinating debate in *Nature* by asking why men of science had a prejudice against using the word "scientist" (Campbell 1924). Richard Gregory, the editor of *Nature*, not content with an editorial on the use of the word, then asked the Chemical Society (and probably other societies) whether it used the word in its publications and received a negative reply: the word was never used. Armstrong was one of the many contributors to a debate for and against the use of the term that took place in *Nature* over the following weeks.[6] He was against the word's use and advocated "sciencer" to harmonize with "do-ers" like bakers, butchers, builders, boxers, and grocers. It would appear that scientists had been deterred from using the word for fear of criticism from literary pedants. In fact, the term entered common use only in the 1940s.

The 1920s saw the arrival of mechanistic organic chemistry and the introduction of the vocabulary and symbolism of reaction mechanisms. The young Christopher Ingold joined the council in 1923 when Armstrong was invited to compose an obituary of James Dewar. Armstrong was not impressed by mechanistic ideas of floating electrons, and at a council meeting on 15 October 1925, he notoriously moved, in the presence of Ingold

---

5. Chemical Society council minutes, 30 March 1922, f. 61.
6. Chemical Society council minutes, 20 November 1924, f. 218 (HEA 1925b). This contribution was a response to correspondence in *Nature* 114 (8 December 1924): 823–25; 115 (20 December 1924): 897–98; 115 (10 January 1925): 50. For earlier debates on the word, see S. Ross 1962.

and Nevil Sidgwick, that a memorandum should be sent to all the officers of the Chemical Society "that the absurd game of chemical noughts and crosses be taboo within the Society's precincts and that, following the practice of the press in ending correspondence, it be an instruction to the officers to give notice—'That no further contributions to the mystics of polarity will be received, considered, or printed by the Society.'" Needless to say, the motion was not even seconded, being quite properly taken as one of Armstrong's jokes.[7]

Nevertheless, he returned to the attack a year later when he urged the Publications Committee to follow the example of the Société Chimique in appointing a committee of three to report on the "inwardness" of the speculations that were appearing "in recent scientific communications dealing with substitution in benzene."[8] The council must have known that this move was again a criticism of Ingold's and Robert Robinson's investigations. The suggestion was duly passed to the Publications Committee, which took no action. Surprisingly, Armstrong and Robinson, who could be equally cantankerous, got on well. They joined forces in April 1926 to urge that the work of the council and its committees needed to be made more efficient, and they suggested that this matter should be investigated by a small committee.[9]

Like other learned societies, the Chemical Society found its membership severely declining in the 1920s. No immediate action was taken and, two months later, on 17 June 1926, Armstrong resigned from the council. He was immediately asked to reconsider, and although his response is not recorded, Armstrong turned up at the next meeting as if nothing had happened.[10] The brief falling out was probably caused by a trenchant series of articles that Armstrong wrote for the *Chemical Age* at this time that more or less claimed that the Chemical Society was "moribund" (HEA 1932–1933). In the first of the six articles, he stated that "the Society has no social coherence; this is a root cause of its difficulties.... The change from its early constitution to the present democratic form has been an entire failure" (HEA 1932–1933, part 1, 491).

It was not until 1932 that a Reconstruction Committee that included Armstrong was to be appointed to implement measures to improve the situation. Drastic reconstruction did then occur, with more provincial

---

7. Chemical Society council minutes, 15 October 1925, item 20, f. 285.

8. Chemical Society council minutes, 18 February 1926, item 13.

9. Chemical Society council minutes, 22 April 1926, 26, item 17. The draft of his resignation letter, 4 May 1926, is found in AP2.180.

10. Chemical Society council minutes, 17 June 1926, 36, item 5.

fellows being elected to the council, a reduction of fees for younger fellows, the revamping of the *Journal*, the delivery of endowed named lectureships in provincial centers as well as in London, and, above all, the arrangement of London meetings around groups of papers that invited discussion. Armstrong had suggested all of these measures on many previous occasions during the 1920s. Despite their disagreements, Armstrong undoubtedly was a persuasive force for change in the Chemical Society. Most of the changes were implemented while Scottish chemist George G. Henderson was president, between 1932 and 1933 (Henderson 1933).

In April 1923, Armstrong attended an international meeting organized by the French Société de Chimie Industrielle, which he praised in *Nature* for its organization and interest (HEA 1923f). The occasion proved an opportunity once again for him to meet Herbert Levinstein (head of the family dyestuffs company and son of its founder Ivan Levinstein) and, ironically, hear him lecture in French on the downfall of the British dyestuffs industry. A group visit to Reims and its cathedral, which had been largely destroyed during the German invasion, made Armstrong understand why the French were so bitter and so determined not to admit German chemists to international conferences.

On the centenary of Stanislao Cannizzaro's birth, Armstrong and William Pope were appointed the Chemical Society's representatives at the seventh meeting of the Congress of Industrial Chemistry at Palermo, from 22 May to 1 June 1926.[11] Two years later, the summer of 1928 saw Armstrong representing the society at the 8th Congress at Strasbourg, followed a month later by his attendance at the 9th International Conference of Pure and Applied Chemistry at the Hague. Such European meetings usually resulted in a lively article for *Nature* or some other publication (HEA 1928d). He observed shrewdly that "whatever the value of the conferences proper, these excursions afford precious opportunities to the delegates of coming into close personal contact." Among the Hague delegates was his American bête noire, Wilder Bancroft. Perhaps due to Armstrong's personal magnetism, they appear to have gotten on well despite their differences over physical chemistry.

Both the Society of Chemical Industry and the Institute of Chemistry had instituted local sections soon after their respective formations. The Chemical Society did not consider following their example until 1925. The issue would, once again, mean a change in the charter and the bylaws. Armstrong pointed out that the opinion of the entire membership should

---

11. Chemical Society council minutes, 18 February 1926, 12, item 7.

be sought before a decision was made.¹² It was not until 1932 that the society appointed representatives in towns outside London and granted them powers to organize local events of interest to members. Meanwhile, Armstrong frequently entertained local groups of the Institute of Chemistry during the 1920s. For example, in a talk entitled "The Nescience of Science and the Conceit of Ignorance," delivered to the Manchester Group in February 1925, he called the chemistry profession "a disunited rabble" and pleaded for the unification of the professional societies (Anon. 1925).

Armstrong continued to attend annual meetings of the British Association for the Advancement of Science in the 1920s, though he no longer gave papers. Instead, he found himself reporting on the proceedings—usually in a critical vein. He was not, however, the only person finding fault with the peripatetic organization. When the association met in Cardiff in August 1920, *Nature*'s local correspondent, Dr. R. V. Stanford, deplored "the apathy of local people of the educated classes to the presence of the Association" (Stanford 1920, 13). Armstrong agreed, and going further, he declared the association "practically defunct" with little hope of resuscitation. He doubted whether it could "ever again fulfil the desires of its earlier promoters, who undoubtedly held its primary function to be that of advancing public appreciation of scientific discovery" (HEA 1920e). When he read the printed reports and speeches in 1921, Armstrong complained additionally of their poor English (HEA 1921b and 1929j). Stanford's article stimulated a considerable debate that went on until mid-October.¹³ Almost everyone agreed that the British Association needed reform. In postwar Britain, Armstrong was keen to see the British Association resume its occasional excursions to meetings in the colonies—he saw this practice as a vital exercise in erecting barriers against the "spurious puritanism" that was sweeping America (a reference to anti-evolutionism), and also threatening Britain and its colonies (HEA 1922g).

The decade ended with an invitation to deliver the opening address of an international meeting of the Société de Chimie Industrielle in Barcelona in October 1929, his first visit to Spain since a brief visit in 1873. Some 270 French and four hundred Spanish chemists, together with chemists from another fourteen countries, attended this Congress and heard Armstrong give his address in French: "Structure Moléculaire: La Vie et la Couleur: Pensées Allégoriques d'un Chimiste en Espagne" (HEA 1929i).

---

12. Chemical Society council minutes, 19 March 1925, item 14, f. 249; Moore and Philip 1947, 139.

13. *Nature* 106 (1920): 69–72, 101–09, 110–13, 144–47, 179–79, 211–12. Contributors included Lodge, Lankester, Pope, Soddy, and Napier Shaw. See Brock 1981a, 89.

During a long journey across Andalusia from Seville to Granada and on to Córdoba, he became engrossed by the olive trees:

> The labour entailed in plucking the fruit must be enormous—the industry can only be possible in a country where the cost of living is very low. Travelling hour by hour through such country, the chemist, if taking notice, can but wonder how so modest looking a tree does the trick of making oil ... Heaven save us, however, from ever manufacturing the oil synthetically and so destroying the peace of mind and rational occupation of multitudes of happy beings engaged upon healthful work consonant with their intelligence. Our modern lust to manufacture must be curbed. (HEA 1929i, 1198)

Leaving aside the eugenicist implications in his language, the fate of the Indian natural indigo industry was obviously still on his mind, and the quality of food was a subject that he took up strongly in his last decade. In Spain, he was also greatly impressed again by the Gothic cathedral that dominated Burgos in northern Spain and that he had visited in 1873. It led him, like Auguste Comte before, to reflect that mankind needed a new calendar of saints and of saints' days, and sculptures that commemorated great figures of science like Darwin, Liebig, Pasteur, and Volta.

Although he never returned to North America after the British Association met in Canada in 1909, he was always interested, mostly in a negative manner, with the way American chemistry had developed. Consequently, he welcomed the opportunity to review the semicentennial history of the American Chemical Society (C. Browne 1926). He admired the way the ACS had been able to maintain a grip on the union between science and practice, so that its annual meetings included both academic and industrial chemists; whereas in Britain the Society for Chemistry and Industry had gone its separate way.

> Our Chemical Society is now a positive danger as it is supposed to be representative of the subject generally whilst, in effect, it is but the preserve of a narrow academic clique, aloof from the world. On the other hand, the Society of Chemical Industry does not sufficiently represent the higher industrial interests. (HEA 1926r, 806)

Noting the book's account of the ACS's development of an abstracting service, while praising *Chemical Abstracts* for its coverage, he concluded that, in general, abstracting had been detrimental for the progress of chemistry. "In my early days," he wrote, we read everything as it appeared: many of us bought at least the leading English and foreign journals. We were therefore constantly learning and constantly training: the voices of the master-workers were continually in our ears. To-day, very few read and scarcely any one maintains a library. Students and even workers glance through this or that section of the abstracts, just as they skim *Tit Bits* in the train. "Tit-bitry," in fact, prevails everywhere, the art of reading is uncultivated and all but unknown, the whole chemist is a fast-disappearing species, the "bit-o-chemist" will soon be sole survivor. (HEA 1926r, 806)

As discussed in chapter 9, the review also enabled him to skewer Wilder Bancroft for one more time. He expressed relief that Bancroft had at last been reading some of the literature and "having swallowed the water-complex in this age of prohibition, will ultimately come to realise what are the essentials of chemical change. It is pitiable that we should have wandered these fifty years past in the wilderness of doubt on such a subject." Armstrong's bark was always worse than his bite, and he always got on personally with his "enemies." Bancroft was welcomed to Armstrong's family home in June 1923, when the American was visiting London.[14]

Another American chemist whom he had met and admired in Geneva in 1892 was Ira Remsen (1846–1927). From 1879 onwards, Remsen had been closely involved in the development of the saccharin industry in America, and since Armstrong had been at Leipzig when Kolbe first synthesized salicylic acid, they had found a common interest. Later, Remsen had chaired a committee to consider whether benzoic acid was a safe food preservative, testing it upon themselves. Armstrong, in a memorial notice of Remsen, judged that the success of Johns Hopkins University, then a unique postgraduate institution, owed everything to Remsen's leadership (HEA 1927f; Hannaway 1991).

The decade ended with an unexpected honor, the award of the Albert Medal by the Royal Society of Arts in June 1930. The award, presented by the Duke of Connaught at Clarence House, was for "his discoveries in chemistry and his services to education"; in other words, it was a lifetime achievement award. The duke made no mention of heurism. In accepting the medal, Armstrong praised Prince Albert's achievements and the foresight of the City and Livery Companies of the City of London who had created Finsbury College and the City and Guilds (Central) College (HEA 1930e).

14. Visitors' book, 5 June 1923, RSC archives. Bancroft would have been attending a meeting of the Faraday Society.

The Lake District Campaign

Armstrong had always been an outdoorsman. When his children were young, he had taken them for regular holidays in Margate, the nearest seaside town to Lewisham and easily reached by train. These journeys had led him to a fortuitous meeting with the local doctor, Arthur Rowe (1858–1926), an amateur geologist, when Armstrong consulted him about what he thought was influenza, which turned out to be a mild case of pneumonia. The two men became friends immediately and had the audacious idea of tramping around the coasts of southern England and the Yorkshire coast to study and photograph chalk-cliff formations. This project led to a fine series of articles illustrated by Armstrong in the *Proceedings of the Geologists' Association* between 1900 and 1908 (Rowe 1900). They remained close friends until Rowe's death in 1926, when Armstrong memorialized him in *Nature* (HEA 1926k). A modest man and a Margate local worthy, Rowe had hoped to produce a chalk atlas together with Armstrong, but nothing came of it. Armstrong had also tried to have Rowe elected to the Royal Society, but Rowe would not have it, preferring only to be honored by the awards of the Geological Society's Wollaston and Lyell Medals in 1901 and 1911 respectively.

Sometime during the 1870s Armstrong first visited the Lake District, and thereafter began to take regular Easter holidays there. Friends and students were invited as well, and Armstrong would tramp with them over the hills and dales. It was there amidst the lakes that he began to read the writings of John Ruskin, who had purchased Brantwood, a house and grounds overlooking Coniston Water, in 1871. Ruskin undoubtedly influenced Armstrong's literary style. The National Trust had been founded by Octavia Hill and others in 1884, and it seems probable that Armstrong became a member during the Edwardian period, when the popularity of the Lakes for walking holidays expanded. At the end of the First World War, he was appalled to learn of proposals to expand the roadway network through the district to allow better access by motorized vehicles.

> I would add my protest to those of the many who seek to preserve this mountain solitude from the intrusion of the dust-raising, stench motor and all the modern evil consequences of a high road through a region marked out by Nature for retirement and the refreshment of the soul-weary walker. (HEA 1919d)

It was not just the motorcar that was disturbing the peace, but also the "discordant motor cyclists." If the road in the Sty Head district were macadamized, as was proposed, the whole region would be ruined.

Coming down upon the pass during the past week from the snow slopes of Great End and from the screes of Great Gable, both heights near upon three thousand feet, in most perfect weather, the incomparable and rugged beauty of the scene has appealed to me, as it always does, but with a new force, as I thought of the threatened interference with the present natural conditions. The stream which flows from the tarn is worthy of worship alone; those who have eyes to see will have noticed the wonderful play of colour—blue, green, grey, red, brown—in its rocky bed, seen through water of crystal clearness, with which the rude path at its side is in complete harmony. Discord must arise if any formal road be introduced into such a situation. (HEA 1919d)

He went on to stress the area's scientific importance as a former volcanic region, as well as its beauty from the modelling of the mountains and the colors of the landscape. "If we were in any way alive to its aesthetic value, we should set it aside as a national reserve; at least let us seek to protect it from defacement at the hands of those who would make it but a mere passage way." When land was put up for sale in the area of the pass in September 1920, he urged the nation to purchase it so that the land would remain inviolate (HEA 1920g). Again in 1926, he urged building restrictions on new homes and road-widening schemes. What appalled him was that the National Trust accepted the necessity of road-widening schemes that destroyed part of the glacial patterning of the landscape (HEA 1926e; 1926n; 1926q; 1928g).

Armstrong's dislike of the motorcar was evident when a so-called new fuel, "ethyl petrol," was put on the market by American oil companies in 1923. The essential additive for this improved "anti-knock" fuel developed by the chemist Thomas Midgley was tetraethyl lead. The question arose whether widespread use of this fuel could in time result in generalized air pollution by lead, a known poison. Investigations in the United States proved negative insofar as that when a petrol can was left standing open, the evaporate contained no lead. In a letter to the *Times*, Armstrong argued that the American tests, which had been reported verbatim by the British government's committee on the safety of ethyl petrol, were valueless (HEA 1928c; Anon. 1928; Anon. 1930b). The Ministry of Health, he argued, was encouraging the use of a known poison while, at the same time, it prohibited the addition of a harmless antiseptic preservative to ice cream. Meanwhile, he pointed out, another government department, the Home Office, prohibited the use of white lead in house paints. The use of the prefix "ethyl" was also a misnomer and misleading. The petroleum industry, he concluded, was "playing fast and loose with our scientific terminology." He went on to complain that he had not been consulted even though he had studied the effects of lead and other poisons for the last twenty years—ever since he had been one of the chief disputants at the League of Nations Labour Conference in Geneva in 1921, where he had defended the use of white lead in paints (HEA 1921i; HEA 1921f). While the presence of tetraethyl lead in petrol might offer no immediate danger, he was worried about its longer-term effects on the cells of the body, especially nervous and fatty tissues.

> I am prepared to let the garage folk and motorists take care of themselves and poison themselves, with lead tetraethide. But I do object to lead being vomited at the public at large. It is a cumulative poison and the stench of cars is already objectionable. To pump more poison into the atmosphere is indefensible even if presently the quantities are only homeopathic. (HEA 1928c)

He feared that as anti-knock additives were improved, the problem would become intolerable. It was best to nip the problem in the bud.

As a result of this letter, Armstrong was asked to give evidence before the committee on the safety of "Ethyl," though he evidently had no influence on the committee's report in 1930, which exonerated ethyl petrol. In a scathing letter to the *Times* when the final report appeared, he observed:

> It has little to do with lead tetraethide, the incriminated constituent. It does not touch the question of the possible poisonous character of the substance specifically. £2,947 15s 3d of public money has been spent in showing what was already known—that we can put up with small amounts of lead. Nothing has been learnt of the effect of the tetraethide upon human beings—which ostensibly was what the Committee was "out for to see." No impartial medical study of the problem has been made. (HEA 1930d)

It was not until the 1970s, when the poisonous effects of lead in the air were proven and catalytic converters became widely available, that governments banned the use of lead anti-knocking agents in petrol; the manufacture of lead paints was only prohibited in 1992. Armstrong had been a voice of truth crying in the wilderness.

Environmental issues now played an increasing role in Armstrong's life. When the zoologist Sir Arthur Shipley noticed while on an ocean voyage that ocean liners and other vessels were frequently discharging their oil tanks at sea and that the oil took a long time to disperse, he speculated that this discharge would be harmful to fish and life in the sea (Shipley 1921). Armstrong agreed and stated that oil was not easily dispersed. Humans were beginning to foul the oceans. Fishing was a major industry. Life had begun in the sea, and a fish diet was man's fundamental food. The threatening menace of oil pollution, he wrote, must be minimized (HEA 1921a). Once again, Armstrong was an early leader of the incipient environmental movement.

Book Reviewing

During the 1920s, Armstrong honed his skills as a book reviewer in *Nature*. Editor Gregory understood that Armstrong always provided great insight into an author's intentions, coupled with excellent prose and outrageous and opinionated polemic. Gregory ensured that Armstrong's reviews always headed the books under review each week. In a long review of Joseph Mellor's silicate chemistry in his "stupendous" *Comprehensive Treatise of Inorganic and Theoretical Chemistry*, Armstrong praised the pottery chemist's single-handed achievement:

> It is impossible to do justice to the greatness of his work—to rate it at its proper value, to appreciate the truly scientific spirit in which throughout the years, he has wrought for us, the excessive modesty of his presentation, his wondrous talent as a compiler, his almost uncanny faculty of unearthing information, his power of curt expression and concise statement, his logical, unbiased attitude. That one man should accomplish so much is more than remarkable. (HEA 1926g)

Chemists and historians of chemistry a century later will all agree that Mellor's one-man survey of inorganic chemistry was an astonishing achievement. The bulk of the essay was then devoted to an interesting comparison between the chemistry of carbon and silicon.

In an unusual commission, which demonstrates *Nature*'s eclectic contents in the 1920s, Armstrong was asked to review H. G. Well's new three-decker novel of 885 pages, *The World of William Clissold*. Although framed as Clissold's autobiography, critics then and since have correctly seen it as a vehicle for Wells's personal philosophy and his aspirations for a worldwide clerisy of intellectuals (including scientists) who would establish a world republic. Armstrong dismissed it as a mere novel (the final book VI on the relations between men and women he ignored completely) and viewed the whole work as about social pathology, his admiration being curtailed to its indictment of "our public-school system and the ancient universities" and to Wells's parody of publishing a scientific paper in the Royal Society's *Philosophical Transactions* (HEA 1926o).

Prior to Armstrong's retirement at the Central, lectures on fuel technology had been given at the Chemistry Department of the RCS by a visitor from the University of Leeds, William Bone. In 1912, Bone was elected to a new chair of Fuel and Refractory Materials at Imperial College. Since fuel was a subject of great interest to Armstrong, and because they shared a love for Richard Wagner's operas, they had become friends and often disputants. Most of Bone's thirty-year-long research on fuels was outlined in his book *Flame and Combustion of Gases* (1927), which he co-authored with his colleague Donald Townend. In a long review of the book in *Nature*, Armstrong, in what Bone described as "a breezy review," disputed several of the authors' explanations and used the review to propose his own explanations in terms of the hydrone theory (HEA 1927i).

In reply, Bone could not understand why Armstrong insisted that the heat of combustion of carbonic oxide (carbon monoxide) was lower than that of hydrogen, meaning that it would not (and could not) be oxidized by steam. Bone protested that if he were wrong, he was in the excellent company of Marcellin Berthelot and Julius Thomsen. But if he were right, it exposed Armstrong's fallacy in supposing that something more complex than hydrone (steam) was formed in flames (Bone and Townend 1927a). Armstrong replied that he was referring to the heat of formation of liquid water, not hydrone, and begged Bone to follow his arguments rather than to dispute. Bone replied that evidently Armstrong was suggesting that water, not steam, was formed in flame (Bone and Townend 1927b). Where was the evidence, demanded Bone and Townend? Armstrong's reply was feeble, turning the argument over to one concerning the meaning (or non-meaning) of the term "catalytic." Quite rightly, Bone and Townend forthwith dismissed the notion that water played any intermediary role in flame (HEA 1927p; Bone and Townend 1927c).

Instead of continuing this argument, Armstrong turned instead to the way crystallographers like William Lawrence Bragg were representing the structure of common salt (HEA 1927j). Quoting Robert Burns's aphorism, "Some books are lies from end to end," Armstrong argued that science seemed also to be going in that direction. In a Royal Institution lecture on the structure of silicates, Bragg had stated: "In sodium chloride there appear to be no molecules represented by NaCl." This claim was repugnant to common sense, said Armstrong:

> It is absurd to the nth degree, not chemical cricket. Chemistry is neither chess nor geometry, whatever X-ray physics may be. Small unjustified

aspersions of the molecular character of our most necessary condiments must not be allowed any longer to pass unchallenged ... It is time that chemists took charge of chemistry once more and protected neophytes against the worship of false gods; at least taught them to ask for something more than chess-board evidence. (HEA 1927j, 478)

It seems unlikely that any other chemist agreed with Armstrong, and Bragg wisely made no comment.

Behind this rhetoric, of course, lay Armstrong's continuing vendetta against the ionists. It is interesting, therefore, to read his review of the collected papers of Spencer Pickering (1858–1920), one of the few British chemists to join Armstrong in deriding Arrhenius's ionic theory (HEA 1928a; Pickering 1927). Pickering had left £1000 to the Royal Society with the request that it "procure the writing and publication in book form of an account of my work in pure science or of such part of it as may seem suitable for such treatment" (HEA 1928a, 48). The papers were collected together and edited by Armstrong's pupil, Thomas Lowry, and Sir John Russell (Pickering's friend at Rothamsted), with the biochemist Arthur Harden providing a biography. Armstrong had first met Pickering at the Chemical Society, when he was its secretary, and supported him in getting several of his papers on solutions into print.

> There was universal disbelief in the validity of Pickering's conclusions and particularly of the method he used. ... Pickering had to fight his way to publication through the by no means encouraging reports of referees chosen from outside the chemical circle and therefore credited with superior authority—but coldly aloof from chemistry, if not unsympathetic. ... The work was so obviously good [experimentally], however, so exact and thorough, that he could not well be denied: still, the treatment he received soured his proud nature; he felt he was not being helped, that no sympathy was accorded him. Moreover, he was up against the great wave of fashion which soon set in through Ostwald's persistent advocacy of Arrhenius. No one was prepared to reason. (HEA 1928a, 50)

In view of such an insight, it seems a great pity that Armstrong had failed to write Pickering's obituary when asked.

Like Armstrong's effort to demonstrate the validity of his hydrone model of water chemistry in his *Essays on the Art and Principles of Chemistry*, Pickering's posthumous effort to reawaken interest in his evidence for hydrate formation in solutions was destined to sink without a trace.

Armstrong noted, with approval, that Lowry's appraisal of Pickering's work did justice to his subject and that Lowry accepted that Pickering had been justified in concluding that "solutions exhibit a multitude of discontinuities." However, to Armstrong's disappointment and disapproval, Lowry would not accept that the discontinuities were evidence of distinct hydrates in solution. This denial showed, concluded Armstrong, that Lowry "has neither imagination nor a free mind, no sense of logic. He is willing to accept any freakish new view as it comes along, such as the chess-board conclusion: that there is no bond of union between the constituents of common salt" (HEA 1928a, 51). One wonders how Lowry took this rebuke from his former teacher and Central colleague? Such syncretism was, of course, the direction chemistry took in favor of dissociation; but Armstrong resisted it for the remainder of his life.

In December 1928, Armstrong gave a long review of the biography of Sir Norman Lockyer, the founder-editor of *Nature* magazine, in which so much of Armstrong's work was published and publicized (HEA 1928f; T. M. and W. L. Lockyer 1928).[15] Armstrong praised Lockyer for creating *Nature* as "the *Times* of science" and, typically, stated that Lockyer's great advantage was not to have had any academic training. By being largely self-taught, he had always been "unhampered by professional prejudice." On the other hand, when Armstrong had first met him at the Solar Physics laboratory in South Kensington in 1870, and Lockyer had started trying to dissociate elements spectroscopically, Armstrong realized that lack of training made him unfit for the task (HEA 1928f, 873; Brock 1969). Consequently, he took pains to correct Winifred Lockyer's interpretation that Lockyer's dissociation hypothesis was revolutionary and "at variance with the general trend of scientific thought." That was not the case: Prout's hypothesis—and the feeling that the so-called elements might not be truly elementary—had been in every chemist's and physicist's mind for some time.

> The view was not a novel one—the possibility was never doubted, the objection taken was rather to Lockyer's "slap-dash" way of proving his case. His interpretation of much of the spectroscopic evidence he advanced was thought to be unsatisfactory, especially was this true of his attempt to establish the existence of lines common to several elements.

---

15. In a later letter to Edward Frankland Armstrong, Richard Gregory revealed that "Lady Lockyer [Lockyer's second wife, Thomasina Mary Lockyer] did not write a word in the book," but left it to Lockyer's daughter, Winifred, who was aided by Herbert Dingle. Gregory to Frank Armstrong, 5 November 1945, AP2.349.

To be plain, much of his work was not trusted; exact workers like Dewar and Liveing and Huggins simply would not listen to him; even his friends Frankland and Roscoe shook their heads. (HEA 1928f, 873)

While Armstrong valuably corrected the historical record, it now seems ironic that the explanation for Lockyer's misinterpretation of the spectroscopic data lay in the concept of ionization when applied to spectra.

In November 1930, Armstrong's former pupil Sir William Pope, himself now a towering figure in British science, gave the Norman Lockyer Memorial Lecture, founded in honor of *Nature*'s editor by the British Science Guild (Pope 1930). Gregory asked Armstrong to comment on it for an editorial that Armstrong used as one last roar at Britain's failure to teach the scientific method (HEA 1930h). Pope had, in fact, the same critical attitude as his master toward the general literary ignorance of young people coming up to university and their lack of knowledge of modern languages. It was pointless to emphasize research, wrote Armstrong, "if we do not bring under effective consideration the methods of training used in school and university leading up to such work."

> So few among us have any use for scientific method that the vast mass perish for want of such knowledge. Knowledge we have—a vast knowledge: only the knowledge how to use knowledge profitably is wanting among us. Our great present need is to appreciate the depth of our ignorance, especially in education. (HEA 1930h, 870)

With one exception, all of the obituaries and memorials that Armstrong composed were of friends or of chemists that he knew and admired. The exception was a request that came from the Liverpool branch of the Royal Institute of Chemistry in December 1928. The division had decided to sponsor a triannual Hurter Memorial Lecture in honor of Anglo-Swiss alkali industry chemist Ferdinand Hurter (1844–98), who was also well known for his technical work on photographic exposure and speed times. Armstrong was invited to give the thirteenth lecture in the series in December 1928. He began with the confession that he had never known Ferdinand Hurter personally, except that the name had been "fixed in my memory, a photographic lodestar, since 1890." But just as his friend Jacques Loeb had been able to tickle unfertilized ova to life (an awkward fact for bishops who still accepted the first book of Genesis), Armstrong would make a good story out of Hurter—mainly on account of his photographic work with Vero Driffield, which provided the only rational and scientific foundation for the study of the behavior of photographic film to the action of light (Hurter and Driffield 1890). As his title, "Our King's Wardrobe: Iconolatries Old and New," implied, Armstrong offered a parable (HEA 1929e): like the famous emperor, contemporary chemistry of the 1920s had no clothes.

> Is this not a dainty dish to set before a king? A perfect and most prophetic picture of our garment of research, at least on the theoretical side; true in almost every detail, of our attitude. In photography, on the practical side, the advance is too wonderful for words; on the theoretical, a multitude of tongues has but produced confusion: A language a shepherd uses is "that awful" that it cannot be reproduced. Present-day doctrine is little more than pre-Hurter doctrine, paraphrased, with electrons and ions, and light-quanta thrown in. . . . My good old friend Arrhenius, I feel sure, was trained by one of those weavers of the emperor's clothes. (HEA 1929e, 644)

And with that wry remark, he was away on his hobbyhorse of anti-ionism. But there was now a serious point—a point that was shared by Oliver Lodge in his opposition to Einstein's general theory of relativity in the same decade—namely that mathematics or, rather, mathematical modelling, had lost touch with reality. Thinking, no doubt, of the Debye-Hückel equation, Armstrong said:

> Complex equations are written with five variable terms, but the solvent is not yet included as an agent. Is any one of the terms a reality—are they not all imaginary—are we not but looking at an empty loom? It is possible to play with algebraic signs till Doomsday—it seems to be an entrancing game—is it more than a game? Can the separate factors operating in a solution be evaluated? Are the results necessarily a mathematical fake? In the end, probably, it is all a question of words, not of facts. . . . Electron worship has now a large following, and, in valency speculations, two dots are taking the place of a single dot everywhere to-day. We are all becoming dotty. (HEA 1929e, 645; Stanley 2020)

Mankind had always followed fashion, and likewise, so had science.

Armstrong's position as a former president and vice president of the Chemical Society meant that he could continue to voice his opinions to its council throughout the 1920s. As the society's most senior member, he was also easily prevailed upon to represent the society at meetings of sister societies in Europe. Meanwhile, holidays in England with members of his family, former pupils, and friends, made him increasingly aware, as it had to Ruskin, of the fragility of landscape. Mass tourism aided by railways, macadamized roads, and the petrol motorcar and motorcycle not only threatened the tranquility of the district for the hiker and hill-walker, it threatened to destroy the geological fabric that had attracted people to the lakes in the first place. The motorcar, with its petrol-driven combustion engine made more efficient and powerful by American anti-knock additives, was something Armstrong would have banned from the Lake District on grounds that it spread poisonous lead fumes into the environment. The threat to the fishing industry by ships fouling the seas was but one aspect of his increasing concerns about fresh food supplies, as opposed to the growing commercialization of the processed-food industry in the 1920s. This subject, and others, will be treated in the next chapter.

✻ 14 ✻
Campaigns Old and New

> *Arm Strong.* That's a fine name for him—more power to his elbow.
> —Dean 1929

Food, Nutrition, and Biomedicine

The chemistry and purity of food interested Armstrong all his life. The dramatic series of discoveries of various essential nutrients necessary in tiny amounts for human health enlivened Armstrong's wartime and retirement years, and he took a strong interest in the research. Armstrong never liked the word "vitamine" and was always careful to use his neologism "advitant" instead, on the grounds that the vital accessory food factors uncovered by Casimir Funk, Frederick Gowland Hopkins, and others were not all amines. He had expressed his preference for "advitants" in his Cantor Lectures on food in 1919, where he characterized the word "vitamin" as a "gross experimental blunder." This point of nomenclature was also made by the retired Anglo-Australian chemist Arthur Liversidge in 1921. His complaint was, however, that Americans were dropping the final "e," leading to mispronunciations. It is curious that Armstrong's more precise and arguably more appropriate coinage, advitant, has not survived—not even in the definitive *Oxford English Dictionary*—and that "vitamin" has prevailed (Liversidge 1921; HEA 1921c).

During the First World War, Armstrong strongly supported Lord Ernle, the wartime minister of agriculture, in his newspaper campaign for preserving the British countryside for agricultural purposes, and for new scientific research to boost farming (Ernle 1921; HEA 1922a). Armstrong highlighted Britain's need to rely less on imported food and warned of the dangers of running out of phosphate fertilizers; echoing Liebig, he condemned the fact that Britain poured phosphorus into its sewers and hence out to sea (HEA 1926c). Agricultural education had to be transformed to meet the needs of the 1920s. This was all the more necessary, he said, after reading the complaint of a schoolteacher, C. M. Mathieson, about the poor quality of school food in English public schools (Mathieson 1922). As we have seen, Armstrong was quick to decry the emphasis on counting calories rather than concentrating on the quality of meals.

> We have learnt but recently that even the value of animal food—of milk, butter and meat—depends largely on the way the animals themselves are fed, whether in the open on green food of not.... From close observation and special study of the problem, with large numbers, during the past seven years, I am satisfied that children have not been fed wisely at school, and that they suffer in development as a consequence. The failure has been in quality rather than quantity. (HEA 1922b)

He considered the quality of milk a serious problem now that pasteurization was being applied after cows were removed from urban centers and fed in the distant countryside. A discussion about the effects of pasteurization on milk quality was planned for the Dairy Show toward the end of 1922. Armstrong was not invited, but he came to the meeting at the Guildhall anyway. At the first opportunity, he got to his feet and denigrated the meeting for not including any scientist in the discussion. As we saw in chapter 11, he was concerned that pasteurization had never been properly assessed. Did heating milk above blood heat destroy more than just pathogenic germs? The research had not been done. All the constituents of milk probably had a nutritious purpose, but there was a possibility that some vital nutritious constituent was being destroyed in the heating process. The cow itself also needed better food and should be grass-fed as well as stall-fed (HEA 1922h).

This topic, as well as the issue of lead in paints and the disinfection of wool to prevent anthrax, was the subject of his sharp lecture to the Nottingham Section of the Society of Chemical Industry in December 1922, which attacked government for interference with industry and the well-being of citizens (HEA 1923a). The subject continued to worry him into the 1930s (HEA 1931c and 1931e).

In the summer of 1923, a Departmental Committee on the Use of Preservatives and Colouring Matters in Food was established by Nevil Chamberlain, who was minister of health at the time; the final report of the department was issued in February 1925 (Cobbold 2020). Its rather lax conclusions were criticized by Armstrong's former pupil, Sir William Pope, in the *Times*; and Armstrong himself published a long discussion of the issue a few days later. He reiterated his concerns about the effects on health of the pasteurization of milk, to which he now added his concerns about the use of chemical preservatives in butter and other consumables. He concluded with a warning:

> The factors at work in life are so numerous, it is so difficult to trace effects to their causes, that valid proof of injury will not easily be found. What we are learning is that the healthy body is built up during earliest childhood and that lost opportunities can never be regained. So, we should take every precaution not only to provide all that is necessary, but to avoid all possible sources of interference. In the case of marasmic [severely malnourished] children, the slightest interference may be fatal. Little as we know, vague as is our knowledge, all tends to show that we shall be wise-hearted to avoid the use of preservatives as far as possible. (HEA 1925c)

He returned to the subject a few weeks later, having done some experiments at home by dousing young and mature shoots of laurel leaves from his garden in solutions of benzoic acid and boric acid commonly used as food preservatives (HEA 1925m). He found that baby leaves were quickly blackened by benzoic acid and concluded that the evidence for the safety of various preservatives had not been adequately explained in the Ministry of Health's recent report. Brazenly, he suggested that he, together with the UCL food scientist Jack Drummond and the manager of the British Dyestuffs Corporation, should lead an experimental inquiry. Nothing came of that suggestion. His further experiments with salicylic acid again suggested that the subject should be more thoroughly investigated, though Armstrong agreed that the healing power of the human body was probably sufficient to allow toleration of many long-accepted food preservatives in moderation—a position that has remained generally accepted into the twenty-first century (HEA 1925p).

News that the Medical Research Council, in league with the American and Canadian governments, proposed to patent Frederick Banting's insulin treatment for diabetes infuriated Armstrong. It was yet another example, he felt, of the commercialization of science by the Medical Council and the DSIR. Even research at Rothamsted was being controlled by government interference, and the "old conception that scientific discovery was its own great reward is all but lost to us." But even worse to him was the fact that the Royal Society, which should be controlling such issues, had abrogated its responsibilities. "Ere long we shall be a mere body of Fellows—concerned only with our own washing; even reduced to patenting our discoveries, to make them appear of account in the public mind" (HEA 1922i). This remark was in response to the news that Banting and his collaborator Charles Best had sold their insulin patent to the University of Toronto. Nor was he pleased to learn that chemical studies on fats—basic research that was the legitimate concern of universities—was being carried out at the National Chemical Laboratory, whose role, in his view, should be restricted to applied chemistry (HEA 1926h).

Despite the satire suggesting to the young that he was an old fogey, Armstrong could demonstrate that he was still very much up to date when he presented the final chemistry paper of his studies on enzymes (part 23) to the Royal Society in June 1922 (HEA 1922k). This paper suggested a way in which Hopkins's recent work on the milk enzyme that oxidized xanthin and hypoxanthin might be explained by Armstrong's model of chemical change and his model of water as hydrone, namely as a hypoxanthine reverse electrolysis. But if the Royal Society's publications committee was willing to allow Armstrong one more attempt to advance his hydrone theory, it drew the line when Armstrong (partnered with William Barlow) presented a paper extolling the Barlow crystallographic model as an explanation of the tetrahedral carbon atom. In July 1928, for the second time in his life, Armstrong faced the ignominy of having his paper rejected. Angrily, he published the paper in *Chemistry and Industry* instead, with the observation: "We submit our speculations, without brag [word play on Bragg], for discussion by those whom they may appeal—knowing full well that at our age we have no right to have and exercise imagination" (HEA 1928e, 892).

Incidentally, Armstrong's old theory that chemical change was electrolytic in origin received an unexpected boost in 1925 from his former pupil Martin Lowry, who was now professor of physical chemistry at Cambridge. Although Lowry had come to accept G. N. Lewis's octet theory and the general idea of the electronic atom, he showed in 1925 how a catalyst could aid the transfer of hydrogen atoms in conjugated chains of carbon atoms from one part of a molecule to another in what were called "prototropic compounds" (Lowry 1925). Organic compounds such as ethyl acetoacetate acted like graphite in their ability to conduct a hydrogen atom from one end of its molecule to the other. Lowry saw this phenomenon as interpretable in terms of Armstrong's belief that all chemical action was reversed electrolysis. Rather than being grateful to his former pupil, Armstrong attacked him ferociously for "putting new labels upon over-labelled bottles." Brutally, he told Lowry to go and read Faraday's three volumes of collected research on electricity (HEA 1925d, 568). To this rude and very public critique of a professor at the University of Cambridge, Lowry wisely chose not to reply.

Science Education

Although the heuristic campaign had long since ended, Armstrong continued to be invited to address audiences about science education. A particularly fine example is an invitation to speak to the London branch of the Institute of Chemistry and Industry in March 1924, at which the chairman praised him for his literary style. Indeed, the lecture was liberally scattered with apposite references to Wordsworth, Dean William Inge, Anatole France, and Ruskin. Armstrong first asked what a chemist was, knowing that in Britain the man in the street immediately thought of a pharmacist, and suggesting that *real* chemists hid themselves from the public. This state of affairs was the fault of chemists themselves, but it was also the fault of their education and use of "indefensible jargon which it cannot itself understand." He reproduced a paragraph of such cant from a recent issue of the *Proceedings* of the Cambridge Philosophical Society, of which he had been a member since 1914, which was rife with such neologisms as "emulsoid," "cationoid," and "lyotropic" (HEA 1924b, 146). He would not advise his sons to attend Cambridge to face such gibberish, he said; rather, better send them to France where there was a respect for language. The tide of science had developed to his advantage at both Finsbury College and the Central Technical College, but after thirty years of trying to develop the art of thinking in students he had concluded that not much could be done. He continued: "The real difficulty lay in securing sympathetic assistants—assistants who did not nullify one's efforts. Colleagues, too, constantly counteract one's efforts." He lamented the way that physical chemistry and mathematics had entered the curriculum. The "let it be granted" assumptions of mathematicians completely wrong-footed chemists whose job was "to reason why, to experiment and explore, to study the ways of materials, of tangible things generally."

> If I were charged to-day to found a school—I have founded two, one of which [Finsbury] is gone for lack of leadership and the other [Central] suppressed, because we dared to attempt to make students, especially engineers, think and be reasoning beings, not automata—I should have no lectures until a large amount of preparation work had been done and intimacy gained with chemicals. Chemistry should be taught as a whole—I would run the school as a factory, not as a forcing house. Students might swear as much as they pleased, in the vernacular, provided no jargon was used, no Ikons worshipped. (HEA 1924b, 150)

Armstrong concluded this magnificent tirade by deploring the way the word "research" had become debased during his lifetime. The word had presumably come from the French *chercher*, to search or inquire. The additional English prefix had debased its real meaning. "Today, the silliest little experiment is termed a research." He admitted that while he had been an enthusiastic preacher of the gospel of research at the turn of the century, the word had at that time signified the opposite of what "research" meant in the 1920s. The changing use of words by the modern generation annoyed him continually.

Through the efforts of the Science Master's Association, the question of teaching biology in schools became an issue again in the 1920s. At Christ's Hospital, the laboratory block that Armstrong had helped design was doubled in size, chiefly to accommodate the teaching of biology alongside physics and chemistry. Thus, when the vice chancellor of Birmingham University drew attention to a dearth of undergraduates taking biological degrees, linking this dearth with the dire need for such graduates to help run the Empire, Armstrong was quick to take up the subject in the *Times* (Robertson 1928; HEA 1929a). Biology, he declared, was only "a higher branch of chemistry, apart from its morphological side." Unfortunately, the arrival of the School Certificate pass/fail examination system in 1918 had destroyed the possibility of schools teaching scientific method adequately and hindered the uptake of biology.

When the pioneer expert on the analysis of paints used by artists, Edinburgh chemist Arthur Pillars Laurie (1861–1949), condemned the use of "experts" rather than men of science in analyzing the probity of artists' works, Armstrong queried how one made the distinction. The term "science" needed to be used "scientifically" (Laurie 1929; HEA 1929b). It seems likely that Armstrong had met Laurie in February 1929 when he was invited to address chemistry students at Glasgow University and their Alchemists' Club. His lecture, which was printed in the student magazine, the *Alchemist*, was introduced by the students' chairman with the following interesting profile:

> H. E. Armstrong—frail of physique, white-haired and bearded, at first sight an unexpected figure in such company. Later impressions efface the first to leave a lasting memory of the clear sparkling eyes, the capable articulate hands, the confident and combatant speech of the born fighter. Life had taught him many things, and he shares his knowledge gladly. It has *not* taught him yet to suffer fools gladly—even fools in authority. Nor has it taught him to abandon his own belief before the other case is proven. It is pathetic in a sense that such simple qualifications should

make a man unique. But Chemistry and Science have grown accustomed to Armstrong standing alone while opposing ideas were advanced, accepted, canonised—and abandoned. The Ionic Theory of Arrhenius is one case very much in point. There have been others. (Dean 1929, 51)[1]

The Glasgow students were entertained with some vintage punch lines. Armstrong claimed that his text, *The Chemistry of Carbon Compounds* of 1874, had given a "well-proportioned skeleton of the subject, carrying sufficient clothing to warrant its appearance in decent society." Although a vast amount of new clothing had since been added, there had been few worthwhile bones added to the skeleton in the past fifty years. A similar book written in 1929, Armstrong asserted, might be published without greatly adding to its length—rather in the way that contemporary women were setting the example of "showing how little clothing it is necessary to wear—how much freely and advantageously the skeletal mechanism can be used when lightly clad." There was no mention of the recent development of physical organic chemistry or the electronic theories of Robert Robinson and Christopher Ingold, both of which now seem such significant developments. Instead, his epistle was that chemistry had become unnecessarily cumbersome because physical chemists had spoiled the elementary teaching of the subject. Tyros were no longer encouraged to think and find out for themselves, with the result that chemists had become badly trained and unfit for industrial employment.

Armstrong was always full of advice, sound and otherwise. When a debate took place in the *Times* about careers for the sons of middle-class parents, he urged them to leave school at seventeen at the latest and spend a couple of years doing something practical, learning to be independent and use their initiative. In these gap years, as we might call them now, the sons could learn a language prior to going abroad to one of Britain's dominions, where intellects were as much in need as were the arms of laborers (HEA 1922d).

Commemorations

The year 1925 saw two important centenary celebrations—of the birth of Thomas Huxley and of Faraday's discovery of benzene. Huxley was com-

---

1. The Alchemical Society, which still exists, was founded by students in 1916; its magazine, the *Alchemist*, ran from 1925 to 1942. Earlier, in March 1925, Armstrong was elected an honorary member of the Royal Philosophical Society in Glasgow (AP1.10. Diplomas).

memorated by a special *Nature* supplement edited by Richard Gregory, who invited nearly two dozen distinguished men to reflect on how Huxley had influenced their specialties. Armstrong was naturally asked to write about Huxley's contribution to science education (HEA 1925f). He was quite explicit: he had not gained anything from attending Huxley's classes at the School of Mines when he was a student at the Royal College of Chemistry. Huxley, he said, was far too didactic, though his ability to draw anatomical details on the blackboard was fascinating.

> I did not come into personal contact with him until 1884–85, when I was translated to South Kensington, one of the small band charged with the working out of a scheme which he had done much to promote. He took no special interest in us [at the Central College].... Doubtless his mind was over-full at the time and his health bad. His outward manner was the more disappointing, as, from the beginning of my career as a teacher, I had been greatly influenced by his writings and was consciously anxious to tread in his footsteps.... Huxley, in the main, was outwardly didactic, with a definite tendency to pontificate; at heart he was ever the inquirer. (HEA 1925f, 743)

Armstrong said that Huxley had thought "science" and "scientific method" were one and the same thing, in that science was nothing but "trained and organised common sense," and he quoted liberally from the several educational essays in the collected works of Huxley (Huxley 1893). Unfortunately, Huxley's messages that science was an important part of human culture had been "delivered to no purpose."

> Herbert Spencer, Huxley and their school over-rated man's educability. They seem not to have grasped the fact that a weapon which is so novel and so all-powerful cannot and does not appeal to the vulgar mind. (HEA 1925f, 747)

He was to expand upon this idea in another lecture on Huxley a decade later.

Armstrong's contribution to the Faraday benzene centennial was much more hands-on than the Huxley celebration (J. Wilson 2012). In a letter to the *Times* in May 1925, he announced that a Committee of the Royal Institution, the Chemical Society, the Society of Chemical Industry, and the Association of Chemical Manufacturers would celebrate Faraday's report of the discovery of "benzol" to the Royal Society on 16 June 1825 (HEA

1925g).[2] The committee, under his chairmanship, included William Henry Bragg, William Pope, and his own son Frank. He saw this event as a golden opportunity to publicize the public role of science, a function much practiced by Faraday. He arranged for William Pope to deliver a Friday Evening Discourse at the RI, to be followed by further talks on the Saturday and a grand dinner accompanied by speeches in the splendid Livery Hall of the Goldsmiths' Company (HEA 1925i). The room would be lit by some 578 candles made possible by M. E. Chevreul's development of the smokeless candle from his research on fats. Half seriously, he suggested that 16 June should become a saint's day in the chemical calendar (HEA 1925i, 870). The public had its "Saint Lubbock's days" (the saint in question being John Lubbock, who had introduced bank holidays), so why shouldn't chemists? Many foreign delegates were invited, but not German chemists.

The event, like many such events involving Armstrong, was not without a farcical element. Armstrong wanted to strike a Faraday medal and, accordingly, he borrowed photographs of William Thomas Brande and Faraday from the Royal Institution so that a benzene centenary medal could be struck. He alone decided that the first recipient should be the dye-maker James Morton (1867–1943), who had devised a green vat dye. Morton had played a considerable role in saving the British dyestuffs industry during the First World War (HEA 1917c). Subsequently, in 1925, his weaving and dyeing firm had been absorbed into the British Dyes Corporation, of which Armstrong's son Edward Frankland Armstrong had become managing director. Morton had thereby become a personal friend of the Armstrong family, as had E. F. Armstrong's assistant, the Welsh chemist John Thomas. When writing an appreciation of Carl Duisberg in 1935, Armstrong picked out both Morton and Thomas as "the nearest this country had yet produced to a Duisberg" (HEA 1935g, 1025; Anon. 1933b). The award had to be kept a secret from Morton, but as the event approached, Armstrong discovered that Morton was not registered to attend because he was on holiday in Italy. After desperate pleas that he return in time for the meeting, Morton obliged, only to find that the medal had not yet been struck and was to be awarded virtually at the meeting. Bizarrely, the medal was not ready until about ten years later in 1935 (Morton 1971). The medal was never issued to anyone else, and the dyes and all the profits from the 1925 event were passed to the Chemical Society to support its occasional Faraday lecture.

---

2. "Benzol" was the form of benzene most familiar to motorists in the 1920s. In 1991, Faraday's sample of benzene in the RI's Faraday Museum was analyzed by British Petroleum chemists and found to be 99.7 percent pure.

Another Armstrongian dustup concerned the John Scopes "monkey trial" in Tennessee, which attracted widespread publicity in Great Britain, just as it did in America. Armstrong read about it in *Nature*, which like other newspapers and periodicals saw the trial of the American teacher as one of "freedom of thought." Inevitably Armstrong saw it rather differently, as just another pertinent example of how all humans failed to adhere to the Pauline injunction: "prove all things, hold fast to that which is good." If only scientists followed that precept, he held, science periodicals might be half their size and save learned societies a great deal of money. His letter to *Nature* on the subject ended with his old hobbyhorse of scientific method and the failure to teach it. But there was also a new voice in his sermonizing: he had discovered Ruskin's *Modern Painters*, "the work of an art critic with a mind of transcendent power, not an obscurantist," and he reiterated W. K. Clifford's message in the 1877 essay "Ethics of Belief" that "it is wrong always, everywhere and for anyone to believe everything on insufficient evidence" (HEA 1925k).

Anti-Physics

The Belgian industrial chemist Ernest Solvay had planned to hold conferences for select members of the international scientific community every four years. The first of these, in physics, was held in 1911, but plans for a meeting of chemists was disrupted by the war and did not take place until 1922 under the chairmanship of Sir William Pope (Solvay 1922; Fauque 2019). Armstrong was not present. A second meeting took place in Brussels in 1925, when the topic for discussion was "Structure and Chemical Affinity." Armstrong was a delegate. He reported on the proceedings in *Nature* and also presented a paper on catalysis that drew on his recent thoughts on enzymes that he had presented to the Royal Society (HEA 1925h; 1925n; 1927a). The conditions of chemical change had been very much on his mind, therefore, when he read a paper by Herbert B. Baker and Margaret Carlton that confirmed what Baker had shown years before: that reactions did not occur under extreme dry conditions (Baker and Carlton 1925). The paper was purely experimental in character and made no reference to Armstrong's views on water. Water (or in Armstrong's terms, an electrolyte) had to be present to drive the reaction (HEA 1925p). Armstrong clearly wanted to seize on any report that seemed to proffer evidence of his model of chemical change. Similarly, after listening to physicists at the Royal Society debating the formation of ozone in the upper atmosphere, he suggested (HEA 1926b; 1926j) how his catalytic theory could offer a chemical explanation (one that is broadly accepted today).

Armstrong published a letter to *Nature* in February 1926 that addressed the closely related question of the cause of lightning in thunderstorms. If Bancroft was Armstrong's American foil for receiving brickbats, Oliver Lodge remained his chief British target. Armstrong had been criticized in a review of a lecture by Lodge at the Institute of Physics by "one J. S. G. Thomas, whose acquaintance I, unfortunately, do not enjoy" for questioning G. C. Simpson's theory of thunderstorms made in 1909, and explaining them by the hydrone theory (HEA 1926a). In the letter he attacked Lodge's well-known spiritualist views:

> Sir Oliver Lodge has a great name and his opinion counts for much with the public, because of his position in science. Surely, we have the right to ask him to dissociate himself from our body [of scientists] for saying in public: "And what of man? If his death is the end of him, the value of his existence may be doubtful. But if, *as I know* (my italics) that is not the end of him, then there may be infinite progress in store." We of the body scientific know that we can very rarely say: *I know*, least of all when speaking of the great problem that has exercised the mind of man throughout the ages. (HEA 1926a, 196)

Lodge replied sarcastically that Armstrong seemed to glory in "uncertainty" and that his sceptical attitude "prevents his own enjoyment of the great discoveries of the present generation, because they do not dance to the *drone* of his water bagpipe—a serviceable instrument but in danger of becoming a fetish" (Lodge 1926). Lodge made a palpable hit and Armstrong was left only with a verse of George Meredith's to defend himself (HEA 1926f).[3]

Armstrong fired another amusing broadside against the interpretation by an American chemist, Dwight C. Bardwell, that hydrogen acted as an anion in metallic hydrides (Bardwell 1922; HEA 1926d). Bardwell's colleague G. N. Lewis, who had first formulated the idea that hydrogen could both behave as a positive metal at the head of Group 1 in the periodic table, and act as a halogen in Group 7 of the table, came to his defense. "We always enjoy Armstrong's attacks and must not complain if we come within his range," Lewis began; but he had to come to the aid of Bardwell's beautifully designed experiment (Lewis 1926). The chemical conclusion was impeccable. Armstrong would have none of this argument. In a long

---

3. "O sir, the truth, the truth! Is't in the skies / Or in the grass or in this heart of ours?" From "A Ballad of Fair Ladies in Revolt," 1876. Ironically, Meredith was arguing for women to be treated equally with men!

reply he accused Lewis, in good humor, of casting aspersions on hydrogen's character and concluding, inevitably, that modern chemists had lost the art of scientific method (HEA 1926i).

Armstrong clearly took pleasure in satirizing recent developments in physics. And it is his scepticism (or what may seem now as "eccentric views" toward contemporary physics) for which he is, consequently, often remembered (Gratzer 1996). For example, when the Americans William D. Harkins and Hugh A. Shaddock improved the Wilson cloud chamber to take three-dimensional photographs of atomic disintegration, he pleaded ignorance.

> I know that I am very ignorant, but old and unrepentant as I am I still live to learn from young people and watch their doings with delight, tho' maybe they are sometimes a little "previous" . . . Probably, therefore, I am more than stupid in being surprised at "the atom (presumably oxygen of mass 17) which is synthesised." Suppose, however, that a poor errant molecule of fair hydrone were the stricken "atom," it might well be "electrolysed" and give OH = 17 + H = 1. (HEA 1927b)

In other words, there were other possible interpretations of the cloud chamber streaks. He knew, he said, that "it was wrong for a poor worm of a chemist to turn and put a common or garden interpretation upon the work of august and fashionable physicists: that they should be regarded as kings who can do no wrong. I have, however, preserved from my youth the memory of a king who wore no clothes and myself still work a little in the garden with not too much regard for weather." He remained implacably contemptuous of physicists (and astronomers like James Jeans) who knew little chemistry.

As the last remark confirms, Armstrong had little sympathy for contemporary physics. In a letter to *Nature* on space and time as described by his old sparring partner, Oliver Lodge, in a reflection on Einstein's lecture at University College, Nottingham, on 6 June 1930, Lodge had suggested that Einstein had only reached conclusions that he (Lodge) had held, namely that "matter is passive" and "space is active." This duality followed from Einstein's famous equation, $E = mc^2$, so that "matter is inert; space is energetic. Matter *does* nothing, though it serves to display the energy of which space is full" (Lodge 1930). Armstrong, quoting extensively from Lewis Carroll's "The Hunting of the Snark," ridiculed Lodge's position.

> "Science" is something that we are endeavouring (perhaps I ought to say, should endeavour) to pass on to the masses. It were time, therefore

that some protest were made against the language used by one of our would-be leaders ... The public is being played upon and utterly misled by the dreamery of the rival mathematical astronomers and physicists—not to mention the clerics—who are touring today and raising the game of notoriety to a fine art. In rivalry to religious mysticism, a scientific pornography is being developed which attracts the more because it is mysterious—apparently the professors are seeking to outrival Mrs [Mary Baker] Eddy. (HEA 1930f)

Lodge, wisely, did not reply, though he might well have taken offense to Armstrong's concluding limerick underlining that matter *was* something: "Or talk to Ol'ver Lodgey / About his great big body."

In 1927, to out-Kolbe Kolbe, Armstrong decided to give wider publicity to his views on the structure of water in opposition to Arrhenius's theory of ionization. The book, *Essays on the Arts and Principles of Chemistry*, was issued by his son-in-law's publisher, Ernest Benn, and comprised long excerpts from his article "Chemistry" for the most recent (twelfth) edition of *Encyclopaedia Britannica* (1922), various prewar essays on the ionic theory from *Science Progress*, and a reprint of his Messel Memorial Lecture of 1923 (HEA 1922j). The book was dedicated to

my seven grandsons and my seven granddaughters, hoping that one or more may develop the individuality to think for him or herself and not bow the knee to authority. (HEA 1927s)

There were very few reviews of the book, and the reaction was entirely negative. Armstrong's explanation of chemical change was too complicated, and, more especially, it could not be translated into mathematical language. The ionic theory and mechanistic organic chemistry had won the day.

He suffered the same fate with his advocacy of the Barlow-Pope modelling of crystals. Although Pope himself had abandoned the theory in favor of the Braggs' model of crystallization, Armstrong continued his advocacy as late as 1928, when he published a joint essay with Barlow to show how the tetrahedral character of carbon's valences affected the shapes of paraffins and polymethylenes (HEA 1928e). It appears that a second part of this essay was never written, or at least never published. He had already adumbrated the model in an essay for a book that celebrated the work of Isaac Newton (Greenstreet 1927, 5). When Barlow died in 1934, Armstrong paid homage to his old friend in the *Times*, calling him "an inspired geometrical genius" (HEA 1934b).

The Golden Anniversary

On 30 August 1927, the Armstrongs celebrated their golden wedding anniversary in the home of their son-in-law and daughter at Haverstock Hill (HEA 1927h). The Armstrongs' six children and eight grandchildren were present, and the guests included H. B. Baker, the current president of the Chemical Society; Sir Richard Gregory, the editor of *Nature*; and Sir William Cope (1870–1946), a Welsh politician who was then junior lord of the Treasury. Armstrong was presented with a striking portrait by the artist Thomas C. Dugdale (1880–1952), which had previously been displayed publicly at the Royal Academy's summer exhibition. After Armstrong's death it was presented to the Royal Institution by his family, where it remains today. In addition, former staff and students at the Central presented the couple with a bound illuminated address (AP2.A9). *Nature* commented on the coincidence that Sir Oliver Lodge and his wife had also celebrated their golden wedding anniversary the week before the Armstrongs:

> Both Sir Oliver and Prof. Armstrong are distinguished by the originality of thought and independence of action; and both are held in affectionate regard by all who have come under their influence. Their different natures—or rather certain factors which characterise them—are represented to some extent in their eldest sons, Sir Oliver's being a life-long student of art and full of imaginative insight, while Prof. Armstrong's is devoted to industrial chemistry, and is managing director of the British Dyestuffs Corporation, Ltd. Heine said that man should be careful in the choice of his parents, and with the Lodges and Armstrongs the children have had the advantage of both wise nurture and noble. (Gregory 1927)

An unsigned editorial by F. E. Hamer (who was present at the celebrations) in the *Chemical Age* paid handsome tribute to Louisa Armstrong:

> A woman as remarkable in her character and sphere as her husband has been in his, belonging to the type of woman who make great careers possible to husbands and son. [The couple were] surrounded by sons all distinguished in one way or another, and by a group of grandchildren that promise to carry on honourably the family tradition.

Hamer's encomium concluded by citing Louisa's final comment (noteworthy considering her husband's views regarding gender relations): she "remarked that while her husband always thought that he was having his own way, she might now safely confess that in fact, she had always had hers" (Anon. 1927).

The 1920s saw Armstrong campaign on behalf of purer foods and attention to the nutrition of both farm animals and humans. He remained wakeful concerning the state of British education in schools and universities and seized the opportunities provided by significant Huxley and Faraday anniversaries, as well as the many lecture invitations he received every year to criticize and advise. In 1925, Macmillan reprinted his educational essays on scientific method with a new preface (HEA 1925s). His sharp criticisms of physical chemists continued and were extended, with amusing literary flourishes on the actions and opinions of contemporary physicists. These campaigns and sharp criticisms were to continue into his last decade of life. As the reports of his golden wedding anniversary demonstrate, by the 1930s the cantankerous chemist had become something of a national treasure.

✳ 15 ✳

The Lewis Carroll of Chemistry

> We of the body scientific are seeking to impose ourselves mentally
> upon the world—to rival the word-painters—having long
> done so in ways mechanical and succeeded: with terrible
> effect in part yet with much benefit in some fields.
> —HEA 1932c

In his retirement years Armstrong enjoyed his role as an elder stateman of British science. On one level, and in some contexts, he was a cantankerous gadfly and pot-stirrer. Although often deadly serious, he also loved being a flamboyant jester and literateur of science. On every level, he continued to be irritating to some, entertaining to many, and often decidedly constructive. The present chapter provides a sampling of his many Fifth Business activities.

Social Responsibility of Science

In retirement, Armstrong continued to attend meetings of the Chemical Society, the Royal Society, the Royal Institution, the Royal Society of Arts, and the occasional meeting of the British Association for the Advancement of Science, even though he expressed frequent annoyance at the way physics had come to dominate the scientific enterprise (HEA 1933h and 1934f). Above all, he continued to write vigorously for newspapers and the scientific press.

There had been a huge growth of international organizations at the end of the nineteenth century toward the control and standardization of measuring units, nomenclature, chemical analysis, and the bibliographic control of scientific literature (Fox 2016b; Hepler-Smith 2015). Among the many organizations that Armstrong became involved in was the International Association of Chemical Societies (IACS), founded by

Wilhelm Ostwald in Paris in 1911. When peace had been restored by 1919, an International Research Council was established to replace the defunct IACS. Following a conference in Paris to which German scientists were not invited, a number of new international unions were created for coordinating the peacetime reorganization of research. One of these was IUPAC, the International Union for Pure and Applied Chemistry. This new organization took over the responsibilities of the earlier IACS (which had concentrated solely on pure chemistry), and it soon came to play a vital role in establishing international agreement on chemical symbols, nomenclature, and atomic weights (Fennell 1994; Van Tiggelen and Fauque 2012). Although Armstrong played no direct role in these restructuring events, he frequently commented on them, or attended meetings as an auditor. He attended the ninth international meeting in Madrid in 1934 with the Manchester natural products chemist Robert Robinson. Similar in personality, they became good friends.

In November 1934, the Maison de la Chimie was opened in Paris in an elegant refurbished eighteenth-century building. Armstrong, together with T. S. Price and C. S. Gibson (treasurer and secretary of the Chemical Society), was there to represent the Chemical Society. The Maison was Paul Otler's and Henri La Fontaine's realization of Wilhelm Ostwald's plan to erect a world center for the subject in the form of an institute in Brussels (Fauque 2011). Besides assembling a first-class library, the Maison would also act as a center for the documentation of chemistry and provide an abstracting service as an expansion of Otler's and La Fontaine's prewar service. Today, it is a chemistry-oriented conference center, in addition to providing offices for various scientific organizations.

On a smaller scale of ambition, in the 1860s the Royal Society had begun a catalogue of all the world's scientific papers published in scientific periodicals since 1800. Armstrong was troubled when in 1922 the society abandoned this effort and a follow-on project, as its continuation had become financially impossible. From then on, Armstrong became highly critical of the Royal Society's management. When Ernest Rutherford was elected president of the society in November 1925, Armstrong warmly congratulated him for reintroducing critical discussions and debate after the formal reading of papers (D. Wilson 1983, 465). Nevertheless, he became increasingly upset by the way the society was being run by an elite of self-perpetuating fellows, a complaint that continued under the presidency of biochemist Frederick Gowland Hopkins, who succeeded Rutherford in 1930. Ironically, this criticism was the same irritated complaint that Armstrong himself had faced as secretary of the Chemical Society fifty years previously.

With the growing breakdown of political and economic certainties in the 1920s and 1930s, many younger fellows shared Armstrong's view that the Royal Society ought to be showing leadership in solving the world's problems, rather than just promoting pure research. However, it was not Armstrong but Rutherford's former radioactivity collaborator Frederick Soddy who became the society's leading critic. He had been elected to a chair at Oxford in 1919 only to face obstruction and indifference in his attempts to improve laboratory facilities and to establish an Oxford school of radioactivity research. In the struggle, he became increasingly bitter and eccentric, more or less abandoning scientific research. In Soddy's view, the Royal Society should have been taking the lead in energizing the scientific community to take their social responsibilities seriously and work out ways in which science could help solve the world's economic problems. Armstrong wholeheartedly agreed.

In the foreword to a collection of essays by a like-minded group of socialist intellectuals, *The Frustration of Science* (1935), Soddy declared that the public should require that

> universities and learned societies should no longer evade their responsibilities and hide under the guise of false humility as the hired servants of the world their work has made possible, but do that for which they are supported in cultural release from routine occupations, and speak the truth though the heavens fall. (Soddy 1935)

Armstrong completely agreed with Soddy's view that contemporary science had become the body of "hopeless idiots" by allowing the public to believe that "science" meant writing fine words "about the stars and the smashing of atoms, to little useful end, whatever the ultimate value may be—no one of us making clear how our work has to do with bread and butter" (HEA 1935j, 345). The fault lay, Armstrong argued, in the Royal Society because it had allowed the bureaucratic Department of Scientific and Industrial Research to take its place and thereby deprived science of any collective voice. However, Soddy's solution—the formation of a scientific clerisy—would be difficult to achieve.

Armstrong reviewed the book in *Science Progress*, noting that Soddy stood "as a forest giant above the low arboreal growth of Oxford, a man who, daring to look abroad, has views of his own and expresses them, forcibly urges scientific workers the need in which they stand of developing a sense of social responsibility" (HEA 1935j, 345). He dismissed the essayists' views as woefully inept. Daniel Hall, the notional editor of the volume, had provided no clear lead on agriculture; James Crowther's view on aviation were complete nonsense ("Flying is probably the greatest curse that 'science' had brought upon mankind" because of its deadly potential in warfare); J. D. Bernal's treatment of science and industry (a foretaste of his *Social Function of Science*) was "high-fallutin" nonsense; and the only worthwhile essay was that by Vincent H. Mottram, professor of physiology at King's College, who had argued that medicine should be concentrating on the prevention of disease rather on its cure. The dismissive review ended with the news that he had recently attended his brother-in-law's funeral in Eastbourne, where he had stayed in Huxley's former house. What would Huxley have said about our failure to organize science for social purposes, he mused? He would have erupted in blazing wrath, in all probability, Armstrong felt.

The book was published shortly after a movement for reform of the Royal Society, led by Soddy and aided by Armstrong, had sputtered and failed. Although the "Soddites" lost their case, as Hughes has shown, W. H. Bragg and the society's official secretary, Henry Lyons, did make subtle reforms—a greater emphasis on applied science and, most notably, keeping in touch with fellows through a periodic circular, *Occasional Notices*. In April 1938, a year after Armstrong's death, this circular became *Notes and Records of the Royal Society*, today a prominent journal for research in the history of science. Frank Armstrong congratulated Soddy: "Your Royal Society crusade is rapidly bearing fruit. What a lot of changes! How interested my father would have been. The trouble with the Royal is that it is mainly a branch of the Civil Service" (Hughes 2010, S111).

The Royal Institution

In his student days at the Royal College of Chemistry and later in the 1870s, Armstrong attended some of the evening lectures of the Royal Institution. In 1908, he mentioned that he had been first introduced there in 1867 by Frankland, presumably by providing him with entrance tickets as a guest. In February 1873, he delivered four Thursday daytime lectures on "the artificial formation of organic substances," presumably at the invitation of John Tyndall, who was then director (Anon. 1873). The lectures were not published. Armstrong was first asked to give a prestigious RI Friday Evening Discourse on Fischer's work on sugars in 1895, but it also was not published (HEA 1895b). Since his initial allegiance was to the *London* Institution, Armstrong may have long felt deterred from joining the *Royal* Institution—though many other LI members were also members of the RI. It was not until 4 December 1899 that he became a subscribing member of the institution, meaning that he could be invited to play a role in its management. Within a year, he had been made one of its overseeing "visitors" for the season 1902–1903.

After his year as a visitor, he served as a manager of the RI on and off between 1905 and 1920, and he gave occasional lectures in May 1905 and February 1911 on nutrition. His most important invitation was to deliver the second and third Hodgkins Trust Lectures in 1908 and 1915 respectively (HEA 1908e). The wealthy British-American philanthropist Thomas George Hodgkins (1803–1892) had given $100,000 to the Royal Institution under the condition that the fund was to be applied "exclusively in investigation of the relations and correlations existing between man and his Creator. It is the belief of [Mr. Hodgkins] that with the present generation God has almost been forgotten and he appeals to the scientific men of the present and coming centuries to whom we look to direct thought to lead us back to the source of all knowledge" (Brock 2011c).

Following discussions between the managers, Fullerian RI professor James Dewar, and the RI's legal advisers it was decided—contrary to Hodgkins's explicit directive—that the new fund could be legitimately used to support Dewar's work on the liquefaction of air, on the condition that it also sponsored a lecture every seven years that explained and publicized Dewar's findings. Under these directions, the first Hodgkins Lecture was given by popular writer on astronomy Agnes Clerke (Clerke 1885). As a devout Irish Roman Catholic, Clerke had been awarded the RI's Actonian Prize in March 1893 for an essay entitled "Astronomy, as Illustrative of the Wisdom and Beneficence of the Almighty." In her Hodgkins Lecture of December 1901, she provided a simple descriptive review of all the papers published by Dewar between 1893 (when the Hodgkins Trust Fund began to be used to support Dewar's research) and 1900, with particular emphasis on Dewar's success in liquefying and solidifying hydrogen, as well as his solidification of oxygen (Clerke 1901).

Clerke died in January 1907, a year before the second lecture was due. She would almost certainly have been asked to deliver it had she lived; instead, the managers turned to Armstrong, because it was well known that he was intimate with Dewar and admired Dewar's work as a magnificent feat of engineering science. He accepted and had a good eighteen months to prepare the review of some thirty-five of Dewar's lectures and publications. His address, which covered Dewar's work between 1900 and 1907, differed little from Clerke's approach, with one exception. Namely, he tried to show how Dewar's investigations supported his own work on structural chemistry; he did so by devoting a good portion of the address to his centric formulae for benzene, naphthalene, and anthracene. Dewar's investigation of chemical interactions at low pressure also gave him an opportunity to refer to his own theory of hydrone and the extraordinary properties of water.

Armstrong attributed Dewar's failure to liquefy helium (in competition with the Dutch physicist Heike Kamerlingh Onnes) to the RI's lack of funds and to a bout of illness that had inhibited Dewar's work at a crucial stage. Deprived of the Nobel Prize, Dewar's work nevertheless (Armstrong argued) had "developed the use of liquefied gases to the solution of the manifold problems involved in the study of the properties of matter at excessively low temperatures." If the Hodgkins Trust's first review by Agnes Clerke had been Dewar's hydrogen period, he continued, the second deserved to be named the "charcoal vacuum period," because Dewar had shown "the marvellous power of charcoals to absorb gases at low temperatures." Armstrong emphasized the significance of the Dewar vacuum vessel, which had "now [founded] a German industry of considerable magnitude," as an illustration of the way in "which the research work of the Royal Institution has conduced to the realisation of the objects of its founder, Count Rumford." In other words, he argued, pure research leads to valuable tools and inventions that aid the prosperity of the general public.

Armstrong concluded this first of his two Hodgkins Trust Lectures with a long series of reflections on the future of scientific research at the RI. He supposed that Hodgkins had made the donation to the RI in recognition of its important place in the history of scientific discovery. Ironically, Hodgkins may well have confused the Royal *Institution* with the Royal *Society* and intended the legacy for the latter (Brock 2011c). For Armstrong, the RI was "a Holy of Holies," but he suspected this feeling was not shared widely because its achievements remained so largely unknown to the outside world. Home to the important research of Davy and Faraday, it was in his view "the birthplace of discoveries and inventions which have revolutionised our civilisation." But both Davy and Faraday had come to scientific research with little formal education. How was it, then, he asked, that the workers now engaged in scientific investigations who had the benefit of far better educations appeared "to lack not only originality and breadth, but also the critical faculty and that sense of proportion which is so eminently characteristic of Faraday's writings?" It had taken philanthropic outsiders like Hodgkins to assist and bring about a better understanding of the value of scientific inquiry.

The problem was, Armstrong familiarly asserted from his hobbyhorse, that although science had been introduced into schools, its instruction was too perfunctory, barren of interest and lacking disciplinary rigor. The universities, likewise, had failed to recognize that scientific method was a necessary preliminary to further education. Faraday had previously asserted all these things, but they needed reasserting. Ludwig Mond, in endowing the RI with the new, modern laboratory facilities of the Davy-Faraday Laboratory, had assumed that many educated in science at public schools and universities would come to the RI to do their research. His expectations of such a demand had not been justified (K. Watson 2002). The reason was Britain's retention of a belief in classical studies in schools and its worship of examinations. British education was still "dominated by the literary and not be the practical type of mind, by men of narrow purview." Armstrong's hope was that the RI would become an expanded research institute with the professor or professors in charge aided by a staff of student assistants. Ironically, this arrangement was never one to which Dewar aspired.

Armstrong continued to act as an RI manager in the years following this lecture and, by 1913, he had risen to the status of being one of the vice presidents. He returned again in December 1915 to deliver the third and last Hodgkins Trust Lecture, this time on Dewar's final decade of low-temperature research (HEA 1916f). Armstrong's task was now a good deal more difficult to hold to the spirit of the Hodgkins bequest. Dewar had largely given up low-temperature work after 1908 and his research was only tenuously linked to investigations of air and gases. Dewar's increasingly bad temper and inability to get on well with colleagues also inhibited his ability to work on cryogenics. Moreover, membership of the RI had declined since 1904 as London audiences for scientific lectures became less popular. Audience numbers would decline even more as the First World War continued. The unfortunate fact was that very little original research had been conducted in the RI's laboratories between 1908 and 1916. Armstrong's recourse was to review the whole of Faraday's and Dewar's scientific careers and to add his own reminiscences at appropriate points. Inevitably, therefore, he went over much of the ground that Agnes Clerke had covered in the first Hodgkins Lecture. He disguised this ruse by calling the years under review as Dewar's "High Vacuum Period." In passing, Armstrong mentioned that he now accepted that the chemical elements were complex, and he even praised the work of Soddy on isotopes (whose existence he had previously denied).

He concluded the final Hodgkins Lecture by again examining the "future of original experimental research," but this time generally, and not specifically at the RI. He reiterated the value to his personal scientific development and career that his attendance at RI lectures had always been. He repeated his detestation of the continuing emphasis in schools on a classical education. The war, he trusted, was bound to change this state of affairs because it had exposed the value of utilizing scientific knowledge. He expressed approval of the government's creation of a special Department of Scientific and Industrial Research, and he likewise approved of the implication that university academic and industrial research would be brought closer together. And he hoped that this convergence promised a future for the RI as "a glowing gem in an otherwise featureless casket." Money had to be given to "competent inquirers almost unconditionally to be of service" (HEA 1916f, 782–85).

Not surprisingly, the RI held no more Hodgkins Trust Lectures. By the 1920s, the financial interest from the original monetary gift was no longer separately itemized in the RI accounts. By the 1990s, the sole clue to the American gift that had supported Dewar's cryogenic researches was a small portrait of Thomas Hodgkins that hangs in one of the RI's many rooms.

Armstrong continued to act as a manager and as a vice president up until the time of Dewar's death in March 1923. As Dewar's closest (and, one might say, only) scientific friend, it fell to Armstrong to deliver a memorial oration on 18 January 1924. Heavily influenced by Lytton Strachey's *Eminent Victorians* (1918), Armstrong adopted Strachey's controversial advice to the modern biographer to attack the subject in unexpected places and "shoot a sudden revealing searchlight into obscure recesses." The result was a masterful portrait of an unloved figure in British science. Armstrong gave no more lectures at the RI during the 1920s and 1930s; it appears that he deliberately cut himself away from its management. Perhaps, without the presence of Dewar, he felt unable to speak; and although he was on friendly terms with William Henry Bragg, the new director, he was uncomfortable with the RI's movement toward physics and physical chemistry.

Obituary Writing

As Armstrong's word-portrait of Dewar demonstrates, Armstrong was skilled in the art of writing obituaries. He published around fifty long and short memoirs of fellow chemists and other friends during his lifetime. He offered the following Stracheyan advice:

> It is only when obituary notices are to be written, that the difficulty of securing any real picture of a man and his achievements becomes the stark, staring certainty it usually is. Then the lights are over-painted and the shadows are usually left out. The *nil nisi* [*bonum*: of the dead say nothing but good] adage has done infinite injury. Later on, when the

general historian sets to work, he finds himself without safe material to go upon—so history becomes a polite and perjured fiction. (HEA 1930c, 809)

During the 1920s, Armstrong became increasingly called on to compose obituaries of his famous contemporaries—a task that he conducted brilliantly. However, as so often happened, Armstrong was dilatory in carrying out the writing. When called upon to compose the notice of Dewar for the Chemical Society he faced a problem, because he had also taken on obituaries of two other intimate friends, Rudolph Messel and Horace Brown.

The Dewar obituary, as we have seen, was first given as a lecture to the Royal Institution in January 1924. But, instead of being published in the RI's *Proceedings*, it was issued immediately as a small monograph by the publisher Ernest Benn. Shorter versions of the lecture were then developed for *Nature* and *JCS* (HEA 1923b; 1924a; 1928i). These first two versions of the obituary were further adapted for the Royal Society (HEA 1926s). The Chemical Society had to wait to the last; its full obituary of Dewar did not appear until four years later, in 1928. Armstrong's excuse was that he wanted to utilize the collected papers of Dewar, which were not published until 1927 (Dewar et al. 1927).

Armstrong's dilatory behavior caused great problems for the secretaries of the Royal Society. In the end, physicist James Jeans took it upon himself to ghostwrite the obituary using as its basis Armstrong's RI monograph. When Jeans consulted Lady Dewar about his sketch, she found fault with several facts, forcing Jeans to make corrections and to make sure Armstrong approved of his amendments. Thus, although signed off as written by Armstrong, it was Jeans who composed the official Royal Society obituary of Dewar (HEA 1926s).[1]

In October 1927, the French celebrated the centenary of the birth of chemist Marcellin Berthelot. Armstrong was asked by the Société de Chimie Industrielle, which was organizing celebratory events, to deliver the keynote address in French. The English text was subsequently published in *Nature* (HEA 1927n). In a week of grand and even lavish ceremonies, the foundation stone of the previously mentioned Maison de la Chimie was laid (Fox 2016a). The man behind the project, French chemist Charles Moureu, did not live to see the Centre opened in 1934, having died in 1929. Armstrong, who had had dealings with him as a member of what had become the International Union of Pure and Applied Chemistry, memorialized him in *Nature*. He pointed out that Moureu had succeeded

---

1. Jeans to Lady Dewar, 22 February 1926, RS archives, NLB/69/285.

to Berthelot's chair at the Collège de France, where he had done sterling work on alkaloid chemistry as well as detecting the presence of noble gases in French springs and waters. He had accomplished much before turning his attention to internationalism and the creation of the Maison de la Chimie (HEA 1929f; Fauque 2011). A month later, in the presence of the French ambassador, Armstrong gave another, entirely different appraisal of Berthelot at the Royal Society of Arts in London (HEA 1927m and 1927q). This time, he stressed the influence on Berthelot of the philosopher Ernest Renan. William Pope and Robert Mond offered further reflections on Berthelot's importance in the discussion after the lecture.

More controversially, Armstrong was asked to give a personal account of the electrical engineer Hertha Ayrton, the second wife of his former Finsbury and Central College colleague, William Edward Ayrton, after her death in August 1923. His essay was decidedly provocative, opening:

> Mrs Ayrton was one of those who aspired to prove that woman can be as man as an original scientific inquirer. Did she succeed? . . . I have but small qualification for the office, yet as she was my colleague's wife and we often met and were in fair sympathy, I was able to take notice of her idiosyncrasies and of the conditions under which she was placed. (HEA 1923d)

His first offense was to portray Ayrton's first wife, Mathilda Chaplin, a pioneering female doctor, as an ethereal Mélisande, the heroine of Debussy's opera. If he were to write an opera portraying his scientific friends, he continued, Mrs. Hertha Ayrton would have to be portrayed as Brünnhilde from Wagner's *Ring des Nibelungen*, who "aspired to be an active companion of scientific heroes." He appreciated the irony that Edith, the daughter of Ayrton's first marriage, had married a Jew, the cultural Zionist author Israel Zangwill; and that Ayrton's second wife, Hertha, was likewise Jewish. "I often told him that he and his wife were an ill-sorted couple: being both enthusiastic and having cognate interests, they constantly worried each other about the work they were doing."

> He should have had a humdrum wife, "an active, useful sort of person," such as Lady Catherine recommended Mr Collins to marry [in Jane Austin's *Pride and Prejudice*], who would have put him into carpet slippers when he came home, fed him well and led him not to worry either himself or other people, especially other people; then he should have lived a longer and a happier life and done far more effective work, I believe. (HEA 1923d, 801)

Did this account paint a picture of his own wife Louisa and their domesticity?

He went on to damn Hertha with faint praise, implying that her husband and their other Finsbury College colleague, the mechanical and electrical engineer John Perry, had tended to exaggerate her talents even though, as he granted, she had been "an indefatigable and skilful worker." A campaign to elect her a fellow of the Royal Society had, he thought, been ill-advised and illegal, and he was pleased that it had failed.

> I never saw reason to believe that she was original in any special degree; indeed, I always thought that she was far more subject to her husband's lead than either she or he imagined. Probably she never had a thorough scientific equipment; though a capable worker, she was a complete specialist and had neither the extent nor depth of knowledge, the penetrative faculty required to give her entire grasp of her subject. (HEA 1923d, 801)

The effort to get Hertha Ayrton into the Royal Society had been organized by none other than Perry, but the society simply argued that married women were ineligible.[2] Perry then campaigned for her to be awarded the Royal Society's Hughes Medal (for "original discoveries in electricity and magnetism") instead.

Reading Armstrong's condescending dismissal of this pioneering woman scientist and inventor understandably raises hackles today; in 2021, a letter to *Nature* urged retraction of Armstrong's obituary (Brandt 2021). It also raised hackles at the time. Letters poured into *Nature*'s offices speaking in her defense and condemning Armstrong for his insensitivity. Among them was one from the Ayrtons' daughter, Barbara Ayrton Gould, but Gregory privately defended Armstrong's portrait as "mild chaff" that "should not be taken too seriously" (Baldwin 2015, 92, 260).

Gregory only published two of these critical letters. One was from the Ayrtons' family doctor: "It is difficult to understand how anyone professing to have been their friend could suggest that they were 'an ill-sorted couple'... I am well qualified to protest against the heartless comments upon the private life of a very noble woman of whose living presence we are so recently bereaved.... Surely for the rest of his life he [Armstrong] will regret not having declined that 'appeal' for an obituary notice" (Mills 1923). In his defense, Armstrong merely said the good doctor had not understood his sense of humor.

2. Perry to James Larmor, 8 January 1902, RS NLB/23/1/418 and NLB/23/2/166.

> When I used to tell [the Ayrtons] that they were "ill-sorted," knowing this full well and knowing me, they did but smile. As did Mrs Ayrton—when, to terminate one of our fruitless discussions on woman as man, I sometimes said: "We will admit you are 'up to us' (apart from being yourselves) when you are regularly engaged as chefs and produce one to go down to posterity with [celebrated 19th-century chef Alexis] Soyer." (HEA 1923e)

The other printed letter was from Thomas Mather, William Ayrton's collaborator on the textbook *Practical Electricity* and colleague at the Central. Mather rigorously denied Armstrong's implication that Mrs. Ayrton had lacked originality, correctly pointing out that she took out eight patents between 1913 and 1918, and that she had published inventions as a single woman before her marriage. It should also be noted that in 1904, Perry's efforts on her behalf were crowned with success: Hertha Ayrton was awarded the Royal Society's prestigious Hughes Medal. She was also the first woman elected a member of the Institute of Electrical Engineers (Fara 2014).

When organic chemist William Henry Perkin Jr. died in September 1929, Gregory inevitably asked Armstrong for a memorial essay. Perkin, said Armstrong, saw the chemical laboratory as the fons et origo of chemical discovery. Until Perkin arrived in Oxford in succession to William Odling, Oxford had been chemically dead; "now it is alive—even exploding; alive too in the other sciences." The failings that Armstrong had criticized in 1904 and 1917 had all been corrected (HEA 1904c; 1917a; 1929h). As usual, Armstrong was fascinated by heredity, the comparison of father and son, and regretted that Perkin Jr.'s marriage had not produced children. There seemed to be "some influence at work promoting sterility in scientific workers. The problem is one needing close attention, especially by the advanced woman: no birth control seems called for, rather the opposite." Referring to Perkin's idiosyncratic practice of starting the day at one end of a clean bench and working along it leaving a trail of dirty apparatus, Armstrong opined that Perkin "was as mad as several *Mad Hatters*."

Perkin, Armstrong believed, had pursued research just for its own sake. "His joy over each beautiful new substance was extreme—he gloried in constructing. This gave him his power but limited his sense of proportion and his influence." He had notably failed to create a chemistry school at Oxford, and graduates from the colleges had not joined him.[3] On the other hand, Armstrong was of two minds about the wisdom of Perkin's most

---

3. Jack Morrell's judgment is that Perkin did create a research school (Morrell 1993).

noteworthy educational act, the establishment of a fourth year to be spent on research for an honors degree.

> It is a grave question whether this be a wise provision. It involves the assumption, that the years previous to the fourth are not years in which the spirit of inquiry is inculcated and rampant. They should be—yet, if they were, it would not be necessary to set aside the fourth for the purpose. To begin in the fourth is too late. (HEA 1929h, 625)

Although he disliked the Nobel Prize, he could not understand why Perkin had never been awarded one.

> The Swedish assessors seem to have no eye for an English chemist— thus, in the past, they have overlooked men such as Crookes and Dewar ... I have heard it said, that this was because these were not nominated by their countrymen: the excuse will not apply to Perkin Jr. If the rules for the race involve this, the sooner they are altered the better: it should rest with the assessors to take all necessary steps to make the award in a scientific spirit, not contingent upon chance nomination. (HEA 1929h, 626)

In fact, both Crookes and Dewar were nominated many times, but Perkin never.

Occasionally—and less controversially—Armstrong profiled the living. Between 1873 and 1938, *Nature* published a total of forty-eight pen-portraits of notable scientists under the general title "Scientific Worthies," most of whose subjects were still living at the time of publication (Barr 1965). Each essay was accompanied by a steel engraving and usually came at the beginning of a new volume in the series. In the summer of 1927, Gregory asked Armstrong to compose the forty-fifth memoir on natural products chemist Richard Willstätter (1872–1942). Armstrong does not appear to have ever met the German Jewish chemist, but he was familiar with, and in admiration of, Willstätter's elucidation of the structures of plant chemicals, particularly of colored materials. He opened the portrait with a literary quotation, as he often did, this one a passage from *Leaves of Grass* in which Walt Whitman asks, "What is grass?"

> The chemist who can teach so much of grass, who can go so far towards answering the question put by the child to which the poet confessedly had no answer, who can also lay bare the secret of colour in flowers, may be placed above even the poet. The poet but deals with the superficial

and with fancies; at best he is a mere painter. The full beauty of Nature, the structure of her wondrous mechanism, is patent only to the chemist. (HEA 1927g, 1)

In this elegant essay, Armstrong also noted the synthetic repercussions of Willstätter's research and drew especial attention to the work of Robert Robinson in Manchester on the preparation of plant anthocyanidins. Always a man to draw a moral, Armstrong concluded:

I perhaps more than anyone can appreciate the value of work so varied [as Willstätter's] yet always in logical connexion—can wonder at the genius displayed and the self-sacrificing devotion of the worker to his task. Only the few among us can be aware what such inquiry means, what it involves—what are its joys—what its pains. (HEA 1927g, 5)

The Chemist in the Colored Waistcoat

The centenary of Michael Faraday's laboratory demonstration of the principle of the dynamo took place in 1931, and large celebrations were planned by electrical engineers all over the world. Armstrong welcomed a new biography of Faraday by the electrician Roll Appleyard in *Nature*, but he hoped that "in near days to come, hastening as we now are the exhaustion of natural resources, through disproportionate use of machinery and greed of gold, [Faraday's] spiritual legacy may be of more value to us in overcoming and directing the crude, untamed human nature within us" (HEA 1931d). An example of our moral failure was the discharge of oil in the world's oceans that not only killed birds but threatened the plankton that sustained so much ocean life (HEA 1931e). While Armstrong could not compete with the electrical engineers in commemorating Faraday, he was determined that "Faraday the chemist" should not be lost sight of in the celebrations in which the Institute of Electrical Engineers planned an exhibition of Faraday's work in the vast space of the Royal Albert Hall. Three stands were given over to Faraday's work in chemistry, its applications in industry, and (via benzene again) to dyestuffs. The latter stand was Armstrong's responsibility. "I have taken charge of benzene and the naphthalene-sulphonic acids and have tried to show what has been made of them, by a long line of ingenious chemists" (HEA 1931j; J. Wilson 2012).

With the exception of wallpaper manufacturers, the chemical industry was overwhelmingly supportive of Armstrong's plans for exhibits. Rather than showing rows of labelled bottles containing colored powders and liquids, Armstrong borrowed and extended the use of dyed banners and cloths that had been a feature of the 1925 celebrations. He also arranged a series of colored railway posters along a screen at the back of the organ gallery to show "how geology may now be studied from the railway platform: color chemistry, too, if you will" (HEA 1932a, 113). And when he spoke in Faraday's lecture theatre at the Royal Institution, he delighted the auditors by appearing in evening dress with a brightly colored waistcoat—a ploy he adopted for the remainder of his active life (HEA 1932d; Rodd 1940, 1437). The event was beautifully described by John Vargas Eyre, who was present:

> With the lecture theatre draped with fabrics of many colours, like an eastern bazaar and the lecturer, Armstrong, wearing an evening waistcoat of sky-blue and pink, the scene was the most dazzling ever staged in that historic lecture theatre. Having unfolded the story of the profound changes colour brought about by the subtle changes in the molecular architecture of the dye, the climax came when an attractive young lady stepped forward as "Miss Hook of Holland" and proceeded to remove coloured petticoat after coloured petticoat in illustration of Armstrong's running commentary on the nature of the colour exhibited and of the chemist's control of colour so that they could produce different colours from the same base. (Eyre 1958, 203)[4]

In giving its own report of the lecture, *Nature* named Armstrong as the "venerable Lewis Carroll of chemistry" and appearing as a "benevolent high priest of colour." It added how ingeniously Armstrong had evaded portraying complicated formulae of straight chains and rings by deploying the backbone of a mackerel and whiting with its tail in its mouth; "also buttons on safety-pins deputised for the unwieldy formulae of the dyes chemist, all used to illustrate the genealogy of benzene." For Armstrong, the hexagon and the tetrahedron were the heraldic symbols of modern chemistry (HEA 1930c, 810).

A striptease, fish, and safety pins had been his props for this extraordinary lecture delivered at the age of eighty-four.

---

4. The model (possibly a granddaughter) has not been identified. "Miss Hook of Holland" was a musical comedy first performed in 1907; it remained popular until the 1950s among amateur groups. The heroine is given some fifteen petticoats by different admirers during the play's action.

The Dyestuffs Industry

Stephen Miall, Armstrong's son-in-law, published his still-useful history of the British chemical industry to celebrate the jubilee of the Society for Chemical Industry in 1931. It described the people, families, and firms that had contributed to all the different sectors of the industry, including the rise and fall of dyestuffs production. Although he did not review it publicly, Armstrong used its appearance to set out his own views on why the British dyeing industry had been overtaken by German competition—a story Miall had told in his book (Miall 1931). In a sharp essay for the *Pharmaceutical Journal*, Armstrong blamed the loss of British domination of the synthetic dyestuffs market on a failure to develop it—just simply through lack of intelligence, especially commercially (HEA 1931g). Interestingly, Armstrong placed much of the blame on W. H. Perkin (Sr.) himself.

> His discovery of mauve was sheer accident: no special credit is due to him on that account, but it is impossible to overrate the wonderful ability he displayed in grasping the opportunity the discovery gave and the marvellous skill he showed in starting up the dyestuff industry—from absolute scratch. . . . Yes, he failed, not in the long run, but in a short one, under twenty years . . . The discovery of rosaniline (by no other name could it dye so sweetly) soon put mauve into the shade, especially when violets, blues and greens were obtained from it. (HEA 1931j, 323)

Meanwhile, Perkin had neglected to build anything on mauve, and while Carl Graebe and Carl Liebermann's identification of the structure of alizarin stirred him to fresh efforts with anthracene as raw material, he allowed himself to be beaten at the patents game, and he failed to take on talented assistants to keep the firm buoyant (HEA 1933b).

He repeated his argument in 1935 when correcting a *Nature* editorial on the dyestuffs industry (HEA 1935g). Here he explicitly mentioned that Heinrich Caro, the effective leader of the German dyestuffs industry, was Jewish. He did this, no doubt, to draw attention to the Jewish contribution to Germany's industrial power at a time when Jews were being persecuted by the Nazi government. Despite his criticism of Perkin, Armstrong happily celebrated the eighty-ninth anniversary of Perkin's patent for mauve (HEA 1936i).

The death of his great German friend, industrialist Carl Duisberg, in March 1935 gave Armstrong the opportunity to compare and examine Duisberg's important role in the rise of German chemical industry; to reassert where British industry had gone wrong; and to argue that Germany was also abandoning the method to which it owed its previous success. He recalled visiting Duisberg's Bayerwerk at Leverkusen in 1890 and thinking that its central laboratory was the best chemical institute in the world (Wetzel 1991, 177). German militancy and commercial determination had been obvious to him when Duisberg and his son were guests of the Armstrong family in the Lake District just before the First World War. The German and English sons had innocently teased that "at no distant date they would meet on the [battle]field," while Duisberg himself had joked that he would like one day to be part-owner of Derwentwater, if Britain ever became a German conquest. "The German is a complex character," Armstrong wrote:

> As chemists, it is essential for us to learn to dissect out the elements of his mentality. Having spent nearly three years as a student in Germany before the War of 1870, I have been witness of some of his interactions. Before '70, they were a primitive, almost pastoral, music-loving nation, wonderfully intelligent and absolutely indefatigable workers, asking only to be well led and ever willing to be led, almost child-like in their simplicity. On the other hand, they were curiously lacking in sense of proportion and without humour; you had not far to go below the surface to meet with unpleasant if not barbaric reactions, such as Wagner has brought out wonderfully in his [operatic] tetralogy—such as the Red Queen saw in Alice: "A nasty vicious temper." (HEA 1935h, 1024)[5]

Final Views on Education

Student societies continued to seek Armstrong as a speaker, knowing (as *Nature* once said) that he "garnished a deal of common sense with trenchant criticism":

> As a chemical criminologist Prof. Armstrong has no equal; for many decades he seems to have watched his fellow chemists and educationists stray from the obvious path of rectitude and sail one by one into every kind of intellectual disgrace. Such unrivalled experience, together with the habit of fearless insistence on the importance of the most important things, serves to equip him admirably to play the double part of detective inspector and prison chaplain. (HEA 1932e)

---

5. See also HEA 1935d, where he hinted that the rising Nazi party threatened a lapse back to barbarism.

This comment arose from lectures that he had given to Imperial College's Chemical Society in January 1932 on the subject of "Waste Chemistry." Armstrong went through a whole itinerary of wasted opportunities in chemical education, the wasted vocabulary of contemporary chemists, and the wasted opportunities caused by Britain's obtuse narrowness of vision in wasting natural resources.

Education was always a strong theme in his invited lectures. When he addressed the West Kent Scientific Society (the successor of the Blackheath and Lewisham Scientific Society) in March 1930 on the "present curse of education," his theme was taken up in a long anonymous editorial in *Nature* (Anon. 1930a). Armstrong's title deliberately echoed that of the book by the British author and journalist, Harold Edward Gorst (1868–1950), whose *The Curse of Education* had shocked the Edwardians (Gorst 1901). In his usual inflammatory style, Armstrong made a vehement attack on contemporary school and university education. His chief target, however, was the damage caused by "the external examiner." Such examiners were "physically, mentally and morally harmful to all concerned, to teachers and taught, to parents and examiners":

> Our bondage comes in large measure from the anti-heuristic, unpractical University of London, and the wicked example it has set—by opening the matriculation to others than those who were proceeding to its degrees. The University of London has lived upon matriculation fees—these are the main support of the large and ever-growing army of bureaucratic officials . . . In our gardens, when plants begin to show signs of healthy growth, we do not take them up to examine their roots. We regard them lovingly, knowing that if we wish them to flourish we do well to let them alone and merely take care that their surroundings be kept normal and healthy: in due course they flower and fruit—we test the fruit by eating it, not by examinations. The bad is obviously bad. Teachers, if and when competent, know full well how far their pupils—to use terms of the phase rule—are reacting towards their environment. (HEA 1930b, 560)

As the anonymous "A.A.E." commented in a *Nature* editorial, this argument was somewhat illogical. On the one hand teachers were incompetent, and on the other, they were competent to judge their pupils without external validation (Anon. 1930a, 560).

In the same month of this talk, Armstrong read Winston Churchill's racy (early) autobiography, *My Early Life*, which he found full of meat for the student of education on account of Churchill's "irresponsible upbringing" (HEA 1930i). He noted Churchill's account of his struggles to learn Latin and his rescue from disgrace by the fine teaching of English and English literature he experienced at Harrow. Armstrong was impressed:

> I would *force* this passage upon the attention of every English teacher and whip him hard until he mastered its meaning and noted its spirit. In one of the schools with which I am connected [Christ's Hospital], in which the curriculum had been entirely classical and mathematical, several years ago the decision was taken, partly owing to my insistence, to substitute English for Latin in the Lower School and then to enforce Latin only upon boys of distinct literary ability. The attempt was a failure because the classical masters would not and could not teach English; so they reverted to their early evil source of general Latin torture. (HEA 1930i, 984)

Imperial College established an annual series of Huxley Memorial Lectures during the centenary of his birth in 1925, the first being delivered by biostatistician Karl Pearson. Armstrong himself had already commemorated the centenary of Huxley's birth in 1925 with some reflections on Huxley as an educator (HEA 1925f). He had also previously lectured on the development of the scientific institutions of South Kensington in a lecture to the Old Students' Association of the Royal College of Science in September 1920 (HEA 1920c). In 1933, Armstrong was invited to deliver the Huxley Memorial Lecture, which he entitled "Our Need to Honour Huxley's Will" (HEA 1933m; Brock 1973, 55).[6] Its chief interest today is its autobiographical content, with his longstanding concerns for the methods of education and forceful criticism of contemporary scientific method. His view of Huxley was certainly not quite what the auditors probably expected. Huxley, he asserted, "was a marvellous exponent—therefore, a bad teacher, as are all who are eloquent . . . a master of fine logic, but encased in hard bones." He had attended many of Huxley's lectures at the Royal School of Mines when a student at the Royal College of Chemistry, but they had failed to hold his interest because Huxley never explained how to find things out. In retrospect, Armstrong regretted that he had not attended Huxley's sixpenny Working Men's Lectures where Huxley told deductive stories, such as how to investigate a piece of chalk. On the other hand, if Huxley's work as a teacher lacked power and inspiration, he had

---

6. The lecture was reported in *Nature* 131 (13 May 1933): 684–85; and *Chemical Age* 28 (13 May 1933): 437.

become a master of education through his writings and public addresses, particularly the third volume of his collected essays, *Science and Education*

In reminiscing about the development of the teaching and research at South Kensington, Armstrong put in a good word for the teaching of physics in the Royal College of Science by fellow Kolbe pupil Frederick Guthrie, who had recently been slandered by H. G. Wells in his autobiography. In a subsequent letter to *Nature* in November 1933, he praised Guthrie as the last of the physical chemists in the real sense of someone investigating real stuff and not mathematical fantasies (HEA 1933j; Brock 1998). In an aside to his Huxley Memorial Lecture, he recalled:

> Good as was the top floor [of the Royal College of Science building], down below in the basement, by the chemist-turned physicist, Guthrie, who developed a logical practical course, on self-help lines, of extraordinary value—long since on the scrapheap, I fear; given its final quietus by the all-pervasive electron. It is little short of shameful that South Kensington has let this go by the board—unrecorded. Real earthly physics is a fast-disappearing art. . . . None of Huxley's colleagues, except Guthrie, had the least interest in educational method. (HEA 1933m).

The Abbey School in Reading was founded as an independent girls' boarding school in 1887 aimed at the daughters of fathers in the professions. It had changed its original name to the Abbey School during WW1 in honor of the fact that Jane Austen had attended a boarding school adjacent to Reading's Abbey. Its curriculum under headmistress Helen Musson was progressive; in May 1934, Armstrong was invited to open the school's new biology laboratory and to address the pupils. His speech was saucily entitled "Silken Stockings" (HEA 1934d).

> In my youth it was accounted dangerous to let a boy into a girls' school; I am not sure that an old boy, with experience of girls, is not more dangerous.

The misogyny and patronizing tone of old remained:

> Man is an irrational being. What woman is no man daresay—the more as she is several in one and he cannot live long enough nor intimately enough with them all to fathom their depths. Fundamentally full of "sweetness and light," she is subject to the strangest aberrations—a

victim of uncontrollable emotions special to her sex, by nature a most jealous slave of fashion, peculiarly obstinate in outlook.

Unfortunately, women's intelligence was not cultivated to play the vital role that women had to play in saving civilization from its commercialism. What Armstrong wanted was a return to "quiet living, honest living":

> We only cultivate that of thoughtless rushing about, quietly killing each other, with almost super-gladiatorial skill. We are as barbaric as we ever were. The Church has lost its power—because it remains irrational and unscientific; it retains an influence, simply because man, in the mass, is irrational and built to obey. In no other condition can societies hold together. Nevertheless, new orders are being formulated; a new Church is coming to the fore; the Church of the Search after Truth, in place of one that assumes Truth; the Church of Scientific Method.

The Abbey School's new laboratory was "an altar to the coming faith." He urged the girls to study how their viscose (rayon) stockings were made from the cellulose of wood that had been the invention of the industrial chemist Charles Frederick Cross. He went on to deplore the way modern women, in seeking independence from male domination, had turned to fashion and beauty parlors.

> I fear the asserted independence of woman is superficial and unreal. She is more and more making herself the tool and slave of man, by undertaking work that is so monotonous and mechanical that he will not submit to it, over any long period; all in order that she may be free to keep company or walk out with his highness in the evening. That the day may come when she will rise up in her wrath and destroy the typewriter is perhaps too much to hope for. Only then will she revert to the work that is worthy of her.

And that work was, of course, domestic—albeit, with a new twist: the preparation of fresh food to ensure that humankind became free from disease and malnutrition.

Friends and Contemporaries

Inevitably, the 1930s were filled with deaths of contemporaries that he was asked to memorialize or for whom he volunteered reminiscences and characterizations. A friend with whom he often (but not always) saw eye to eye was the American Harvey Washington Wiley (1844–1930). They had first met in Minneapolis in 1897, when Armstrong was lecturing on behalf of the Lawes Agricultural Trust at one of America's agricultural research stations. Wiley, who then headed the Bureau of Chemistry of the US Department of Agriculture, "was the life and soul of a large meeting; ever full of resource."

> Thus, on one excursion, in a dry town on a very hot day, displaying a surprising geographical instinct, he took some of us poor sufferers to a pharmacy and tendered prescriptions on our behalf: the medicine we got passed all the Brer Rabbit tests for good ale and no doubt saved our lives. (HEA 1930g)

They soon became great friends and later exchanged letters.[7] To Armstrong's chagrin, however, after his marriage, Wiley became a teetotaller, causing Armstrong to reflect that "my old friend became a first-class humbug in the matter of drink: probably he was never a man of really balanced, scientific judgement." His obituary of Wiley turned into a diatribe against American Prohibition. He found Prohibition extremely ironic given the vast quantities of caffeine that Americans imbibed not only in coffee but in their obsession with soft drinks laced with kola extract. Ironically, he noted, Wiley had actually warned Americans against drinking too much caffeine in 1912. The two friends also came to disagree over the purity of processed food. Wiley was adamant that preservatives like benzoic acid and boric acid should never be added to foodstuffs. This stance had particularly angered Armstrong, because in his view the prohibition of such preservatives had nearly destroyed the American cream industry.

By the 1930s, Armstrong had become a past master at portraying his male friends. "A FRIEND cannot be defined," he wrote:

> He is never made: he comes, when or how who shall say? Only where the wind listeth. He cannot be a woman: subtle, homosexual harmonies tie the relationship. He is the greatest and rarest of discoveries: the inestimable loss. The intensity of friendship may vary greatly: waiting as it does upon opportunity for its upgrowth, ripening with time, its character is of instant determination: at least, you know at once who are the people you will like.

---

7. I have not been able to consult Wiley's extensive correspondence in the Library of Congress.

That quotation headed his *Nature* obituary of American writer and former chemist Ellwood Hendrick (1861–1930) (HEA 1931a).[8] Educated in chemistry at the University of Heidelberg, Hendrick had become manager of the Aniline Dye Works at Albany, New York, at the age of twenty. The works soon failed, and he turned to selling insurance for thirty years before, in 1917, joining Arthur D. Little's chemical consultants' company in Cambridge, Massachusetts. All Hendrick's spare time, however, was devoted to writing. Armstrong had met him at an unidentified chemists' conference in the early 1920s, and although they never met again, they became devoted correspondents. It was not chemistry that cemented their friendship, but a joint delight in humorous and whimsical writing—a style that Armstrong happily adopted in his later journalism. Largely forgotten today except by a few Americans who have read his letters to the strange Greek-Irish journalist Lafcadio Hearn, Hendrick's best-known work was *Percolator Papers* (1919), a set of whimsical essays, many of which had been written when he edited the monthly bulletin of the Chemists' Club of New York (Hendrick 1917; 1919a; 1919b).

What strikes the reader about these commemorations of friends is not so much the details of a life but the way Armstrong finds a moral in their lives. This tendency is best illustrated in his essay on "The Monds and Chemical Industry: a Study in Heredity," which was prompted by the death of Sir Alfred Mond, the first Baron Melchett, industrialist, financier, and politician, in December 1930 (HEA 1931b).[9] In this essay, he drew on Hilaire Belloc's remark in his recently portrait of Cardinal Wolsey: "I propose to present his character and story with this object: to cite as the great example of those who do mightily yet cannot see what they are doing and who stand on the edge of doom with no vision of its approach" (Belloc 1930). Armstrong believed biographers should pay far more attention to hereditary character. He had met Ludwig Mond in the early 1870s, when the latter's son Alfred was barely four years old, and watched him emerge from an industrial environment and evolve into a politician.

> I merely wish to claim, in all modesty, that biography should be the recognised province of the structural chemist: he alone can appreciate

---

8. Hendrick lacks an entry in the *American National Biography*. He was president of the Chemists' Club of New York, 1918–1920. There are three letters from Hendrick to Armstrong in AP2.372–74.

9. See also Sir Alfred Mond to Armstrong, 12 January 1914; Mrs. Frida Mond to Armstrong, 20 September 1915; and Robert Ludwig Mond to Armstrong, 3 March 1928 (AP1.318–19; AP1.323; AP2.471).

the complete interdependence of character and structure. The supreme interest of chemistry comes from the fact, that it is the study of character as affected by structure. Our senses seem to be but distortions of structure and electrical ripples—purely physical emotions. (HEA 1931b, 238)

In other words, Armstrong's aim was to present character and motive, faults and merits, very much in the spirit of Lytton Strachey's *Famous Victorians*, and as he had done brilliantly in his memoir of Dewar. He continued to puzzle about hereditary talent in families until the end of his life (HEA 1937b).

Given his Galtonian interest in heredity, it must have shocked him severely when his grandson, Kenneth Frankland Armstrong, was killed on 2 January 1935, at the age of only twenty-five, in a mountaineering accident. Armstrong had been disappointed that his other sons, Clifford, Richard, and Harold, had not pursued chemical careers as Frank had done. He must, therefore, have been hugely distressed that his grandson from Frank had shown great promise in research only for him to die at such a young age. Frank Armstrong had educated Kenneth with Sanderson at Oundle School from 1922 to 1926, after which (despite his grandfather's distaste for Oxford education) he had read chemistry at Magdalen College, graduating with honors in 1930 and a BSc by research in 1931. The BSc awarded him a Henry Fund scholarship to spend two years with J. B. Conant at Harvard. There he had worked on the structure of chlorophyll, a subject of interest to Oxford's newly appointed professor, Robert Robinson. Awarded a Harmsworth Senior Scholarship at Merton College, Oxford, Kenneth Armstrong joined Robinson in the Dyson Perrins Laboratory in 1934. During the Christmas vacation of 1935, Kenneth and an old Oundle School friend, John Howard, had gone climbing in the Austrian Tyrol—no doubt with Robinson's approval, since Robinson and his wife were expert mountaineers. Both the young chemists were killed when an avalanche occurred as they ascended a hillside near Vent. "The loss of these two brilliant young chemists," Robinson wrote, "must surely be one of the most poignant tragedies on record in these columns" (Robinson 1935; Anon. 1935a).

Armstrong continued his active interest in the progress of chemistry into his eighties, though he also continued to attack physical chemists for thinking that mathematizing chemical phenomena really explained chemical reactions. Along with Frederick Soddy, he believed that the Royal Society no longer offered scientific leadership. He publicized the chemist's craft often by wearing vividly colored waistcoats to lectures and public events. He continued to express anxieties about diets, the supply and use of coal, and above all, the failure of British education to train people to think. Finally, as the venerable old man of chemistry, it often fell to him to honor former friends and colleagues—a task he took extremely seriously, believing that there were moral principles to be learned from the lives of those who had contributed to society's enjoyment and betterment. Such activities were to continue into the final years of his long life.

✦ 16 ✦
The Final Years

> The fact is that physical chemists never use their eyes and are most lamentably lacking in chemical culture. It is essential to cast out from our midst, root and branch, this physical element and return to our laboratories.
> —HEA 1936k

By 1930, Armstrong was eighty-two years of age. Two years later, he and his wife celebrated fifty-five years of marriage in a day spent quietly with a few members of their family. Louisa celebrated her ninetieth birthday in October 1933 but was too frail to attend the celebratory dinner that had been arranged by the family. She died on Christmas Day 1935. Armstrong himself remained in excellent health; he was able to take long hikes without feeling exhausted and retained his wide range of interests.

Armstrong's comments on the science of his day have frequently made him appear to be an eccentric conservative who had completely lost touch with modern science (Gratzer 1996). On the contrary, despite his continuous opposition to what he viewed as the over-mathematization of twentieth-century physics and chemistry, he closely followed what was going on and was fully capable of summarizing such developments—as is clearly shown in his final chemistry entries for *Encyclopaedia Britannica* in 1922 and 1927.

Book reviewing occupied much of Armstrong's final decade, chemically speaking the most important being the review, considered in a previous chapter, of Richard Anschütz's biography of August Kekulé (HEA 1930c). His wide culture was demonstrated in literary reviews of Sir Edmund Gosse's *Life and Letters* (1931) and the Hibbert lectures given by Bengali sage Rabindranath Tagore in 1931 (HEA 1932c). Later, he reviewed the work of another Bengali writer, the autobiography of Prafulla Chandra Rāy, for *Nature*, seeing it as having special value as "a presentation of a complex personality, unique in character, range of ability and experience"

(HEA 1933g). He had met Rāy in Calcutta in 1914, lectured to his department, and visited his chemical and pharmaceutical plant in Bengal.

The Lakes District continued to inspire him, and he also remained a spirited defender of Dorset's coastline, which, he argued, needed protection from intrusive building projects. He recalled his walking holidays with Arthur Rowe in which he had photographed every chalk outcrop. The pristine landscape was now being spoiled. Armstrong pleaded that the coast from the Poole estuary to Weymouth be made a National Reserve and protected from building within sight of the shore. It was not to be, although the National Trust subsequently acquired certain parts of the coastline that forms the Jurassic Coast World Heritage site.

Armstrong's interest in farming and the countryside prompted him to join the Farmers' Club in Whitehall Court, which offered facilities for visiting landowners and farmers conducting business in London. He was its after-dinner speaker at its annual dinner in 1933, where he spoke of agriculture as "a great national industry supported by public funds" (HEA 1936d). In 1935 he had met and heard R. D. Stapledon, the director of the Welsh Plant Breeding Station in Aberystwyth, talk about the importance of grassland agriculture—a subject close to his heart for the production of healthy milk from well-fed cattle. Consequently, when there was protest about Weymouth Town Council's plans to build a sewage works in the charming Jordan Valley in Dorset, Armstrong urged a return to Liebig's plans for the recovery of animal and human sewage for the purposes of agriculture (HEA 1935e), rather than the building of costly and unsightly sewage works whose product would then be dumped at sea. His idea was to collect sewage into reservoirs, use water plants to extract phosphates and other vital minerals, and then collect these and compost them down for farmyard manure to replace that from horses.

In 1933, Armstrong had been honored to become the first president of the newly founded Edward Frankland Lancastrian Society (Anon. 1933a). He delivered the first Frankland Memorial Oration in Lancaster in January 1934. He opened by pointing out that it had been seventy years since he had been Frankland's pupil, when he would have little thought both his eldest son and his eldest grandson would bear Frankland's forename (HEA 1934c). The address followed Frankland's career and training closely, with sarcastic comment on how much better Frankland's self-education was than the formal systems of training that had developed since.

He closed his term of office in January 1936 with a presidential address: "The Coming Religion of Natural Knowledge" (HEA 1936a). This address was Armstrong's last major piece of writing, in which he sketched "the immensity of our progress during the period [of his lifetime]," and called attention to some of its shortcomings (HEA 1936a: 753). One shortcoming was mankind's profligate use of coal, supplies of which he feared were fast running out. "Our future efforts should be directed, not to employing more miners, but to economising the use of coal as far as possible. That coal is our food is clear from the sudden great increase in our population following its [introduction during the industrial revolution]." Taking eugenics to heart, Armstrong believed that "civilisation is fast being developed, if not organised, to promote the survival of the unfit." Man must either work with Nature or "fall back and degenerate" through the operation of ignorance. Cinemas, radio broadcasting, aimless smoking, women flying solo to Australia, and the motorcar were all signs of laziness and "men no longer called upon to use their hands." The mordant reflections ended on the subject of education, with a lament that even his ideal Christ's Hospital had capitulated by teaching Latin on a boy's entry instead of deferring it to later in the curriculum.

For the rest of his life, Armstrong kept up his demand that coal should be better used and mined (HEA 1931f), even denouncing the way that production of synthetic ammonia had disrupted the economic stability of nations and led farmers to use it instead of the ammonia from town gas or the natural process of fallowing land (1931h). Nevertheless, when Haber died in 1933, Armstrong acknowledged him as a friend and accepted that the Haber-Bosch process had helped farmers (HEA 1934a). The pasteurization of milk also continued to obsess him (HEA 1931c; 1931h; 1932f), and when the Ministry of Agriculture and Fisheries reported in 1933 that it was perfectly safe and was inhibiting the transmission of tuberculosis to humans, Armstrong let rip in a tirade against the practice, as well as against the way food generally was becoming industrialized, tinned in "sealed coffins" and losing vital nutrients; and the way land was no longer being nurtured to produce fresh food and healthy cattle (*Report* 1933; HEA 1933d; HEA 1933f).

Agricultural Experiments at Christ's Hospital

Armstrong had previously given an enthusiastic welcome to the Cantor Lectures on vitamins that the biochemist Jack Drummond gave at the Royal Society of Arts in April 1932 (Drummond 1932), though it worried him that science had become too beholden to commerce and industry and less concerned with public welfare (HEA 1932b). Drummond's lectures inspired Armstrong, in his capacity as chairman of Christ's Hospital's Education Committee, to experiment with a group of young boys at the school during the winter of 1932, "with the object of determining whether milk, produced with the aid of summer-cut grass, be of special food value compared with the ordinarily 'stall-fed' milk." In this trial, six cows from a herd of thirty-six were fed on summer-grown grass that had been cut and dried with special care. The experiment was made under the auspices of Sir Frederick Keeble at the Jealott's Hill Farm in Berkshire, which since 1927 had been run by Imperial Chemical Industries as an agricultural research station.

> Throughout the winter, the six remained sleek and gave milk of summer character, whilst the rest of the herd [at Horsham] fell off as usual, giving an ordinary poor winter milk. Similar results have since been obtained elsewhere. (HEA 1935m, 566)

He was astounded by the way that the cattle had fattened and the carotene content of their milk had increased. "Who shall say what other components may not be equally affected?" These experiments were conducted on pupils and cattle on the farmlands of Christ's Hospital, as well as with Keeble in Berkshire (HEA 1933f, 606). The final results were not published by Armstrong; instead, they were incorporated in the dietary conclusions of Dr. Gerald E. Friend, who was the school's medical officer from 1913 to 1947.

Friend had spent twenty years improving the school's diet, sometimes with the advice of Armstrong in his capacity of school governor. In 1933, Armstrong tipped off Drummond (who was then professor of biochemistry at University College London) that Friend had valuable information about the nutrition of schoolboys. In the following year, Drummond and Armstrong's former pupil Vargas Eyre visited Christ's Hospital and learned from Friend that boys' fractures during sporting events had risen during WW1, when margarine had been substituted for butter in the canteen, but had fallen after 1922, when butter had been restored. Drummond was excited by this primary evidence of the importance of fat and vitamins in the diet and persuaded Friend to publish his findings (Young 1954; Pemberton 2003). Friend's twenty-year-long study *The Schoolboy: A Study of His Nutrition, Physical Development and Health* appeared in 1935. Armstrong chose to review it in the quarterly *Wine and Food* journal, whence it received publicity in *Nature* (Friend 1935; HEA 1935m; Anon. 1935b). There he urged boarding schools to serve wholemeal bread, fresh and raw vegetables, sausages containing every kind of "innards," and fresh milk produced by cattle reared on lime-treated pastures. The schoolboy's "tummy" was far more important than sport and house games and the study of dead languages. School farms should be compulsory. It was a school's duty to ensure health, not by fighting diseases with thermometers and antiseptics but with decent food and sunlight to build bodies resistant to infection.

Armstrong also warned readers of the *Times* that the Metropolitan Water Board's continuing reliance on drawing water from the polluted Thames was dangerous; more reservoirs for supplying expanding London were needed (HEA 1935f). At the Farmers' Club in 1936, Armstrong entered vigorously into the discussion of a paper on public health and agriculture given by Scottish nutritionist Sir John Boyd Orr. His remarks were republished by the Royal Society of Arts to give them more publicity (HEA 1936d). Armstrong listed nearly a dozen things that needed to be achieved during the new reign of King Edward VIII, including Britain's self-sufficiency in food production, the widespread availability of fresh non-pasteurized milk and whole-grain bread, better feeding of animals, improved education that included outdoor study, the recirculation of human sewage, and a policy of positive eugenics (!).

Armstrong was made an honorary member of the Pharmaceutical Society in 1933 and asked to deliver an address when the society's School of Pharmacy inaugurated new pharmacological laboratories under the leadership of Joshua Harold Burn (1892–1981). Armstrong's lecture was partly autobiographical—he recalled the many pharmaceutical friendships he had made at the London Institution, for example—but mainly a vehicle for his current obsession with the way the health of the world's population was changing. The secret of good health, he argued, lay in good husbandry. He envisaged a future in which pharmacists would become grocers attached to farming and selling nutritious food rather than drugs (HEA 1933i). Much of the speech was a repeat of suggestions made the previous March in the Sir Jesse Boot Foundation Lecture he had given at the University College, Nottingham (HEA 1933e). In this speech he had revealed that his father had been an inveterate pill-popper. Consequently:

> I have never bought a box of pills in my life, having early made up my mind to be independent of family example. I therefore also never smoked nor did I drink spirits—both drugs. At most, very seldom, I have swallowed a blue pill to stir up my liver [Carter's Little Liver Pills], torpid through lack of exercise, owing to some stupidity of my own such as sitting for days worrying out an address . . . To my own example I may add that of my wife—my senior and daughter of a druggist: she has therefore never willingly taken drugs of any kind and would keep all medicines and medicine men from her sight if possible. (HEA 1933e)

What the World Was Coming To

By 1935, Armstrong was the last of the former Central College's four founding professors. Consequently, he was the lion of the evening when the City and Guilds celebrated the college's semicentennial jubilee on 4 February 1935 (HEA 1935b; 1935c, 151; Whitworth 1985; AP2.A10). Following a reception during which Armstrong was made an honorary fellow of the college, he delivered his reminiscences at the microphone so that the speech could be relayed into other rooms, such was the size of the gathering. The speech emphasized the college's importance for training Britain's engineers but questioned whether the college had considered its graduates' "spiritual needs." This phrasing was not a reference to religion, however, but to their appreciation of nature, including the study of the natural sciences, which Armstrong felt was sadly lacking in contemporary science curricula.

Armstrong attended his final meeting of the British Association for the Advancement of Science at Norwich in 1935. He had scarcely missed a meeting since his first at Liverpool in 1870. He had been commissioned by *Chemical Age* to review the proceedings, at which the president was Imperial College geologist William Whitehead Watts. Armstrong approved of his speech explaining and discussing Alfred Wegener's hypothesis that the great continents had once formed one huge landmass. (The theory of continental drift was not generally accepted until the advent of plate tectonics in the 1960s.) Armstrong's long and caustic review of the proceedings, serialized over three weeks' issues, concluded that the BA's annual meetings had passed their glory and that without changes the association should be wound up (HEA 1935i).

Although he appears not to have attended the meeting in Blackpool in September 1936, he read and enthusiastically endorsed Sir Richard Livingstone's presidential address to the Education Section. Livingstone had argued that secondary education for all should be extended to the age of eighteen and that it should entail realistic education that fitted young people for the marketplace. That, wrote Armstrong, was the crux of education today—education for life and the needs of the day, and teachers who could lead. What was required was a great leader, like Lord Fisher had been for the British navy (HEA 1936j). He remained convinced that the heuristic method of instruction remained the best system of education, and in two articles for *Chemical Age*, written in his eighty-eighth year, he reiterated how learning by doing in workshop schools would solve Britain's future problems as a nation. He had just read H. G. Wells's latest pessimistic solution to mankind's problems in *The Anatomy of Frustration* (1936)—certainly one of Wells's most peculiar publications. Wells was "made of old-man stuff," Armstrong declared, because he had no real understanding of scientific method. New man stuff was *doing*, not *talking* (HEA 1936m, 453). This was Armstrong's last word on education.

The End

One of London's most memorable buildings and landmarks, the Crystal Palace, was destroyed by fire during the night of 30 November 1936. The disaster of its loss, both as a memorial of the Great Exhibition of 1851 and as a wonderful venue for concerts and entertainment, occasioned a flurry of reminiscences in the letters column of the *Times*. Armstrong, wearing his historian's hat, recalled the palace as a symbol of Britain's greatness and decline. Once, it had made other nations jealous as a symbol of Britain's superior technology; but, having shown that they were not fools, the British had gone to sleep and allowed Germany to develop the industries they had brought into being (HEA 1936p). After the fire, initially he had merely thought that some permanent memorial to the 1851 exhibition should be erected at Sydenham, whence the palace had been moved after the exhibition; but upon further reflection he came up with the suggestion that since Prince Albert had turned South Kensington into a scientific center, the Sydenham site should be turned over to scientific agriculture (HEA 1936q). Of course, nothing came of this reflection and today the area is once more parkland and a sports ground.

The last meeting of the council of the Chemical Society that Armstrong attended was on 21 November 1935. His wife, Louisa, was by then near death. Comforted and cosseted by his two daughters, who were still living with him in Lewisham, at the end of February 1936 he took a sea voyage on *HMS Carthage* to Tangier in Morocco. He likely did so on the advice of his doctor since, by this time, Armstrong was showing signs of urinary disease. On board, he found companionship with a much younger, Oxford-educated physical chemist, Harold Hartley (1878–1972), who happened to be taking the same voyage. Hartley shared Armstrong's interest in science teaching and education and was deeply interested in the history of chemistry. The two men spent the voyage talking about the roles of the German (Liebig, Kolbe, Kekulé), French (Laurent, Gerhardt, Wurtz) and English (Hofmann, Frankland, Williamson) schools in developing structural organic chemistry. Hartley was later to recall these conversations in the Society of Chemical Industry's First Armstrong Memorial Lecture, which he delivered in November 1945 (Hartley 1971, 195).

The sparkle had not faded during 1936. There were a dozen letters to the *Times*, several book reviews in *Nature*, and one final attack on "ionomania" in *Chemistry and Industry* (HEA 1936k). The fate of British agriculture and British diets remained primary concerns (D. Smith 1997). He was appalled that the government was prepared to spend nearly three million pounds subsidizing the sugar beet industry.

> Having allowed this unfortunate industry to be forced into being, we are bound at last to let it down gently; the measure is obviously only one of political expediency; in reality, just a dole to the farming community. . . . Scientific opinion, I believe, is against making the production of sugar an English industry—the more as we thereby throw our West Indian Colonies out of action. It is definitely a substance to be made in the tropics. . . . If we are ever to be rational and apportion the regions of our earth to their most suitable uses, sugar beet must give way to sugar cane. The production of sugar from beet is everywhere a subsidized industry—uneconomic. (HEA 1936b)

And he went on to dismiss sugar as just a fuel and of little dietary value, and even harmful. He took for support a recent book by agriculturist Reginald George Stapledon (1882–1960), which he was reviewing favorably in *Nature* (HEA 1936e).

He had no sympathy for farmers' growing use of artificial (inorganic) insecticides, which to his mind was "nothing short of alarming."

> Everything in farming is becoming artificial. Cultivation is mechanized and made soulless: the loving art and care of the skilled ploughman is being lost to the land . . . As fertilizer we mostly use chemicals straight from the factory: the intensity of the harm these are now doing to our soils is little considered . . . Research is on wrong lines. In farming, as in medicine, we need to develop methods for securing healthy, resistant growth rather than of curing disease. These insect attacks are probably, in no slight measure, due to faults in cultivation. (HEA 1936h, 16)

Armstrong was well aware of the growing militarism of Hitler's Germany and the fact that the University of Heidelberg had dismissed nearly fifty of its staff on racial, religious, or political grounds in 1933. When the University issued invitations to celebrate its 550th anniversary in 1936, British academics were placed in a quandary, as a flurry of correspondence in the *Times* reveals. Oxford decided to send a congratulatory message in Latin but not to send a delegate to the celebrations. This was also the decision of Birmingham University. Ever perverse, Armstrong tried to see the situation from the point of view of the Heidelberg University authorities (HEA 1936c). He doubted that a boycott of Heidelberg's celebrations would be of any consequence for the future, and indeed he thought that it might possibly add to an "incitement of war by arms—which all seem to be fermenting at the moment—in addition to the even more dangerous commercial war now universal." Armstrong's appeasing solution was that British academics should not be separating from German colleagues but should be trying to support them by showing that "we understand the terrible plight in which they are placed."

A few weeks after receiving the congratulations of the Deutsche Chemische Gesellschaft as their oldest member (Anon. 1937), he received a similar acknowledgment from the Royal Society on his sixty years of membership and for having become the society's senior fellow. *Nature* added its own congratulations for his stimulating and provocative contributions to its columns (Anon. 1936). Armstrong returned the compliment when *Nature*'s publisher, Frederick Macmillan, died in June 1936, expressing the indebtedness of the scientific community to him for supporting the journal. "The position taken by the journal, throughout the cultured world, is indeed unique. NATURE is a holy thing. May it continue to be so regarded and be kept whole: no light task" (Macmillan 1936; HEA 1936f).

He remained an avid reader of contemporary scientific papers. His natural-history interests led him to query whether natural selection really was the mechanism by which insects evolved color and camouflage. "No chemist can believe in change by any process of direct mimicry . . . Nature is held under strict enzymic control." He was promptly corrected by Oxford's professor of zoology and expert entomologist, G. D. Hale Carpenter: insect mimicry was an accepted zoological phenomenon solely explicable by natural selection (HEA 1936g).

He corrected his final proof, "Ammonolatry: The Life Element," for *Nature* the week before his death; a review of two recent books, it appeared posthumously. Gregory wrote, in annotation:

> The article represents . . . the final expression of the frank and critical views which he held upon the training of chemists and subjects of research. Whatever significance may be attached to these views, the fact that, while on his dying bed, he desired to make the two volumes [under review] the subject of a contribution to NATURE, is a remarkable tribute to the active attention to scientific subjects right to the last. (HEA 1937e: 134)

The article reviewed *Nitrogen System of Compounds*, written by American chemist Edward Curtis Franklin, whom Armstrong had met in 1906 at Stanford University in Palo Alto, California, soon after the San Francisco earthquake and when lecturing on agricultural chemistry for the Lawes Agricultural Trust (Franklin 1935; Elsey 1991). Franklin was then beginning a lifelong study of the properties of liquid ammonia and its use as a solvent in comparison with water, a subject that naturally interested Armstrong because of his objection to the theory of ionization. He also revealed the long-forgotten fact that Birmingham electrochemist George Gore (1826–1908) had pioneered the investigation of the solvent properties of liquid ammonia, as well as hydrogen fluoride, only to be frustrated by his lack of research funding. "I had the privilege of being made his confidant and always regretted that he had so little encouragement for the Institute of Scientific Research he had founded in 1880" (HEA 1937e, 134).

Armstrong contrasted Franklin's detailed comparison between the compounds of oxygen and nitrogen with the treatment of nitrogen in the second book under review, Nevil Sidgwick's *Organic Chemistry of Nitrogen*, the first edition of which the author had presented to Armstrong when it appeared in 1910. A new edition had just appeared but appeared to have been edited by Oxford associates rather than Sidgwick himself, and the book was now "devoted to the fashionable morganatic game of crab catching, an appropriate exercise for the upper reaches of the Thames." (Armstrong's strange metaphor referred to a section on chelated orthonitrophenol derivatives.) Compared with the original edition, Armstrong felt, the new one was useless as a teaching aid and of no practical value for students. Ignoring the fact that the text was now aimed at postgraduates, Armstrong then launched into his familiar criticisms of Oxford. Nor could he endorse Sidgwick's deployment of resonance structures based upon "the obscure doctrine of wave mechanics." In a vintage Armstrong polemic, he dismissed the idea of resonance as

> a state of suspended animation in a molecular system with a shortened waist. Is this "Much Ado about Nothing"? In what way it is an advance upon van't Hoff is not clear. Kolbe was angered that van't Hoff should introduce metaphysics into chemistry—what he would have said to Sidgwick passes expression. (HEA 1937e, 135)

He continued in vintage form by satirizing Sidgwick's notion of bivalent hydrogen in what was later known as hydrogen bonding. (Ironically, such bonding would have provided Armstrong with an explanation for his hydrone theory.) Whatever "resonance hybrids" might be, he concluded, "they have little practical significance." None of this was the chemistry he had practiced for seventy years. "Chemistry at present is a sick man—'mostly conventional signs.' Something less Snarkian is needed than 'to measure the value of an idea in terms of incomprehensibility.'"

Channelling his mentor Kolbe, this announcement was Armstrong's final word—except that his son discovered the unpublished typescript of the Horace Brown Memorial Lecture that his father had given in 1927 and never completed because, in his opinion, the chemistry of starch was still unresolved. Frank Armstrong immediately offered it to the *Journal of the Institute of Brewing* and the lecture appeared posthumously (HEA 1937f).

Armstrong was confined to his house and bed for the last eight months of his life. He remained in cheerful spirits and, until the very last days, his memory stayed as fresh as it had always been. He died on the afternoon of 13 July 1837. Apart from natural old age, the cause of death was registered as a urinary bladder condition, probably cancer. One of the last visitors to his bedside in May 1937 was Joji Sakurai, the Japanese pupil of Williamson's who had become a family friend as a member of the former International Research Council and of the IUPAC. Sakurai too was a dying man, but both men conversed for hours over half a bottle of champagne (Donnan 1939).

Thus ended the life of "a veteran of the chemical world [and] probably the most prominent figure in British chemistry," observed the *Chemical Age*. Its editor noted that Armstrong's last letter to the weekly journal had appeared in January 1937 under the typical eye-catching heading: "An Appeal to ICI to Open a Lunatic Asylum" (HEA 1937d). This missive was a vintage Armstrong protest against slovenly, jargon-filled chemical articles. Armstrong was a crusader with many missions ranging from attacks on physical chemistry and school and university examinations to the quality of milk and food or the taste in evening dress. His son-in-law provided an appropriate one-sentence epitaph. Contemporary chemists "would think of him with affectionate remembrance and regret that there is no one left who can do so much to give life and vigour to chemical discussion, to instruct and to divert chemists, and enliven their minds" (Miall 1937).

Conclusions

> Whether one agrees with Prof. H. E. Armstrong or not—and frequently one does not—there is always to be found in his utterances much that is arresting, and in his writings much that stimulates.
> —Anon. 1930a

In his final two years as a widower, Armstrong was assisted by his unmarried daughter, Nora, in Lewisham, and by his physician son, Robin, who lived less than a mile away at Blackheath. It was they, together with their elder brother Frank, who arranged a Christian funeral at St. Dunstan's-in-the-East, despite the fact that their father was not a churchgoer and had always been a rationalist and agnostic in the tradition of T. H. Huxley and Edward Clodd. Following the service, Armstrong was cremated at the Golder's Green crematorium. Over 160 people attended the funeral, many representing the various associations, schools, and societies to which Armstrong had belonged. Armstrong's surviving pupils and colleagues were there in strength. The family was represented by his sons Frank, Robin, Harold, and Clifford and daughters Edith, Annie, and Nora, as well as their spouses and several of the fourteen adult grandchildren. His chief executor was his eldest son Frank, who, later in sorting through his father's papers, presented the Chemical Society with a number of his parents' memorabilia, including the home visitors' book. The estate was valued at £16,368, a sum worth over a million pounds at this writing in 2022.

As a brilliant author of vivid obituaries, Armstrong deserved a good one himself. There were handsome tributes in the *Times* and other daily newspapers, and thoughtful notices in *Nature* (by W. P. Wynne), *Chemistry and Industry* (by his son-in-law Stephen Miall), the *Journal of the Institute of Brewing*, and the *Proceedings of the Royal Society of Edinburgh* (by his old adversary James Kendall). Both the French and German Chemical Societies noted his passing and acknowledged his originality as a chemist and controversialist. A chemist of "the tribe [of] sciencers," as he preferred to be known, the Chemical Society wanted his most famous pupil, Sir William Pope, to write the obituary; Pope gladly began the task but was forced to abandon it by the illness that led to his death in October 1939. It was therefore left to Armstrong's pupil, Ernest Rodd, to build a fine obituary from the notes that Pope had compiled (Rodd 1940). The Royal Society was similarly hampered in its search for a memorialist. The war was well underway by the time the society was able to persuade botanist Sir Frederick W. Keeble to produce a delightful portrait of his intimate friend, whom Keeble regarded as "one of the great personalities of his time" (Keeble 1941, 245).

In June 1938, Frank Armstrong organized a special issue of the *Central* (E. F. Armstrong 1938a and 1938b) in commemoration of his father. Its appearance led to the proposal in March 1939 of a Henry Armstrong Memorial Trust Fund to "honour a man of singleness of purpose, interested in so much that really matters: in education, chemistry, biology, agriculture, geology, music, art and literature, and above all, with a fine appreciation and human understanding of his fellow-men" (A. Wilson 1939). Initiated by a group of Armstrong's former pupils, this trust aimed to raise £3000, which, among other purposes, was to be used to erect a bronze bust of Armstrong by sculptor Estcourt James Clack and presented to the City and Guilds College, together with a copy for Christ's Hospital in Horsham. It also subsidized the publication of J. Vargas Eyre's biography of Armstrong in 1958. The trust was finally dissolved in the 1970s.

Meanwhile, in November 1938, the council of the Society of Chemical Industry, at the suggestion of food chemist Leslie Lampitt, instituted what was intended to be an annual series of "Armstrong Lectures." The Blitz put paid to any immediate lecture, and it was not until February 1941 that, appropriately, the first lecture was given by his eldest son (E. F. Armstrong 1941). The continuing war made any annual successor impossible, and it was not until 1945 that the lectures were resumed as explicit "Armstrong Memorial Lectures," the first of which was given by Sir Harold Hartley in 1945 (Hartley 1971). Frederick Keeble gave variations of his Royal Society obituary in 1947, and one of Armstrong's last pupils at the Central, Richard Colgate (who had become a food chemist), completed the initial sequence in 1953. The SCI had by then run out of lecturers who had personally known the old man. Instead, the lectures have continued intermittently to the present day, with invited talks by specialists in the field of chemical engineering and industry. Not surprisingly, Christ's Hospital school continues to proudly extol its science teaching by tracing its origins to the influence of Armstrong and the appointment of science teachers like Charles Browne and Gordon Van Praagh, who employed a heuristic approach to the science curriculum (Rodd 1968; Van Praagh 1973; Van Praagh 1992; Van Praagh 2001).

Like many who survive to old age, Armstrong found some of the changes in society uncomfortable. Chief among them was his attitude toward women, which today would be unacceptable. He was a man of his times, brought up in the Victorian era, when "a woman's place was in the home"—a social concept that for Armstrong was reinforced by certain ideas taken from evolutionary biology and eugenics. He seems to have realized with disappointment that the case for women's emancipation from the home had continually strengthened by the 1930s. For the octogenarian Armstrong, it was not that the grass was greener in another country (though Germany did have its merits); rather, he was, in Horace's words, a laudator temporis acti, one who believes things were better in the past.

His conservatism extended to his science. In contrast to his contemporary jousting partner, Oliver Lodge, whose classical physics was altered by the direct confrontation with modern physics of relativity and quantum theory, Armstrong stayed resolutely a classical organic chemist. Also unlike Lodge, Armstrong chose not to write popular books that explored the modern chemistry of electrons, atomic structure, and ions with their faults and difficulties, but instead expressed his sceptical opinions in public lectures and letters to newspapers or *Nature*. And once more in contrast to Lodge, he failed to grasp the power that radio and broadcasting had in reaching the common man. He was fastidious to such a degree that he would not allow anything for publication that he thought was second-best. This perfectionism affected his refereeing activities, and it also explains why he frequently got into trouble for failing to deliver articles on time.

On the other hand, like Lodge for physics in the physics community, between 1900 and 1937 Armstrong was probably the best-known British chemist, with a reputation that extended into the chemical communities of Europe and America. By no means was Armstrong always on the wrong track. His "centric benzene" was empirically and retrospectively justified by resonance theories. Heuristic teaching has again become en vogue, with our novel "flipped classrooms" and other pedagogical innovations. Even his "ionomania" campaign had a real point, in showing some real deficiencies in the ionic theory of solutions that needed correcting.

Above all, Armstrong stood out amongst contemporary chemists as an inveterate iconoclast. His opposition to the physical chemistry that Ostwald and others argued had to be the basis of general or theoretical chemistry was total. That meant that, as the subject developed after the 1890s, Armstrong had to develop his own model of chemistry, which he based upon the idea of electrolysis that Faraday had introduced successfully in the 1830s, along with a rigorous determination not to allow the physical chemical school to interfere with organic chemistry and its theory of structure. (Of course, many physical chemists saw themselves as likewise following in the footsteps of Faraday, an ideal model for the first physical chemist.) Armstrong's "associationist" theories of hydrates formed between solvents and solutes eventually proved to be very much on the right track, but his model was neither mathematizable nor quantitative. In contrast, the physical chemists that Armstrong so despised were able to generate formulae and equations that generated useful information about acidity and conductivity—examples of their heurism!—even while physical chemists were well aware that their mathematical models were oversimplified. Most notably, their models ignored the probable role of a solvent in electrolysis or, more generally, in reactions carried out in the liquid phases.

Unfortunately for Armstrong, the physical chemists' campaign coincided with the physicists' discovery of radioactivity and isotopy. Here, older ideas from Prout's hypothesis played a role in Armstrong's thinking. However, once Armstrong was forced to concede that helium was elementary, he strove to formulate purely chemical explanations for the phenomenon of radioactivity. He ignored the evidence for the existence of the electron for many years, and it was not until the 1920s that he came to terms with the physicists' model of the atom. As with his association theory of chemical change, Armstrong's explanations of radioactivity and the periodic pattern of the properties of the elements were too complex to be acceptable, and so they were silently ignored. By contrast, most of his contemporaries quickly accepted the "new physics."

If Armstrong had been able earlier to accept that the electron played a role in chemistry, and that its partial separation from an atom in an organic molecule under the influence of a neighboring molecule's coulombic attraction would lead to its polarization and hence vulnerability to forming a new structure, his role in the history of modern chemistry might have been very different. His pupil Arthur Lapworth was able to see the electron's significance; Lapworth's Manchester colleague Robert Robinson developed a model of chemical change that involved electrons moving from place to place within a molecule when placed in the presence of a suitable neighboring molecule. In the hands of Robinson's rival, Christopher Ingold, this model provided the understanding of chemical change for which Armstrong had been searching (Nye 1994).

One of Armstrong's convictions illustrates the importance of looking at the work of scientists whose work has been forgotten or simply absorbed into the great tree of knowledge; this conviction stemmed from his belief that the easiest and best approach to teaching and understanding a scientific problem is to examine it from a historical point of view. Even if Armstrong's chemical work is now of little interest to practicing chemists, the remark underlines why that work remains of value to the historian in understanding the pathway to contemporary chemistry.

True historicism for historians of science requires understanding the contributions, interactions, and importance of protagonists across the entire range of prominence, from our familiar retrospective heroes all the way down to "invisible technicians" and other less visible but essential players. Historians have come a good way recently toward respecting and incorporating the latter, and they have begun redressing imbalances, for instance by telling the true and hitherto neglected stories of women in science. But the "middle level" of prominence, a station that Armstrong retrospectively has occupied, has been mostly terra incognita. In Armstrong's case, historical neglect is made even more problematical by the fact that his contemporary significance and influence were in fact far greater than has been retrospectively considered.

When we regard Armstrong as a character playing "Fifth Business," we can understand his vital role in the shaping of British scientific culture generally in the first half of the twentieth century. Most of us play Fifth Business in the development of society. We are not heroes or heroines; nevertheless, our roles are crucial to the unfolding of our social and cultural histories. Armstrong played to perfection the Fifth Business character in the development of twentieth-century attitudes toward physics and chemistry, British science and industry in the European context, and the stewardship of the countryside, environment, and public health.

Acknowledgments

Because this book has gestated in my mind for virtually fifty years, my debts to other scholars are legion and far too many to identify individually. It would be invidious, however, not to acknowledge my debt to the late Jeanne Pingree (1925–2016), an unforgettable character to all historians of science of my generation who explored and used archives at Imperial College. The current archivist at Imperial, Anne Barrett, graciously provided the photographic portraits. I was given generous assistance at the following archives: the Royal Society of London; the Royal Society of Chemistry; the City and Guilds archive in the London Metropolitan Archives; the Meldola papers in the Passmore Edwards Museum in Stratford; and the Rothamsted Research Library in Harpenden. I am enormously appreciative of the kind assistance and sterling advice of Karen Merikangas Darling at the University of Chicago Press, and I am deeply obliged for the first-rate copy editing of Jessica Wilson. Two outstanding anonymous reviewers provided advice that substantially improved the final manuscript. My thanks are also due to Richard E. Rice in Montana for his valuable help in compiling a list of Armstrong's obscurer publications when we thought of publishing an anthology of Armstrong's writings—alas, a project that ended in failure. Finally, I owe profound gratitude to Alan Rocke for editing the manuscript when I became ill. His kindness and generosity in taking this project on has been immensely appreciated by the Brock family.

As always, I am indebted to my long-suffering wife, Elvina, for allowing me to live a "life of Armstrong" rather than "the life of Riley" in retirement for which she had hoped.

Archives Consulted

City and Guilds archives, London Metropolitan Archives, London
Imperial College Archives, London
Meldola papers, Newham Museum Service, Passmore Edwards Museum, Stratford
Rothamsted Research Library archives, Harpenden
Royal Society of Chemistry archives, London
Royal Society of London archives, London

Cited Works by Henry Edward Armstrong

1868 (with Edward Frankland) "On the Analysis of Potable Water." *JCS* 6: 77–108.
1870a *Zur Geschichte des Schwfelsäure-Anhydrids: Seine Einwirkung auf einige Chlor- und Schwefel-Verbindungen* (Leipzig: n.p.).
1870b "Contributions to the History of the Acids of the Sulphur Series. 1. On the Action of Sulphuric Anhydride on Several Chlorine and Sulphur Compounds." *PRS* 18: 503–13.
1871 "On the Action of Sulphuric Acid on the Natural Alkaloids." *JCS* 9: 56–60.
1874a *Introduction to the Study of Organic Chemistry: The Chemistry of Carbon and Its Compounds* (London: Longmans Green).
1874b (with T. E. Thorpe) "Report on Isomeric Cresols and Their Derivatives." *Brit. Ass. Reports* 1874: 73–74.
1875a (with T. E. Thorpe) "Report on Isomeric Cresols and Their Derivatives." *Brit. Ass. Reports* 1875: 112–14.
1875b "Note on Isometric Change in the Phenol Series." *JCS* 28: 520–23.
1875c "On Food." *Manchester Science Lectures for the People* (Manchester: Heywood).
1876a Abstract-translation of Wilhelm Körner, "Researches on Isomerism amongst the So-Called Aromatic Substances Containing Six Atoms of Carbon," *JCS* 29: 204–41. Original: "Studi sull'isomeria delle così dette sostanze aromatiche a sei atomi di carbonio," *Gazzetta chimica italiana* 4 (1874): 305–441.
1876b (with George Harrow) "Note on the Action of Nitric Acid on Tribromophenol." *JCS* 29: 474.
1876c (with George Harrow) "On the Action of Potassic Sulphite on the Haloid Derivatives of Phenol." *JCS* 29: 474–77.
1876d (with George Harrow) "Note on the Action of Nitric Acid on Tribromophenol." *JCS* 29: 477–78.
1876e "Inorganic Chemistry." *Encyclopaedia Britannica*, 9th ed. (Edinburgh: Black): vol. 5, 467–544.
1879 "The Dissociation of Chlorine." *Nature* 20 (14 August): 257–58.
1880a "The Dissociation of Chlorine, Bromine and Iodine." *Nature* 21 (18 March): 461–62.
1880b "Ideal Chemistry." *Nature* 23 (16 December): 141–42.
1880c (with C. E. Groves and W. A. Miller) *Organic Chemistry*, 5th ed., part 3, *Chemistry of Carbon Compounds, or Organic Chemistry* (London: Longmans Green).

1880d *Introduction to the Study of Organic Chemistry* (London: Longmans Green).
1881 (with N. C. Graham) "Research on the Laws of Substitution in the Naphthalene Series." *JCS* 39: 138–43.
1883a (with A. K. Miller) "Zur Kenntniss des Metaisopropylmethylbenzenes." *Berichte der Deutschen Chemischen Gesellschaft* 16: 2748–50.
1883b (with A. K. Miller) "Zur Kenntniss des Camphers." *Berichte der Deutschen Chemischen Gesellschaft* 16: 2255–61.
1884 "On the Teaching of Natural Science as a Part of the Ordinary School Course, and on the Method of Teaching Chemistry in Science Classes, School and Colleges." *Proceedings of the International Conference on Education*, edited by Richard Cowper, 2: 69–82. Reprinted in HEA, *The Teaching of Scientific Method and Other Papers on Education* (London: Macmillan, 1903, 1910): 219–34.
1885 "Presidential Address to the Chemical Section of the British Association" (10 September). *Brit. Ass. Reports* 1885: 945–64.
1886a "Chemical Physics" [review of J. P. Cooke, *Elements of Chemical Physics*, 4th ed. (London: Macmillan)]. *Nature* 34 (2 September): 405–06.
1886b "Electrolytic Conduction in Relation to Molecular Composition, Valency and the Nature of Chemical Change: Being an Attempt to Apply a Theory of Residual Affinity." *PRS* 40: 268–91.
1887a "The Past and the Possible Future of Our Association: Presidential Lecture." *Proceedings of the Lewisham and Blackheath Scientific Association* 1887: 14–28.
1887b "The Determination of the Constitution of Carbon Compounds from Thermochemical Data." *Philosophical* 23: 73–109.
1887c "Comparison between the Views of Dr Arrhenius and Professor Armstrong on Electrolysis: Reply to Professor Lodge's Criticism." *Brit. Ass. Reports* 1887: 354–58.
1887d "An Explanation of the Laws Which Govern Substitution in the Case of Benzenoid Compounds." *JCS* 51: 258–68 and 583.
1887e "The Alkaloids: The Present State of Knowledge Concerning Them and the Methods Employed in Their Investigation." *J. Soc. Chem. Industry* 6: 482–90.
1887f Anonymous review of Norman Lockyer, *Chemistry of the Sun* (London: Macmillan, 1887), in *Chemical News* 56 (23 December): 267.
1888a Review of M. M. Pattison Muir and Douglas Carnegie, *Practical Chemistry: A Course of Laboratory Work* (Cambridge: Cambridge University Press, 1887). *Nature* 37 (19 January): 265–68, with reply by Muir, *Nature* 37 (2 February): 318–19.
1888b "Report of Committee Inquiring into the Teaching of Science in Elementary Schools." *Brit. Ass. Reports* 1888: 164–72.
1888c "The Origins of Colour and the Constitution of Colouring Matters [I]." *PCS* 4: 27–31.
1889 "Suggestions for a Course of Elementary Instruction in Physical Science." *Brit. Ass. Reports* 1889: 229–50. Reprinted in HEA, *The Teaching of Scientific Method and Other Papers on Education* (London: Macmillan, 1903, 1910): 300–44.
1890a "The Terminology of Hydrolysis, Especially as Effected by Ferments." *JCS* 57: 528–31; and *Nature* 42 (21 August): 406–07.
1890b "British Association Procedures." *Nature* 42 (28 August): 414.
1890c "Exercises Illustrative of an Elementary Course of Instruction in Experimental Science." *Brit. Ass. Reports* 1890: 299–309. Reprinted in HEA, *The*

Teaching of Scientific Method and Other Papers on Education (London: Macmillan, 1903, 1910): 345–66.

1890d "The Structure of Cycloid Hydrocarbons." *PCS* 6: 101–05.

1891a "The Teaching of Scientific Method" [lecture to the College of Preceptors, May 1891]. Reprinted in HEA, *The Teaching of Scientific Method and Other Papers on Education* (London: Macmillan, 1903; 1910): 1–10.

1891b "Science v. Art at South Kensington." *Times*, 20 May: 13.

1891c (with W. J. Pope) "Studies of the Terpenes and Allied Compounds: Sobrerol, a Product of the Oxidation of Terebenthene (Oil of Turpentine) in Sunlight." *JCS* 59: 315–20, 1118.

1892a "The International Conference on Chemical Nomenclature." *Nature* 46 (19 May): 56–59.

1892b "British Association Procedure." *Nature* 46 (28 July): 291–92.

1892c (with J. F. Briggs) "The Relative Orientating Effect of Chlorine and Bromine I: The Constitution of Para-Brom- and Para-Chloraniline-Sulphonic Acids." *PCS* 8: 40.

1892d "Contributions to an International System of Nomenclature: The Nomenclature of Cycloids." *PCS* 8: 127–31.

1893a "The Conditions Determinative of Chemical Change." *Nature* 48 (6 July): 237–38.

1893b "Flame." *Nature* 49 (30 November): 100–01, 149–50 (Arthur Smithells), 172 (HEA), 198 (Smithells).

1893c "Notes on Hofmann's Scientific Work." *JCS* 69: 637–732; reprinted in *Chemical Society Lectures 1893–1900* (London: Chemical Society, 1901).

1894a "Scientific Method in Board Schools." *Nature* 50 (25 October): 631–34.

1894b "Technical Training under the London County Council." *Times*, 6 December: 15.

1894c "Presidential Address to Chemical Society." *JCS* 66: 336–78.

1895a "Research in Education." *Nature* 51 (14 March): 463–67.

1895b "The Structure of the Sugars and their Artificial Production." *PRI* 14 (29 March): 521

1895c "The Royal Commission on Secondary Education." *Nature* 53 (28 November): 79–82.

1895d "The Conditions Determinative of Chemical Change." *JCS* 67 (1895): 145–48.

1896a "Osmotic Pressure and Ionic Dissociation." *Nature* 55 (26 November): 78–79.

1896b "Kekulé" [brief paragraph for Lord Lister to read in presidential address to Royal Society]. *PRS* 60: 301.

1896c [editor] *The Jubilee of the Chemical Society* (London: Chemical Society).

1897a "The Need of Organising Scientific Opinion." *Nature* 55 (4 and 11 March): 409–11 and 433–35.

1897b "Letters on Chemical Society Controversy over Election of Dewar as President." *Chemical News* 75 (16 March): 154–55; (9 April): 178; (15 April): 190.

1897c "Heuristic Instruction in Physical Science." *International Congress on Technical Education* (London: Trounce): 8–13.

1897d "Chemistry," in *Chapters on the Aims and Practice of Teaching*, edited by Frederic Spencer (Cambridge: Cambridge University Press): 222–59.

1897e "The School Board and Practical Science." *Times*, 25 November: 12.

1898 "The Heuristic Method, or the Art of Making Children Discover Things for Themselves." *Board of Education, Special Report on Educational Subjects* 2: 389–433.

Reprinted in HEA, *The Teaching of Scientific Method and Other Papers on Education* (London: Macmillan, 1903, 1910): 235–99.

1899a "Chemists and Chemical Industries." *Nature* 59 (9 March): 438.

1899b "Indigo." *Times*, 31 October: 4.

1899c "An Explanation of the Laws Which Govern Substitution in the Case of Benzenoid Compounds." *PCS* 15: 176–78.

1899d "Discussion [with Others]: The Laws of Substitution, Especially in Benzenoid Compounds." *Brit. Ass. Reports* 1899: 683–87.

1900a "Public Schools and National Defence." *Times*, 5 July: 12.

1900b "Public Schools and National Defence." *Times*, 21 July: 4c.

1900c "Isomorphous Derivatives of Benzene" [report of a committee led by HEA, H. A. Miers, and W. P. Wynne]. *Brit. Ass. Reports* 1900: 167–70.

1900d (with W. Berry) "Metasulphonation of Aniline." *PCS* 16: 159.

1901a "The Downfall of Natural Indigo." *Times*, 15 April: 13–14. Reprinted in HEA, *The Teaching of Scientific Method and Other Papers on Education* (London: Macmillan, 1903, 1910): 144–52.

1901b "Inquiry into Foreign Methods of Commercial and Industrial Education." *Times*, 19 December: 7.

1901c "Science in Education: The Need of Practical Studies." In *National Education: Essays towards a Constructive Policy*, edited by Laurie Magnus (London: Murray): 103–27. Reprinted in HEA, *The Teaching of Scientific Method and Other Papers on Education* (London: Macmillan, 1903, 1910): 153–76.

1901d "Frankland Memorial Lecture." *JCS* 81: 193–96.

1902a "Chemistry." In *The New Volumes of the Encyclopaedia Britannica, Constituting, in Combination with the Existing Volumes of the Ninth Edition, the Tenth Edition* (Edinburgh: Black): vol. 2, 708–46.

1902b "The Classification of the Elements." *PRS* 70A (20 March): 86–94, 459–66.

1902c "The Military Education Report." *Times*, 5 August: 10.

1902d Opening address to Section L (Education). *Brit. Ass. Reports* 1902: 820–44; reprinted in HEA, *The Teaching of Scientific Method and Other Papers on Education* (London: Macmillan, 1903, 1910): 35–96.

1902e "The Conditions Determinative of Chemical Change and of Electrical Conduction in Gases, and on the Phenomena of Luminosity." *PRS* 70A: 99–109.

1903a "The Assumed Radio-Activity of Ordinary Materials." *Nature* 67 (5 March): 414.

1903b *The Teaching of Scientific Method and Other Papers on Education* (London: Macmillan).

1903c "The Application of Science to Industry." *Times*, 2 September: 12.

1903d "Army Training." *Times*, 19 November: 15.

1903e (with E. F. Armstrong) "Studies on Enzyme Action. I. The Correlation of the Stereoisomeric αβ-Glucosides with the Corresponding Glucoses." *JCS* 83: 1305–13.

1903f "The Basis of a Rational Curriculum." *Brit. Ass. Reports* 1903: 883–84.

1904a "Report on American Education." In *Mosely Education Commission Report*, edited anonymously by HEA (London: Cooperative Printing Society): 7–25.

1904b "Common Sense in Education." *Times*, 12 January: 9.

1904c "Oxford on the Upgrade" [review of W. Warde Fowler, *An Oxford Correspondence of 1903*]. *Nature* 70 (16 June): 145–47.

1904d (with E. F. Armstrong) "Studies on Enzymes Action. II. The Rate of Change Conditioned by Sucro-Clastic Enzymes, and Its Bearing on the Law of Mass Action." *PRS* 73: 500–16.

1904e (with T. Martin Lowry) "The Phenomena of Luminosity and their Possible Correlation with Radioactivity." *PRS* 72A: 258–64.

1905a "Studies of Enzyme Action. VII. The Synthetic Action of Acids Contrasted with That of Enzymes: Syntheses of Maltose and Isomaltose." *PRS* 76B: 592–99.

1905b "Studies of Enzyme Action. VI. Lipase (1)." *PRS* 76B: 606–8.

1905c (with W. Robertson) "The Significance of Optical Properties as Connoting Structure. Camphorquinone-Hydrazones-Oximes-Diazo-Derivatives: A Contribution to the Theory of the Origin of Colour and to the Chemistry of Nitrogen." *JCS* 87: 272–97.

1906a "Radium." *Times*, 10 August: 6.

1906b (with E. F. Armstrong) "The Origin of Osmotic Effects. I." *PRS* 73A: 264–67.

1908a "Co-Education." *Times*, 2 May: 6.

1908b "Studies of the Processes Operative in Solutions. VI. Hydrolysis: Hydrolation and Hydronation as Determinants of the Properties of Aqueous Solutions." *PRS* 81A: 80–95.

1908c "The Training of Faculty." *School World* (December): 477–78.

1908d "Scientific Control of Fuel Consumption" [address at the autumn meeting of the Iron and Steel Institute, Middlesborough]. *Journal of the Iron and Steel Institute* 3: 234–62.

1908e "Low Temperature Research at the Royal Institution of Great Britain, London, 1900–1907: The Hydrogen Period" [Hodgkin Trust Lecture], *PRI* 19: 354–412.

1908f "The Outlook: A Grand Experiment in Education." *Nature* 78 (9 September): 618.

1908g "Report of the Sub-Committee on Elementary Experimental Science." *Brit. Ass. Reports* 1908: 501–25.

1909a "A Dream of Fair Hydrone (A Chemical Idyll)." *Science Progress* 3 (January): 484–99. Reprinted in HEA, *Essays on the Art and Principles of Chemistry* (London: Ernest Benn Ltd., 1927): 117–38.

1909b "The Thirst of Salted Water, or The Ions Overboard." *Science Progress* 3 (April): 638–56. Reprinted in HEA, *Essays on the Art and Principles of Chemistry* (London: Ernest Benn Ltd., 1927): 139–64.

1909c "Opening Address (Abridged)" [Presidential address to Chemistry Section B, Winnipeg, August]. *Brit. Ass. Reports* 1909: 420–54.

1909d "The Revolt of Women." *Daily Chronicle*, September.

1909e (with E. F. Armstrong) "The Origin of Osmotic Effects. II. Differential Septa." *PRS* 81B: 94–96.

1910a *The Teaching of Scientific Method and Other Papers on Education*, 2nd ed. (London: Macmillan). New preface dated May 1910.

1910b "The Meaning of 'Ionisation.'" *Nature* 82 (17 February): 458.

1910c "Research Chemists." *Times*, 9 March: 15.

1910d (with E. F. Armstrong) "The Origin of Osmotic Effects. III. The Function of Hormones in Stimulating Enzymic Change in Relation to Narcosis and the Phenomena of Degenerative and Regenerative Change in Living Structures." *PRS* 82B: 588–602.

1910e "Morphological Studies of Benzene Derivatives. I. Introductory." *JCS* 97: 1578–84.
1911a (with E. F. Armstrong) "The Origins of Osmotic Effects. IV. Note on the Differential Septa in Plants with Reference to the Translocation of Nutritive Materials." *PRS* 84B: 226–29.
1911b (with F. P. Worley) "Studies of the Processes Operative in Solution. XII." *JCS* 99: 349–71.
1911c (with F. P. Worley) "Studies on the Processes Operative in Solutions. XIII. The Depression of the Hydrolytic Activity of Acids by Paraffinoid Alcohols and Acids." *Chemical News* 103: 145–46.
1911d "Studies of the Processes Operative in Solutions. XIX. The Complexity of the Phenomena Afforded by Solutions: A Retrospect with an Addendum on Non-Aqueous Electrolytes." *Chemical News* 103: 97–111.
1912a Review of George Senter, *A Text-Book of Inorganic Chemistry* (London: Methuen, 1911), in *Science Progress* 6: 502–03, with reply by Senter on 698–99.
1912b "Some Consequences of Graham's Work. The Nature of Elements: The Diffusion of Liquids" [Graham Memorial Lecture, Glasgow, 16 January]. *Science Progress* 6: 584–614; also in *Proceedings of the Royal Philosophical Society of Glasgow* 43: 67–96.
1912c "Tuberculosis and the Milk Supply: Raw or Heated Milk." *Times*, 28 September: 10.
1912d "The Stimulation of Plant Growth." *Journal of the Royal Horticultural Society* 38: 17–21.
1912e (with E. Horton) "Studies on Enzyme Action. XV. Urease: A Selective Enzyme." *PRS* 85B: 109–27.
1913a "The Mystery of Radioactivity." *Science Progress* 7: 648–55.
1913b "Gas and the Art of Leader Writing." *Chemical World* 2 (13 November): 339–41.
1913c "The Properties of Alcohol in Relation to Its Physiological Effect." *J. Inst. Brewing* 19 (December): 518–45.
1914a "The Significance of Optical Properties." *Chemical World* 3 (January): 3–4.
1914b "Sir Oliver Lodge, Intolerant, Infallible." *Bedrock, a Quarterly Review of Scientific Thought* 2 (January): 411–22.
1914c "A Move towards Scientific Socialism." *Chemical World* 3 (March): 67–71.
1914d "The Place of Wisdom (Science) in the State and in Education" [address to Section L of BAAS in Melbourne, 14 August]. *Nature* 94: 213–19.
1914e "Remarks on Recent Physics." *Brit. Ass. Reports* 1914: 294.
1915a "A Letter to the Editor." *Morning Post*, 15 March, cited in *Nature* 95: 119.
1915b "The Organisation of Science." *Times*, 22 June: 9.
1915c "The Organisation of Science." *Times*, 15 July: 7.
1915d "The Development and Control of Industry by Public Influences." *J. Soc. Chem. Industry* 34 (31 July): 765–69.
1915e "The Organisation of Science." *Times*, 9 August: 9.
1915f "Where Are We Going?" *Times*, 10 December: 9.
1915g "The Extension of British Trade." *Morning Post*, 21 April.
1916a "Chemistry in Relation to Engineering" [vice-presidential address given to the Junior Engineering Society on 29 February].
1916b "Unorganised Science." *Times*, 27 March: 9.
1916c "Fuel Economy: A National Policy Required." *J. Soc. Chem. Industry* 35: 765–67.

1916d "The Problem of Coal, with Reference to the Complete and Providential Utilisation of the Supplies and Fuels Generally: A Preliminary Discussion and Scheme." *J. Soc. Chem. Industry* 35: 220–71.

1916e Vote of thanks to Horace T. Brown for his lecture: "Reminiscences of Fifty Years' Experience of the Application of Scientific Method to Brewing Practice." *J. Inst. Brewing* 22 (September–October): 267–348; HEA's remarks on 348–52.

1916f "Low Temperature Research at the Royal Institution of Great Britain. No. 3, 1907–1914: The Charcoal Period." *PRI* 21: 735–85.

1916g "Personal Notes on the Origin and Development of the Chemical School at the Central." *Central* 13: 84–96, with an appendix of notes on the careers of his students.

1917a "Science at Oxford." *Times*, 1 February: 7.

1917b "A Chemist's Views." In *Liberty*, edited by Ernest E. Williams (London: Eveleigh Nash): 42–60.

1917c "Outlook of the British Chemical Industry." *Times Trade Supplement: Coal Tar Products*: 1.

1917d "Hugo Müller." *JCS* 111: 572–88. Reprinted in "Obituary Notices of Fellows of the Royal Society," *PRS* 95A (1919): xii–xxv.

1918a "The Breeding of Pigs; Fat in Human Diet." *Times*, 4 February: 4.

1918b "Pig Breeding: The Production of Fats." *Times*, 18 February: 4.

1918c "Chestnuts versus Oranges; Variety in Diet." *Times*, 6 March: 9.

1918d "Natural Indigo: Prospects of Revival." *Times*, 6 July: 7.

1918e "Coal before Gas: An Unequal Exchange." *Times*, 13 September: 9.

1919a "Milk or Wheat?" *Times*, 2 January: 8.

1919b "Milk or Wheat?" *Times*, 6 January: 7.

1919c "Method and Substance of Science Teaching." *Nature* 104 (15 January): 521.

1919d "Styhead Pass: A National Sanctuary." *Times*, 25 April: 8.

1919e "Wasteful Gas Burning." *Times*, 19 July: 7.

1919f "Problems of Food and our Economic Policy." *Journal of the Royal Society of Arts* 67 (5 September): 653–62; 67 (12 September): 667–76; and 67 (19 September): 681–92.

1919g "Lord Fisher's Faith in Oil: The Future Fuel of the Navy, Invention and Supply." *Times*, 22 September: 6.

1919h "The Future of Oil." *Times*, 27 September: 8.

1919i "The Learned Societies." *Times*, 6 November: 10.

1919j "London University." *Times*, 23 December: 6.

1920a "The Wine Tax; French and English Interests: A Lost Opportunity." *Times*, 24 April: 10.

1920b "Economy in Oil: An Essential Insurance." *Times*, 17 June: 8.

1920c *Pre-Kensington History of the Royal College of Science and the University Problem* (London: Old Students' Association of the Royal College of Science, London).

1920d "The University Problem in London." *Nature* 106 (23 September): 129–31.

1920e "The British Association." *Nature* 106 (23 September): 109–10.

1920f "Our Economic Reserve: Wasteful Use of Coal." *Times*, 4 December: 6.

1920g "Sty Head Pass." *Times*, 20 September: 6.

1921a "Oil at Sea: A Nuisance to Fisheries." *Times*, 25 January: 6.

1921b "Letter to the Editor on British Association Reports of the 1920 Meeting." *Chemical Age* 4 (19 February): 214.

1921c "The Designation of Vitamines." *Nature* 107 (17 March): 72–73.
1921d "Detective Work in the Potbank: The Art of Systematised Inquiry (Research)." *Transactions of the British Ceramic Society* 19: 56–72.
1921e "Electricity Supply." *Times*, 4 July: 8.
1921f "Paints, Painting and Painters, with Reference to Technical Problems, Public Interest and Health." *Journal of the Royal Society of Arts* 69: 655–85.
1921g "Vote of Thanks and Toast to Sir William Pope on his Speech to a Dinner of the Society of Chemical Industry," 7 October.
1921h "Is Scientific Inquiry a Criminal Occupation?" *Nature* 108 (20 October): 241.
1921i "Lead Poisoning and the International Labour Conference." *British Medical Journal* (17 December): 1042–44.
1921j "Relativity and the Problems of Coal, Low Temperature Carbonization and Smokeless Fuel." *Journal of the Royal Society of Arts* 69: 385–407.
1922a "The Land and the Nation; Importance of Research." *Times*, 6 January: 6.
1922b "The Feeding of Schoolboys; More Replies to My Masters: Unwisdom of the Present Diet." *Times*, 25 January: 6.
1922c "Natural Indigo: A Setback to Research." *Times*, 20 February: 6.
1922d "What Can We Do with Our Sons? Some Answers: Dr H. E. Armstrong and Late Schooling." *Times*, 30 March: 7.
1922e "The Indigo Situation in India." *Journal of the Royal Society of Arts* 70: 409–29.
1922f "Natural Indigo: Importance to Indian Industry." *Times*, 25 March: 6.
1922g "The British Association." *Nature* 110 (9 September): 341.
1922h "The Perils of Milk" *Nature* 110 (11 November): 648.
1922i "Medical Research; Changed Scientific Outlook; Position of the Royal Society." *Times*, 21 November: 14.
1922j "Rhapsodies Culled from the Thionic Epos: Chemical Change and Catalysis." *J. Soc. Chem. Industry* 41: 15, 253–70. Reprinted in HEA, *Essays on the Art and Principles of Chemistry* (London: Ernest Benn Ltd., 1927): 189–257.
1922k "Studies on Enzyme Action: XXIII. Homo- and Hetero-Lytic Enzymes." *PRS* 94B: 132–33.
1923a "Legislative and Departmental Interference with Industry and the Common Weal." *J. Soc. Chem. Industry* 42 (5 and 12 January; 2 February): 23–24, 91–92.
1923b "Sir James Dewar, FRS." *Nature* 111 (7 April): 472–74.
1923c Introduction to B. Lagueur [i.e., Stephen Miall], *The Problem of Solution: A Tavern Talk between Certain Chymists and Others* (London: Ernest Benn); also in *Chemical Age* 7 (2 September): 308–11; (16 September): 379–80; (23 September): 408; (30 September): 450.
1923d "Mrs Hertha Ayrton." *Nature* 112 (1 December): 800–01.
1923e "Mrs Hertha Ayrton." *Nature* 112 (15 December): 865.
1923f "Congress of the French Society of Chemical Industry." *Nature* 112 (15 December): 879–80.
1923g "Electrolytic Conduction: Sequel to an Attempt (1886) to Apply a Theory of Residual Affinity." *PRS* 103A: 619–21. Reprinted in HEA, *Essays on the Art and Principles of Chemistry* (London: Ernest Benn Ltd., 1927): 273–76.
1923h "The Origin of Osmotic Effects. IV. Hydrodynamic Change in Aqueous Solutions." *PRS* 103: 610–18. Reprinted (with small changes) in HEA, *Essays on the Art and Principles of Chemistry* (London: Ernest Benn Ltd., 1927): 259–72.

1924a *James Dewar, 1842–1923: A Friday Evening Lecture to Members of the Royal Institution* (London: Ernest Benn).
1924b "The Education of the Chemist." *Proceedings of the Institute of Chemistry* 1924, part 2: 139–56.
1924c "Lucerne Crops." *Times*, 20 June: 21.
1924d "Why Was I Born So Soon?" *J. Soc. Chem. Industry* 43 (22 August): 845–48.
1924e "The Making of the Compleat Chymist." *J. Soc. Chem. Industry* 43 (7 November): 1077–88, 1100–03.
1925a "The Life of Lord Rayleigh." *Nature* 115 (10 January): 47.
1925b "The Word 'Scientist' or Its Substitute." *Nature* 115 (17 January): 50.
1925c "Preservatives in Food; Antiseptic Tests: Effect of Alcohol." *Times*, 12 February: 13–14.
1925d "A Course of Faraday." *Nature* 115 (18 April): 568–69.
1925e "Some Colour Problems." *Journal of the Society of Dyers and Colorists* 41 (May): 161–65.
1925f "Huxley's Message in Education." *Nature* 115 (9 May): 743–47.
1925g "The Discovery of Benzol: Faraday the Chemist: A Centennial Celebration." *Times*, 16 May: 15.
1925h "Institut International de Chimie Solvay." *Nature* 115 (23 May): 817–18.
1925i "The Faraday Benzene Centenary." *Nature* 115 (6 June): 870.
1925j "More Verses from Wilder D. B." *Industrial and Engineering Chemistry* 17 (July): 762.
1925k "Science and Intellectual Freedom." *Nature* 116 (1 August): 172.
1925m "Preservatives in Food: Expert Opinion Needed." *Times*, 19 August: 8.
1925n "Catalysis and Oxidation." *Nature* 116 (22 August): 294–97.
1925p "Food Preservatives: Prof. Armstrong's Experiments." *Times*, 22 August: 6.
1925q "The First Epistle of Henry the Chemist to the Uesanians." *Journal of Chemical Education* 2 (September): 731–36.
1925r "The Conditions of Chemical Change." *Nature* 116 (10 October): 537.
1925s *The Teaching of Scientific Method and Other Papers on Education* (London: Macmillan). Reprint of the 2nd ed. of 1910, with a new preface.
1926a "Do We—Don't We—What Do We—Know?" *Nature* 117 (6 February): 195–96.
1926b "Ozone and the Upper Atmosphere." *Nature* 117 (27 March): 452.
1926c "Chemistry of Apples: Variations in Soils." *Times*, 29 March: 20.
1926d "Bigamous Hydrogen—a Protest." *Nature* 117 (17 April): 553–54.
1926e "Lakeland Scenery: The Duty of Preservation." *Times*, 23 April: 10.
1926f "What We Know." *Nature* 117 (24 April): 590.
1926g "Carbon and Silicon: The Two Foundation Stones of the World" [review of J. W. Mellor, *A Comprehensive Treatise on Inorganic and Theoretical Chemistry*, vol. 6 (London: Longmans Green, 1925)]. *Nature* 117 (15 May): 683–86.
1926h "Wasteful Research?" *Nature* 117 (12 June): 823–24.
1926i "Hydrogen as Anion." *Nature* 118 (3 July): 13.
1926j "The Reduction of Carbonic Oxide." *Nature* 118 (21 August): 265.
1926k "Arthur Walton Rowe." *Nature* 118 (16 October): 561–62.
1926m "Our Bookshelf: Review of *Prof. Dr. Phil. Dr. Jr. H.C. Ludwig Darmstaedter*." *Nature* 118 (13 November): 690.

1926n "Lakeland Scenery: New Roads and Tree-Planting. Borrowdale Imperilled." *Times*, 18 November: 15.

1926p "Education, Science and Mr H. G. Wells" [review of H. G. Wells, *The World of William Clissold: A Novel at a New Angle* (New York: Doran, 1926)]. *Nature* 118 (20 November): 723-24.

1926q "Lakeland Scenery: Road Widening at Borrowdale." *Times*, 3 December: 10.

1926r "A Half-Century of Chemistry in America, 1876-1926." *Nature* 118 (4 December): 806-08.

1926s "Sir James Dewar." *PRS* 111A: xiii-xxiii.

1927a "Review of *Institut International de Chimie Solvay. Deuxième Conseil . . . Bruxelles* (Paris: Gauthier-Villar)." *Nature* 119 (1 January): 8.

1927b "Oxygen = 17.0.ϕ." *Nature* 119 (8 January): 51.

1927c "A Message from Old Leipzig through London (to Germany)." *J. Soc. Chem. Industry* 46 (4 March): 185-88. Translation of "Persönliche Erinnerungen und Gedanken," *Chemische Zeitung* 51: 114-16.

1927d "Jacques Loeb: A Chemist's Homage to the Work of a Biologist." *Journal of General Physiology* 8 (June): 653-57.

1927e "Prof. E. H. Rennie." *Nature* 119 (19 March): 431-32.

1927f "Prof. Ira Remsen." *Nature* 119 (23 April): 608-09.

1927g "Scientific Worthies: XLV. Richard Willstätter." *Nature* 120 (2 July): 1-5.

1927h "Golden Wedding." *Times*, 31 August: 13. Also *Chemical Age* 17 (3 September): 211, 216-17.

1927i "Flame and Combustion" [review of William A. Bone and Donald T. A. Townsend, *Flame and Combustion in Gases* (London: Longmans, 1927)]. *Nature* 120 (24 September): 431-35.

1927j "Poor Common Salt!" *Nature* 120 (1 October): 478.

1927k "Flame and Combustion." *Nature* 120 (22 October): 586.

1927m "Marcelin Berthelot." *Times*, 24 October: 15.

1927n "Marcelin Berthelot." *Nature* 120 (5 November): 659-63.

1927p "Flame and Combustion." *Nature* 120 (3 December): 806-07.

1927q "Marcellin Berthelot and Synthetic Chemistry." *J. Roy. Soc. Arts* 76 (30 December): 145-71.

1927r "Chemistry." *Encyclopaedia Britannica*, 13th ed. (London: Britannica Co.). Reprinted in HEA, *Essays on the Art and Principles of Chemistry* (London: Ernest Benn Ltd., 1927): 1-116.

1927s *Essays on the Art and Principles of Chemistry* (London: Ernest Benn Ltd.).

1927t "The Forms of Carbon and Chemical Affinity." In *Isaac Newton 1602-1727: A Memorial Volume*, edited by W. J. Greenstreet (Glasgow: Bell): 5-15.

1928a "The Nature of Solutions" [review of *The Scientific Work of the Late Spencer Pickering, FRS*, edited by T. M. Lowry and Sir John Russell (London: Royal Society, 1927)]. *Nature* 121 (14 January): 48-51.

1928b "Sir Edward Frankland: A Great Lancastrian." *J. Soc. Chem. Industry* 47 (April): 408-10.

1928c "Ethyl Petrol: The Departmental Inquiry." *Times*, 14 August: 6.

1928d "The Ninth International Conference of Pure and Applied Chemistry." *Nature* 47 (31 August): 889-92.

1928e "The Modes in which Valency is Exercised. I. (with W. Barlow) The Tetrahedral Carbon Atom, Paraffins and Polymethylenes. II. The Structure of Graphite and of Black Carbon." *J. Soc. Chem. Industry* 47 (31 August): 892–97.

1928f "Norman Lockyer's Work and Influence" [review of T. Mary Lockyer and Winnifred L. Lockyer, *Life and Work of Sir Norman Lockyer* (London: Macmillan, 1928)]. *Nature* 122 (8 December): 870–74.

1928g "Motor Roads in Lakeland: Protest and Defence." *Times*, 17 December: 10.

1928h "Horace Brown, 1848–1925." *JCS* 1928: 1061–66.

1928i "Sir James Dewar, 1842–1923." *JCS* 1928: 1066–76.

1929a "Biology for the Empire: Need of Scientific Method." *Times*, 12 January: 8.

1929b "Experts and Men of Science." *Times*, 21 February: 10.

1929c "Dr H. J. H. Fenton." *Nature* 123 (2 March): 317.

1929d "Solutions and Heat Engines." *Nature* 123 (9 March): 346–47.

1929e "Our King's Wardrobe: Iconolatries—Old and New." *J. Soc. Chem. Industry* 48 (28 June): 643–47.

1929f "Prof. Charles Moureu." *Nature* 124 (10 August): 238–39.

1929g "The Riddle of Benzene: August Kekulé." *J. Soc. Chem. Industry* 7 (13 September): 914–18.

1929h "Prof. W. H. Perkin, Jun. FRS." *Nature* 124 (19 October): 623–27.

1929i "After-Thoughts on the Neuvième Congrès de Chimie Industrielle, Barcelona, October 13–19, 1929." *J. Soc. Chem. Industry* 48 (13 December): 1198–202.

1929j "An Anxious Onlooker: What of the Wochenals?" *Chemical Age* 21 (21 December): 565.

1930a "Discussion on Brewing Following Paper by Arthur R. Ling, 'Brewing as a Branch of Science.'" *J. Roy. Soc. Arts* 78 (5 March): 671–83 and 687–89.

1930b "School Science and Educational Values." *Nature* 125 (12 April): 560.

1930c "The Doctrine of Atomic Valency" [review of Richard Anschütz, *August Kekulé*, 2 vols. (Berlin: Verlag Chemie, 1929)]. *Nature* 125 (31 May): 807–10.

1930d "Ethyl Petrol: A Plea for Medical Evidence." *Times*, 6 June: 12.

1930e "Speech on Award of Albert Medal by Duke of Connaught." *Journal of the Royal Society of Arts* 78 (13 June): 906–07.

1930f "Space and Matter." *Nature* 126 (23 August): 275–76.

1930g "Prof. H. W. Wiley." *Nature* 126 (20 September): 444–45.

1930h "The Neglect of Scientific Method." *Nature* 126 (6 December): 869–71.

1930i "Mr Winston Churchill on Miseducation." *Nature* 126 (27 December): 983–85.

1931a "Dr Ellwood Hendrick." *Nature* 127 (3 January): 28–29.

1931b "The Monds and Chemical Industry: A Study in Heredity." *Nature* 127 (14 February): 238–40.

1931c "Pure Milk: Feeding and Breeding of Cattle." *Times*, 23 February: 18.

1931d "Prof. Otto Wallach." *Nature* 127 (18 April): 601–02.

1931e "Plankton Changes on the Sea Coast of Ecuador." *Nature* 127 (16 May): 743.

1931f "Living on Coal: Uses of National Resources." *Times*, 27 August: 6.

1931g "The Dyestuff Industry." *Pharmaceutical Journal* 50 (29 August): 322.

1931h "Let Cabbages Be Kings: A Great Synthetic Factory." *Times*, 29 August: 6.

1931i "Faraday Celebrations." *J. Soc. Chem. Industry* 50 (18 September): 774–76.

1931j "At the Sign of the Hexagon, Albert Hall." *J. Soc. Chem. Industry* 50 (25 September): 793–94.

1932a "Back to the Land." *Nature* 129 (23 January): 112–15.
1932b "The Oncoming of Internecine Strife in Science." *J. Soc. Chem. Industry* 51 (6 May): 411.
1932c "The Mind Judicial." *Nature* 129 (21 May): 739–742.
1932d "Faraday and Benzene." *Nature* 129 (28 May): 787.
1932e "Waste: In Chemistry and Education." *Nature* 129 (18 June): 896.
1932f "Safer Milk." *Times*, 20 December: 8.
1932–1933. "The Institutes of Chemistry [Series of Essays]." *Chemical Age* 27 (29 October 1932): 491–92; part 2, 27 (12 November 1932): 449; part 3, 27 (19 November 1932): 472; part 4, 27 (10 December 1932): 543–44; part 5, 27 (17 December 1932): 567; part 6, 28 (7 January 1933): 4–6.
1933a "Dear Little Buttercup: An Arcadian Gilbertian Rhapsody." *J. Soc. Chem. Industry* 52 (20 January): 53
1933b "Dr. John Thomas." *Times*, 26 January: 14.
1933c "Mr C. M. Stuart." *Nature* 131 (11 February): 194–95.
1933d "Our Milk Muddle." *J. Soc. Chem. Industry* 52 (24 February): 157.
1933e "Pharmacy, Food, Farmer and the Future." Offprinted from the *Nottingham Citizen*.
1933f "Agriculture and Milk Supply." *Nature* 131 (29 April): 605–06.
1933g "An Indian Sage." *Nature* 131 (13 May): 672–74.
1933h "The British Association Meeting 1933 [Leicester]." *Chemical Age* 29 (30 September): 291–92, 317–18.
1933i "Pharmacy of the Future. A Phantasy in Greens." *Pharmaceutical Journal* 131 (7 October): 432–35.
1933j "Frederick Guthrie." *Nature* 132 (4 November): 714.
1933k "Charles Maddock Stuart." *JCS* 1933: 469–71.
1933m *Our Need to Honour Huxley's Will*. Huxley Memorial Lecture, Imperial College (London: Macmillan).
1934a "Professor Haber: A Cheap Nitrogen for Agriculture." *Times*, 6 February: 19.
1934b "Mr William Barlow: A Geometrical Genius." *Times*, 5 March: 14.
1934c "First Frankland Memorial Oration to the Lancastrian Frankland Society." *J. Soc. Chem. Industry* 53 (25 May): 459–66.
1934d "Silken Stockings." *Journal of Education* 66 (June 1934): 386–89.
1934e "The Beginnings of Finsbury and the Central." *Central* 31: 1–15.
1934f [Pseudonymously as "Cheshire Cat"]: "At the Sign of the Cheshire Cat: James Jeans and Co.'s Entire and Wavy Grins." *Chemical Age* 31 (22 September).
1935a "On Active Chlorine." *J. Soc. Chem. Industry* 154 (4 January): 15–16.
1935b "City & Guilds College. Celebrating the Jubilee." *Times*, 5 February: 11.
1935c "City & Guilds College." *J. Soc. Chem. Industry* 54 (15 February): 151.
1935d "Dr Carl Duisberg: An Appreciation." *Times*, 27 March: 16.
1935e "Disposal of Sewage." *Times*, 8 April: 20.
1935f "Water Supply of London." *Times*, 28 May: 12.
1935g "The Dyestuff Industry." *Nature* 135 (1 June): 907.
1935h "Chemical Industry and Carl Duisberg." *Nature* 135 (22 June): 1021–25; also in *Chemical Age* 33 (27 July and 3 August): 81–82, 104–05.
1935i "The British Association Meeting [Norwich] 1935." *Chemical Age* 33 (21 September, 28 September, 5 October): 251–52, 271–72, 310–11.
1935j "On Blind Neglect of Knowledge." *Science Progress* 30 (October): 345–45.

1935k "Food, Farmer, and Future." *Nature* 136 (12 October): 565–67.
1935m "The Schoolboy: A Study of Health at Christ's Hospital." *Journal of the Wine and Food Society* 7 (Autumn).
1936a "The Coming Religion of Natural Knowledge." *Nineteenth Century and After* 119: 752–62.
1936b "Sugar Beet in Agriculture: An Uneconomic Crop." *Times*, 11 February: 10.
1936c "The Celebration at Heidelberg: Plea for Cooperation." *Times*, 28 February: 12.
1936d "Public Health and Agriculture." *J. Roy. Soc. Arts* 84 (24 April): 620–23. Reprinted from *Journal of the Farmers' Club*.
1936e "In Search of Truth of Earth." *Nature* 137 (6 June): 923–26.
1936f "A Tribute to *Nature*." *Nature* 137 (27 June): 1074.
1936g "Insect Coloration." *Nature* 138 (8 August): 242.
1936h "Insect Pests: Dangers of Artificial Farming." *Times*, 10 August: 16.
1936i "Dyestuff Manufacturer's Anniversary." *Times*, 28 August: 15.
1936j "Real Education." *Times*, 21 September: 8.
1936k "Ionomania in Extremis." *J. Soc. Chem. Industry* 55 (13 November): 916–17.
1936m "Mr Wells Will Never See It Through: A Revolution Called for in Education." *Chemical Age* 35 (28 November): 453–54.
1936n "The Workshop Schools of the Future: Learning by Doing." *Chemical Age* 35 (3 December): 477–78.
1936p "Crystal Palace Site: Typical Suggestions: A Memorial of 1851." *Times*, 4 December: 17.
1936q "Crystal Palace: An Empire Palace of Agriculture?" *Times*, 14 December: 10.
1937a "Professor A. R. Ling." *Times*, 19 May: 9.
1937b "Professor A. G. Perkin." *Times*, 2 June: 21.
1937c "Weed Accountancy." *J. Soc. Chem. Industry* 56 (12 June): 545.
1937d "An Appeal to ICI to Open a Lunatic Asylum." *Chemical Age* 36 (24 July): 46.
1937e "Ammonolatry: The Life Element." *Nature* 140 (24 July): 134–38.
1937f "Horace Brown Memorial Lecture." *J. Inst. Brewing* 43 (September): 375–86.

General Bibliography

Abney, William (1903). "Presidential Address" to the Education Section. *Brit. Ass. Reports* 1903: 865–75.
Amundsen, Roald (1913). *The South Pole*, 2 vols. (New York: Lee Kendick).
Anon. (1873). "Announcement of Spring 1873 Lectures." *PRI* 6: 559.
Anon. (1879). "Editorial." *Proceedings of the Lewisham and Blackheath Scientific Association*: 5–7.
Anon. (1889a). "William Tite." *Builder* 56: 34.
Anon. (1889b). "The International Chemical Congress." *Nature* 40 (15 August): 369–71.
Anon. (1909). "Problem of Woman." *[Evening] Standard*, 27 August: 7.
Anon. (1913). "The Central Transfers to Imperial College." *Chemical World* 2: 2–12.
Anon. (1914). "Narrative and Itinerary of British Association Meeting in Australia." *Brit. Ass. Reports* 1914: 678–719.
Anon. (1916). "Report on Special Issue of *The Central*." *Nature* 98 (9 November): 195.
Anon. (1917). "Obituary of Alfred Mosely." *Times*, 24 July: 5.
Anon. (1919). "The British Scientific Products Exhibition." *Nature* 103 (10 July): 374–76.
Anon. (1920). "Education at the British Association." *Nature* 104 (15 January): 521–22.
Anon. (1922). "F[rederick] D. Brown Obituary." *Nature* 110 (7 October): 490.
Anon. (1925). "Professor Armstrong Lashes Out." *Chemical Age* 14: 205.
Anon. (1927). "The Armstrong Celebrations." *Chemical Age* 17 (3 September): 211, 216–17.
Anon. (1928). "Professor Armstrong on Ethyl Petrol." *Chemical Age* 19 (8 December): 534.
Anon. (1929). "The Heuristic Method: A Symposium." *School Science Review* 11: 65–70.
Anon. (1930a) "School Science and Educational Values." *Nature* 125 (8 March): 341–43.
Anon. (1930b). "Report of the Departmental Committee on Ethyl Petrol." *Nature* 125 (10 May): 710–12.
Anon. (1933a). "Professor Armstrong's Tribute to Sir Edward Frankland." *Chemical Age* 28 (28 January): 79.

Anon. (1933b). "Dr John Thomas, A Pioneer of the Dye Industry." *Times*, 26 June: 14.

Anon. (1933c). "Modern Women. Whacking Advocated. Old Professor Cynical. Wants to Drive Them Home." *Belfast Telegraph*, 17 October: 11; and *Portsmouth Evening News*, 17 October: 6.

Anon. (1935a). "Mr Kenneth Frankland Armstrong." *Nature* 135 (2 February): 175.

Anon. (1935b). "School Dietaries." *Nature* 136 (9 November): 749.

Anon. (1935c). "William R. E. Hodgkinson (1851–1935)." *Journal and Proceedings of the Institute of Chemistry* 1935: 274–75.

Anon. (1936). "Prof. H. E. Armstrong: Doyen of the Royal Society." *Nature* 137 (27 June): 1064.

Anon. (1937). "Henry Edward Armstrong." *Berichte der Deutschen Chemischen Gesellschaft* 70A (11 October): 148–49.

Anon. (1948–1949). "Obituary of Oliver H. Latter." *School Science Review* 30: 244.

APSSM (1917). *Report of the General Meeting 1916, with List of Members and Rules* (Harrow: Harrow School).

Armstrong, Edward Frankland (1910). *The Simple Carbohydrates and the Glucosides* (London: Longman Green).

Armstrong, Edward Frankland (1938a). "The City and Guilds of London Institute: Its Origin and Development." *Central* 35: 14–45.

Armstrong, Edward Frankland (1938b). "Henry Edward Armstrong's Autobiographical Notes." *Central* 35: 1–14.

Armstrong, Edward Frankland (1939). "William Alfred Davis." *JCS* 1939: 1225–26.

Armstrong, Edward Frankland (1941). "Henry Edward Armstrong: The First Armstrong Memorial Lecture." *Nature* 147 (29 March): 373–77.

Armstrong, H. Clifford, and C. V. Lewis (1935). *Practical Boiler Firing* (London: Charles Griffin).

Armstrong, Henry E. (1912). "Raw and Heated Milk." *Times*, 4 October: 10. (This author, medical officer for Newcastle-upon-Tyne, was not related to the chemist Henry Edward Armstrong.)

Armstrong, Richard Robins (1913). "The Mechanism of Infection in Tuberculosis." *Science Progress* 7: 335–55.

Armytage, W. H. G. (1957). *Sir Richard Gregory* (London: Macmillan).

Arrhenius, Svante (1904). "The Development of the Theory of Electrolysis." *PRI* 17: 552–65.

Atkinson, W. J. (1909). "Admission of Women to the Geological Society." *Nature* 79 (25 February): 488.

Ayres, Peter (2020). *Women and the Natural Sciences in Edwardian Britain: In Search of Fellowship* (London: Palgrave Macmillan).

Ayrton, W. A. (1887). "The Technical Training at the Central Institution." *Memorandum of the Proceedings of a Drawing Room Meeting for the Promotion of Technical Education Held at the House of Mr E. C. Robins. 5 March 1887* (privately printed).

Ayrton, W. A. (1892). "Electrotechnics." *Journal of the Institute of Electrical Engineers* 21: 5–36.

Badash, Lawrence (1979). "British and American Views of the German Menace in World War 1." *Notes and Records of the Royal Society* 34: 91–121.

Baker, H. B., and Margaret Carlton (1925). "The Effect of Ultra-Violet Light on Dried Hydrogen and Oxygen." *JCS* 127: 1990–92.

Baldwin, Melinda (2015). *Inventing Nature* (Chicago: University of Chicago Press).
Bancroft, W. D. (1924). "The Electrolytic Theory of Corrosion." *Journal of Physical Chemistry* 28: 785–871.
Bancroft, W. D. (1925). "Corrosion in Aqueous Solutions." *Industrial and Engineering Chemistry* 17: 336–38.
Barbusinski, Krzysztof (2009). "Henry John Horstman Fenton." *Chemia, Dydaktyka, Ekologia, Metrologia* 2009: 101–03.
Bardwell, Dwight C. (1922). "Hydrogen as a Halogen in Metallic Hydrides." *Journal of the American Chemical Society* 44: 2499–504.
Barr, E. Scott (1965). "*Nature's* 'Scientific Worthies.'" *Isis* 56: 354–56.
Barton, Ruth (2018). *The X-Club: Power and Authority in Victorian Britain* (Chicago: University of Chicago Press).
Basset, A. B. (1914). Letter to the editor. *Morning Post*, 22 October: 11.
Bayliss, Robert A. (1983). "Henry Edward Armstrong and Domestic Science." *Journal of Consumer Studies and Home Economics* 7: 299–305.
Bazarov, Aleksander (1868). "Direkte Darstellung des Harnstoffs aus Kohlensäure." (PhD diss., University of Leipzig).
Belloc, Hilaire (1930). *Wolsey* (London: Cassell).
Ben-David, Joseph (1971). *The Scientist's Role in Society: A Comparative Study* (Englewood Cliffs, NJ: Prentice-Hall).
Bentley, Jonathan (1970). "The Chemical Department of the Royal School of Mines: Its Origin and Development under A. W. Hofmann." *Ambix* 17: 153–81.
Berry, A. J. (1946). *Modern Chemistry: Some Sketches of its Historical Development* (Cambridge: Cambridge University Press).
Black, R. D. Collison (1973). *Papers and Correspondence of William Stanley Jevons* (London: Macmillan).
Bone, William A. (1913). "The Uses of Gas." *Brit. Ass. Reports* 1913: 440–44.
Bone, William A., and Godfrey W. Himus (1936). *Coal and its Constitution and Uses* (London: Longmans).
Bone, William A., and Donald Townend (1927a). "Flame and Combustion." *Nature* 120 (22 October): 586.
Bone, William A., and Donald Townend (1927b). "Flame and Combustion." *Nature* 120 (12 November): 694.
Bone, William A., and Donald Townend (1927c). "Flame and Combustion." *Nature* 120 (17 December): 880.
Bousfield, W. R., and Martin Lowry (1910). "Liquid Water a Ternary Mixture: Solution Volumes in Aqueous Solutions." *Transactions of the Faraday Society* 6: 88.
Bragg, W. H., and W. L. Bragg (1913). "The Reflection of X-Rays by Crystals." *PRS* 88A: 246–48, 428–38.
Brandt, Danita (2021). "Hertha Ayrton's *Nature* Obituary: A Monument to Sexism in Science." *Nature* 590 (23 February): 551.
Brock, W. H. (1969). "Lockyer and the Chemists: The First Dissociation Hypothesis." *Ambix* 16: 81–99.
Brock, W. H. (1973). *H. E. Armstrong and the Teaching of Science 1880–1930* (Cambridge: Cambridge University Press).

Brock, W. H. (1981a). "Advancing Science: The British Association and the Professional Practice of Science." In *The Parliament of Science*, edited by Roy MacLeod and Peter Collins (Northwood: Science Reviews): 89–117.

Brock, W. H. (1981b). "The Japanese Connexion: Engineering in Tokyo, London and Glasgow." *BJHS* 14: 227–43.

Brock, W. H. (1989). "Building England's First Technical College." In *The Development of the Laboratory*, edited by Frank James (London: Macmillan): 155–70.

Brock, W. H. (1996). *Science for All: Studies in the History of Victorian Science and Education* (Aldershot: Ashgate).

Brock, W. H. (1998). "The Chemical Origins of Practical Physics." *Bulletin for the History of Chemistry* 21: 1–11.

Brock, W. H. (2004a). "John Henry Pepper." In *Dictionary of Nineteenth-Century British Scientists*, 4 vols., edited by B. Lightman (Bristol: Thoemmes): 3: 1572–73.

Brock, W. H. (2004b). "Richard Wormell (1838–1914)." *OxfordDNB*, online (published 2004).

Brock, W. H. (2008). *William Crookes (1832–1919) and the Commercialization of Science* (Aldershot: Ashgate).

Brock, W. H. (2011a). *The Case of the Poisonous Socks* (London: Royal Society of Chemistry).

Brock, W. H. (2011b). "Edith Hilda Usherwood (1898–1988) and the Ingold Partnership." In Brock, *The Case of the Poisonous Socks* (London: Royal Society of Chemistry): chapter 28.

Brock, W. H. (2011c). "Thomas George Hodgkins: The Future of Research at the Royal Institution (London) and the Smithsonian Institution (Washington)." In *The Case of the Poisonous Socks* (London: Royal Society of Chemistry, 2011), chapter 5.

Brock, W. H. (2013). "Bunsen's British Students." *Ambix* 60: 203–33.

Brock, W. H., O. T. Benfey, and S. Stark (1991). "Hofmann's Benzene Tree." *Journal of Chemical Education* 68: 887–88.

Brock, W. H., and Edgar W. Jenkins (2014). "Frederick W. Westaway and Science Education: An Endless Quest." In *International Handbook of Research in History, Philosophy and Science Teaching*, vol. 3, edited by M. R. Matthews (Dordrecht: Springer): 2359–82.

Brock, W. H., and Michael H. Price (1980). "Squared Paper in the Nineteenth Century." *Educational Studies in Mathematics* 11: 365–81.

Broiméil, Úna Ni (2015). "A Tinge of Effeminacy: Masculinity and Natural Manhood in the Mosely Report, 1904." *Paedagogica Historica* 51: 335–49.

Brooks, Ron (2004). "Joseph John Findlay (1860–1940)." *OxfordDNB* online.

Brown, H. Taberer (1916). "Reminiscences of 50 Years' Experience of the Application of Scientific Methods to Brewing." *J. Inst. Brewing* 22: 267–354.

Brown, John R., and John L. Thornton (1955). "William James Russell (1830–1909) and Investigations of London Fog." *Annals of Science* 11: 331–36.

Brown, Robert. *Science for All* (1877–1881), 5 vols. (London: Cassell); further eds. 1883–1888, 1890–1894, 1895–1897, and 1899–1900.

Browne, Charles (1926). *A Half-Century of Chemistry in America, 1876–1928* (Philadelphia, PA: ACS).

Browne, Charles (1954). *Henry Edward Armstrong* (privately printed).

[Browne, Charles] (1962). Obituary of Charles Browne. *School Science Review* 43: 187.

Browne, Sir Thomas (1716). *Christian Morals* (Cambridge: Cambridge University Press, 1927)

Bryant, E. G. (1908). "Letter from the Cape Colony on Women Chemists." *Chemical News* 98: 172–73.

Bryant, C. L. (1950). "Fifty Years On." *School Science Review* 32: 140–45.

Burt, Cyril (1938). "Formal Training." *School Science Review* 20: 653–66.

Campbell, Norman R. (1910). "The Meaning of 'Ionisation.'" *Nature* 83 (10 March): 36.

Campbell, Norman R. (1924). "The Word 'Scientist' or its Substitute." *Nature* 114 (29 November): 788.

Cane, B. S. (1959). "Scientific and Technical Subjects in the Curriculum of English Secondary Schools at the Turn of the Century." *British Journal of Educational Studies* 8: 52–64.

Cardwell, Donald S. L. (1957). *The Organisation of Science in England* (London: Heinemann).

Chadwick, Edwin (1847). *Report of the Health of Towns Association* (London: Hatton and Co.).

Chapman, A. Chaston (1919). "The Inter-Allied Federal Council of Pure and Applied Chemistry." *Analyst* 44: 221–23.

Chemical Society (1901). *Memorial Lectures Delivered before the Chemical Society* (London: Chemical Society).

City and Guilds of London Institute (1880). *Report of the Governors for the Year Ending March 10th 1880* (London: City and Guilds of London Institute).

Clay, Felix (1902). *Modern School Buildings* (London: Batsford).

Clerke, Agnes M. (1885). *A Popular History of Astronomy during the Nineteenth Century* (London: Adam and Charles Black).

Clerke, Agnes M. (1901). "Low Temperature Research at the Royal Institution, 1893–1900." *PRI* 16: 699–718.

Cobbold, Carolyn (2020). "The Introduction of Chemical Dyes into Food in the Nineteenth Century." *Osiris* 35: 142–61.

Coulson, E. H. (1970). "Nuffield Advanced Science Chemistry: An Account of Stewardship." *School Science Review* 52: 261–71.

Cowles, Henry (2020). *The Scientific Method: An Evolution of Thinking from Darwin to Dewey* (Cambridge, MA: Harvard University Press).

Creese, Mary (1991). "British Women of the 19th and Early 20th Centuries Who Contributed to Research in the Chemical Sciences." *BJHS* 24: 275–305.

Cremin, L. A. (1962). *The Transformation of the School: Progressivism in American Education, 1876–1957* (New York: Knopf).

Crompton, H. (1888). "An Extension of Mendeléef's Theory of Solution to the Discussion of the Electrical Conductivity of Aqueous Solutions." *JCS* 53 (1888):116–25.

Crookes, William (1911). "Report of the Perkin Dinner." *Chemical News* 103 (2 June): 253–54.

Crosland, Maurice P. (1962). *Historical Studies in the Language of Chemistry* (London: Heinemann).

Crum Brown, A., and J. Gibson (1892). "A Rule for Determining Whether a Given Benzene Mono-Derivative Shall Give a Meta-Di-Derivative or a Mixture of Ortho- and Para-Di-Derivatives." *JCS* 61: 367–69.

Csiszar, Alex (2018). *The Scientific Journal* (Chicago: University of Chicago Press).

Cutler, Janet (1976). "The London Institution, 1805–1933." (PhD diss., University of Leicester).

Daniell, G. F. (1912). "Educational Conferences Considered in Relation to Science in Public Schools." *Nature* 88 (18 January): 393–94.

Davis, W. A. (1913). "Profile of Professor H. E. Armstrong." *Chemical World* 2: 13–17.

Davis, W. A. (1918). "The Present Position and Future Prospects of the Natural Indigo Industry." *Agricultural Journal of India* 8. Reviewed in *Nature* 101 (18 July): 388–89.

Davis, W. A. (1924). "Some Pages from the Story of Indigo: A Lecture Given to Chemists at Port Sunlight, 28 January 1924." *J. Soc. Chem. Industry* 43: 266–71 and 303–08.

Dean, I. O. (1929). "Introduction to Armstrong." *Alchemist* 4: 51.

Desmond, Adrian (1997). *Huxley: Evolution's High Priest* (London: Michael Joseph).

Dewar, Lady, J. D. Hamilton Dickson, H. M. Ross, and E. C. Scott Dickson, eds. (1927). *The Collected Papers of Sir James Dewar*, 2 vols. (Cambridge: Cambridge University Press).

Dixon, H. B. (1884). "The Conditions of Change in Gases." *Philosophical Transactions of the Royal Society* 175: 617–84.

Dixon, H. B. (1927). "The Gentle Art of Chemical Controversy." *Nature* 120 (9 July): 26–27.

Dolby, R. G. A. (1976). "Debates over the Theory of Solution: A Study of Dissent in Physical Chemistry in the English-Speaking World in the Late Nineteenth and Early Twentieth Centuries." *Historical Studies in the Physical Sciences* 7: 297–404.

Donnan, Frederick (1939). "Baron Joji Sakurai." *Nature* 144 (5 August): 234–35.

Donnelly, James F. (1987). *Chemical Education and the Chemical Industry in England from the Mid-Nineteenth Century to the Early Twentieth Century* (Ph.D. diss., School of Education, University of Leeds).

Donnelly, James F. (1989). "The Origins of the Technical Curriculum in England during the 19th and 20th Centuries." *Studies in Science Education* 16: 123–61.

Drummond, J. C. (1932). "Recent Researches on the Nature and Function of Vitamins." *Journal of the Royal Society of Arts* 80: 949–57, 959–65, 974–80, 983–90.

Duff, James (1953). "H. Bompas Smith (1867–1953)." *Nature* 365 (31 October): 793–94.

Dyhouse, Carol (1976). "Social Darwinistic Ideas and the Development of Women's Education in England." *Journal of the History of Education Society* 5: 41–58.

Eggar, W. D. (1920). "Science for All." *School Science Review* 2: 197–212.

Elsey, Howard M. (1991). "E. C. Franklin." *Biographical Memoirs, National Academy of Sciences* 60: 67–79.

Emmens, Stephen (1899). *Argentaurana, or Some Contributions to the History of Science* (Bristol: du Boistel).

Ernle, Lord (1921). "Land and the Nation: Towards a New Policy, Saving by Science." *Times*, 21 December: 13; and 22 December: 11.

Evans, Clare de B. (1897). "Studies on the Chemistry of Nitrogen: Enantiomorphous Forms of Ethylpropylpiperidinium Iodide." *JCS* 71: 522–26.

Evans, John Castell (1884). "Science Teaching." *Proceedings of the International Conference on Health* 2: 90.

Evans, John Castell (1892). *A New Course of Experimental Chemistry, Including the Principles of Qualitative and Quantitative Analysis, Being a Systematic Series of Experiments and Problems for the Laboratory and Classroom, together with a Separate Key for Instructors* (London: T. Murby).

Eyre, J. Vargas (1916). "Armstrong's Investigations of Solutions, Osmosis and Enzymes." *Central* 13.

Eyre, J. Vargas (1958). *Henry Edward Armstrong 1848–1937: The Doyen of British Chemists and Pioneer of Technical Education* (London: Butterworths).

Fara, Patricia (2014). "Women in Science: A Temporary Liberation." *Nature* 511 (2 July): 25–27.

Fara, Patricia (2018). *A Lab of One's Own: Science and Suffrage in the First World War* (Oxford: Oxford University Press).

Fauque, Danielle (2011). "French Chemists and the International Reorganisation of Chemistry after World War 1." *Ambix* 58: 116–35.

Fauque, Danielle (2019). "1919–1939: IUPAC—The First Life of the Union." *Chemistry International* (July–September): 2–6.

Fennell, Robert (1994). *History of IUPAC, 1919–1997* (Oxford: Blackwell Science).

Findlay, J. J. (1914). "Educational Science in Australia—and Elsewhere." *Educational Times* 67: 539–42.

Fisher, J. A. (1919a). "Lord Fisher on the Navy: A Reply to Critics." *Times*, 26 September: 6.

Fisher, J. A. (1919b). *Memories: By the Admiral of the Fleet* (London: Hodder and Stoughton).

FitzGerald, G. F. (1896). "Helmholtz Memorial Lecture." *JCS* 69: 885–912.

Fleming, Ambrose (1915a). "Organization of Science." *Times*, 17 July: 7.

Fleming, Ambrose (1915b). "Science in the War and after the War." *Nature* 96 (14 October and 17 February): 180–85 and 692–94.

Fleming, Donald (1973). "Jacques Loeb." *DNB* 8: 445–47; *OxfordDNB* online.

Foden, Frank (1962). "Sir Philip Magnus and the City and Guilds of London Institute." *Vocational Aspect of Secondary and Further Education* 14: 102–16.

Foden, Frank (1970). *Philip Magnus: Victorian Educational Pioneer* (London: Valentine Mitchell).

Forster, Martin O. (1914). "Scientific Socialism or Syndicalism." *Chemical World* 3: 103–04.

Foster, George Carey (1910–1911). "William James Russell." *PRS* 84A: xx–xxi.

Fox, Robert (2016a). "Science, Celebrity, Diplomacy: The Marcelin Berthelot Centenary, 1927." *Revue d'histoire des sciences* 69: 77–115.

Fox, Robert (2016b). *Science without Frontiers: Cosmopolitanism and National Interests in the World of Learning, 1870–1940* (Corvallis: Oregon State University Press).

Frankland, Edward (1868). "On a Simple Apparatus for Determining the Gases Incident to Water Analysis." *JCS* 21: 109–20.

Franklin, E. C. (1935). *The Nitrogen System of Compounds* (New York: Reinhold).

Freund, Ida (1920). *The Experimental Basis of Chemistry: Suggestions for a Series of Experiments Illustrative of the Fundamental Principles of Chemistry* (Cambridge: Cambridge University Press).

Friend, Gerald E. (1935). *The Schoolboy: A Study of His Nutrition, Physical Development and Health* (Cambridge: Heffer).

Fry, G. C. (1902). "The Amazing Heurist." *Journal of Education* 24: 320.

Fyfe, Aileen (2020). "Editors, Referees, and Committees: Distributing Editorial Work at the Royal Society Journals in the Late 19th and 20th Centuries." *Centaurus* 62: 125–40.

Garforth, F. W. (1966). *John Dewey: Selected Educational Writings* (London: Heinemann).

Gay, Hannah (2000). "Association and Practice: The City & Guilds of London Institute for the Advancement of Technical Education." *Annals of Science* 57: 369–98.

Gay, Hannah (2007). *The History of Imperial College London 1907–2007* (London: Imperial College Press).

Gay, Hannah, and John Gay (1997). "Brothers in Science: Science and Fraternal Culture in Nineteenth-Century Britain." *History of Science* 35: 425–53.

Gay, Hannah, and William P. Griffith (2017). *The Chemistry Department at Imperial College* (London: World Scientific).

Gibson, C. S., and T. P. Hilditch (1948). "Edward Frankland Armstrong." *Biographical Memoirs of Fellows of the Royal Society* 5: 619–33.

Goodwin, T. W. (1987). *History of the Biochemical Society 1911–1986* (London: Biochemical Society).

Gordon, Hugh (1893). *Elementary Course of Practical Science* (London: Macmillan).

Gorst, Harold Edward (1901). *The Curse of Education* (London: Grant Richards).

Graebe, C., and C. Liebermann (1868). "Ueber den Zusammenhang zwischen Molekularconstitution und Farbe bei organischen Verbindungen." *Berichte der Deutschen Chemischen Gesellschaft* 1: 106–08.

Gratzer, Walter (1996). *A Bedside Nature: Genius and Eccentricity in Science 1869–1953* (London: Macmillan Magazines).

Gray, Herbert B. (1913). *The Public Schools and the Empire* (London: Williams and Norgate).

Greenstreet, W. J. (1927). *Isaac Newton* (London: George Bell).

Gregory, Richard (1908a). "Women and the Fellowship of the Chemical Society." *Nature* 78 (9 July): 226–28.

Gregory, Richard (1908b). "The Sequence of Studies in the Science Section of the Curriculum." *Brit. Ass. Reports* 1908: 526–35.

Gregory, Richard (1909). "Women and the Fellowship of the Chemical Society." *Nature* 79 (11 February): 429–30.

Gregory, Richard (1917). "Science in Secondary Schools." *Brit. Ass. Reports* 1917: 123–207.

Gregory, Richard (1919). "Substance and Method in Science Teaching." *Brit. Ass. Reports* 1919: 354–55.

Gregory, Richard (1920). "Science for All." *School Science Review* 1: 93–94.

Gregory, Richard (1925). Editorial. *Nature* 116 (5 December): 827.

Gregory, Richard (1927). "Armstrong's Golden Wedding." *Nature* 120 (27 August): 307.

Hamlin, Christopher (1990). *A Science of Impurity: Water Analysis in Nineteenth-Century Britain* (Bristol: Adam Hilger).

Hannaway, Owen (1991). "The German Model of Chemical Education in America: Ira Remsen at Johns Hopkins (1876–1913)." *Ambix* 38: 145–64.

Harcourt, Augustus Vernon (1896). "Presidential Address." *JCS* 69: 563–71.

Harrison, Andrew J. (1988). "Scientific Naturalists and the Government of the Royal Society 1850–1900" (PhD diss., Open University).

Harte, N. B. (1986). *The University of London, 1836–1986* (London: Athlone).

Hartley, Harold (1971). "Henry Armstrong (1848–1937) and Some of the Great Figures of 19th-Century Organic Chemistry." In Harold Hartley, *Studies in the History of Chemistry* (Oxford: Clarendon Press): 195–222.

Hayward, F. H. (1904). *The Secret of Herbart: An Essay on the Science of Education* (London: Sonnenschein).

Hayward, F. H. (1908). "The Dogma of Formal or Faculty Training and Its Downfall." *School World* (November): 417–19.

Hayward, F. H., and Arnold Freeman (1919). *The Spiritual Foundations of Reconstruction: A Plea for New Educational Methods* (London: P. S. King).

Heller, William M. (1932). "The Advancement of Science in Schools." *Brit. Ass. Reports* 1932: 209–28.

Heller, William M. (1933). "General Science in Schools." *Brit. Ass. Reports* 1933: 312–30.

Heller, William M., and Edwin George Ingold (1905). *Elementary Experimental Chemistry: An Introduction to Scientific Method* (London: Blackie).

Henderson, G. G. (1933). "Presidential Address." *JCS* 1933: 463–67.

Hendrick, Ellwood (1917). *Everyman's Chemistry* (New York: Harper).

Hendrick, Ellwood (1919a). *The Percolator Papers* (New York: Harper).

Hendrick, Ellwood (1919b). *Opportunities in Chemistry* (New York: Harper).

Hepler-Smith, Evan (2015). "'Just as the Structural Formula Does': Names, Diagrams, and the Structure of Organic Compounds in the 1892 Geneva Nomenclature Congress." *Ambix* 62: 1–28.

Hofmann, August Wilhelm, and J. Blyth (1856). *Report to the President of the General Board of Health on the Metropolis Water Supply* (London: General Board of Health).

Hughes, Jeff (2010). "'Divine Right' or Democracy? The Royal Society 'Revolt' of 1935." *Notes and Records of the Royal Society* 64: S101–17.

Humphrey, H. A. (1935). "The Training of Engineers as Seen in Retrospect by Old Centralians." *Central* 32: 105.

Hurter, F., and V. C. Driffield. (1890). "A New Method of Determining the Sensitiveness of Photographic Plates." *J. Soc. Chem. Industry* 9: 455–69.

Hutchinson, C. T., and W. H. Mills (1929). "Dr H. J. H. Fenton." *Nature* 123 (16 February): 248–49.

Huxley, Leonard (1914). "The Scott Expedition." *Times*, 15 January: 5.

Huxley, T. H. (1893). "Science and Education." *Collected Essays*, vol. 3 (London: Macmillan).

Inhelder, B., and J. J. Piaget (1958). *The Growth of Logical Thinking* (London: Routledge and Kegan Paul).

Jackson, Herbert, and W. W. Watts (1923). "The Conjoint Board." *Nature* 111 (26 May): 706.

Japp, F. R. (1901). "The Kekulé Memorial Lecture." In *Memorial Lectures Delivered before the Chemical Society* (London: Chemical Society).

Jenkins, Edgar W. (2019). *Science for All: The Struggle to Establish School Science in England* (London: UCL/IOE Press).
Johnson, Jeffrey (1985). "Academic Chemistry in Imperial Germany." *Isis* 76: 500–24.
Jones, G. C. (1937). "Professor Henry Edward Armstrong, FRS." *J. Inst. Brewing* 43: 361–63.
Jowett, Hooper (1937). "Hooper A. D. Jowett (1870–1936)." *JCS* 1937: 1328–29.
Kauffman, George B., and Priebe, Paul M. (1978). "The Discovery of Saccharin: A Centennial Retrospect." *Ambix* 25: 191–207.
Keeble, F. W. (1941). "Henry Edward Armstrong." *Obituary Notices of Fellows of the Royal Society* 3: 228–45.
Kelvin, Lord (1906). "Radium." *Times*, 9 August: 3.
Kendall, James (1937). "Introduction to the Symposium on Pure Liquids." *Transactions of the Faraday Society* 33: 2–7.
Kendall, James (1939a). *Young Chemists and Great Discoveries* (London: George Bell).
Kendall, James (1939b). "Henry Edward Armstrong." *Proceedings of the Royal Society of Edinburgh* 59: 259.
Kimmins, C. W. (1902). "The Subjects to Be Taught as Science in Schools." *Brit. Ass. Reports* 1902: 844–45.
Kingsley, Charles (1872). *Town Geology* (London: Dalby Isbister).
Kipping, Frederick (1938). "H. E. Armstrong." *Central* 35: 59.
Koenigsberger, Leo (1906). *Hermann von Helmholtz*. Translated by Frances Welby (Oxford: Clarendon Press).
Koerner, Guglielmo (Wilhelm Körner) (1874). "Studi sull'isomeria delle così dette sostanze aromatiche a sei atomi di carbone." *Gazzetta chimica italiana* 4: 305–446.
Kragh, Helge (2016). *Julius Thomsen: A Life in Chemistry and Beyond* (Copenhagen: Royal Danish Academy).
Kumar, Prakash (2012). *Indigo Plantations and Science in Colonial India* (Cambridge: Cambridge University Press).
Kurzer, Friedrich (2001). "Chemistry and Chemists at the London Institution 1807–1912." *Annals of Science* 58: 163–201.
Lagueur, B. (1923). See Stephen Miall.
Lang, Jennifer (1978). *City & Guilds of London Institute Centenary: An Historical Commentary* (London: City and Guilds).
Laurie, A. P. (1929). "Tests for Old Masters: Experts and Men of Science." *Times*, 19 February: 10.
Layton, David (1981). "The Schooling of Science in England, 1854–1939." In *The Parliament of Science: The British Association for the Advancement of Science 1831–1981*, edited by Roy MacLeod and Peter Collins (Northwood: Science Reviews): 188–210.
Leonard, J. H. (1904). "Specialization in Science Teaching in Secondary Schools." *Brit. Ass. Reports* 1904: 844.
Lewis, G. N. (1926). "Hydrogen as Anion." *Nature* 117 (12 June): 824.
Liversidge, A. (1921). "The Designation of Vitamins." *Nature* 107 (10 March): 45.
Lockyer, T. Mary, and Winifred L. Lockyer (1928). *The Life and Work of Sir Norman Lockyer* (London: Macmillan).
Lodge, Oliver (1886). "Abstract and Translation of Arrhenius." *Brit. Ass. Reports* 1886: 362–88.

Lodge, Oliver (1896). "The Theory of Dissociation into Ions." *Nature* 55 (17 December): 150–51.
Lodge, Oliver (1913a). "Professor Armstrong and Atomic Constitution." *Nature* 91 (28 August): 672.
Lodge, Oliver (1913b). "Atomic Theory and Radioactivity." *Science Progress* 8: 197–220.
Lodge, Oliver (1926). "Uncertainty." *Nature* 117 (27 March): 453.
Lodge, Oliver (1930). "Space and Matter." *Nature* 126 (5 July): 9.
Lodge, Oliver (1931). *Past Years* (London: Hodder and Stoughton).
Lonsdale, K. Y. (1929). "The Structure of the Benzene Ring." *PRS* 123A: 494–515.
Lowndes, A. S. (1940). "Economy in Education." *Nature* 145 (1 June): 863; and 146 (17 July): 133, 1023.
Lowry, T. M. (1915). *Historical Introduction to Chemistry* (London: Macmillan).
Lowry, T. M. (1925). "Graphitic Conduction in Conjugated Chain of Carbon Atoms: A Contribution to Armstrong's Theory of Chemical Change." *Nature* 115 (14 March): 376–77.
Lowry, T. M. (1927). "The Theory of Strong Electrolytes." *Nature* 119 (7 May): 676–78.
Lupton, M. C. (1964). "The Mosely Education Commission to the United States, 1903." *Vocational Aspect* 16: 36–49.
Macdonald, B., and R. Walker (1976). *Changing the Curriculum* (London: Open Books).
MacLeod, Roy M. (1971). "The Royal Society and the Government Grant: Notes on the Administration of Scientific Research, 1849–1914." *Historical Journal* 14: 323–58.
Macmillan, F. (1936). "Sir Frederick Macmillan, C.V.O." *Nature* 137 (6 June): 937.
Magnus, Laurie (1901). *National Education: Essays towards a Constructive Policy* (London: John Murray).
Magnus, Philip (1883). *Technical Instruction* (London: Gresham College Headquarters, 1883); reprinted in Magnus, *Industrial Education* (London: Kegan Paul, 1888): 231–68.
Magnus, Philip (1920). "The Imperial College: Degrees and Diplomas." *Times*, 11 June: 12.
Mansell, A. L. (1976a). "The Influence of Medicine on Science Education in England." *History of Education* 5: 155–68.
Mansell, A. L. (1976b). "Science for All?" *School Science Review* 57: 579–85.
Mason, Joan (1991a). "A Forty Years' War." *Chemistry in Britain* 27: 233–38.
Mason, Joan (1991b). "Hertha Ayrton (1854–1923) and the Admission of Women to the Royal Society of London." *Notes and Records of the Royal Society* 45: 201–20.
Mathieson, C. M. (1922). "The Feeding of Schoolboys: A Matter for Experts." *Times*, 17 January: 6.
Mayer, Anna-K. (2005). "Reluctant Technocrats: Science Promotion in the Neglect-of-Science Debate of 1916–1918." *History of Science* 43: 139–59.
McKenzie, Frederick A. (1901). *The American Invaders, Their Plans, Tactics and Progress* (London: Grant Richards).
Meadows, A. J. (1973). "Specialization: The Recurring Debate." *Chemistry in Britain* 9: 504–06.

Meadows, A. J. (1974). "The Rise of the Scientific Journal." In A. J. Meadows, *Communication in Science* (London: Butterworths): chapter 3.
Meiklejohn, J. M. D. (1876). *Inaugural Address* (Edinburgh: Cameron).
Meiklejohn, J. M. D. (1884). "Professorships and Lectureships on Education." *International Conference on Education* 4: 97–120; abstracted in *Educational Times* 37: 287–88.
Mellor, J. W. (1904). *Chemical Statics and Dynamics* (London: Longmans).
Meyer, Lothar (1883). *Die modernen Theorien der Chemie*, 4th ed. (Breslau: Maruschke and Berendt).
Miall, Stephen (1922). "The Compleat Chymist." *Chemical Age* 7: 308–11, with replies on 379–80, 408, 450.
Miall, Stephen [under pseudonym "B. Lagueur"] (1923). *The Problem of Solution: A Tavern Talk between Certain Chymists and Others* (London: Ernest Benn).
Miall, Stephen (1931). *History of the British Chemical Industry* (London: Ernest Benn).
Miall, Stephen (1932). *Poets at Play: Anthology of Parodies and Light Verse* (London: London).
Miall, Stephen (1937). "Professor H. E. Armstrong." *J. Soc. Chem. Industry* 56: 650.
Miller, A. K. (1946). "Obituary of Alexander Kenneth Miller (1856–1945)." *JCS* 1946: 67–68.
Miller, William Allen (1867). *Elements of Chemistry, Theoretical and Practical, Part III: Organic Chemistry*, 3rd ed. (London: Parker).
Mills, Dr. H. H. (1923). "Mrs Hertha Ayrton." *Nature* 112 (15 December): 865.
Mitchell, Ada (1909). "Professor Armstrong and the Place of Women." *Yorkshire Post*, 2 September: 4.
Moore, T. S., and J. C. Philip (1947). *The Chemical Society 1841–1941* (London: Chemical Society).
Morrell, J. B. (1993). "W. H. Perkin Jr. at Manchester and Oxford: From Irwell to Isis." *Osiris* 8: 104–26.
Morton, Jocelyn (1971). *Three Generations in a Family Textile Firm* (London: Routledge).
Muir, M. M. Pattison (1887). *Elementary Chemistry*, with companion vol. *Practical Chemistry: A Course of Laboratory Work* (Cambridge: Cambridge University Press).
Muir, M. M. Pattison (1888). "Reply to Armstrong's Criticism." *Nature* 37 (2 February): 318–19.
Munby, Allen E. (1921). *Laboratories: Their Planning and Fitting* (London: George Bell).
Murrell, J. N. (1964). "Colour and Chemical Constitution." *Advancement of Science* 20: 489–96.
Nelson, Muriel (1909). "Women's Higher Education." *Yorkshire Post*, 1 September 1909: 4.
Nicol, W. W. J. (1883). "The Nature of Solution." *Philosophical* 15: 91–101.
Norrish, Ronald (1923). "Studies of Electrovalency. III. The Catalytic Activation of Molecules and the Reaction between Ethylene and Bromine." *JCS* 123: 3006–18.
Nye, Mary Jo (1993). "National Styles? French and English Chemistry in the Nineteenth and Early Twentieth Centuries." *Osiris* 8: 30–49.

Nye, Mary Jo (1994). *From Chemical Philosophy to Theoretical Chemistry: Dynamics of Matter and Dynamics of Disciplines, 1800–1950* (Berkeley: University of California Press).
Odling, William (1896). "On the Development of Chemical Theory since the Foundation of the Society." *Jubilee of the Chemical Society* (London: Chemical Society): 26–32.
Page, T. E. (1906). "The Curriculum of Secondary Schools." *Brit. Ass. Reports* 1906: 787–88.
Partington, J. R. (1964). *A History of Chemistry*, vol. 4 (London: Macmillan).
Pemberton, John (2003). "The Diet at Christ's Hospital School in the 1920s and the Work of a Pioneer School Medical Officer, Dr G. E. Friend." *Journal of Medical Biography* 11: 10–13.
Pepper, J. H. (1860). *Boys' Play-Book of Science* (London: Routledge).
Perry, John (1897). *The Calculus for Engineers* (London: Edward Arnold).
Pickering, S. U. (1885). "Atomic Valency." *PCS* 1: 122–25.
Pickering, S. U. (1927). *The Scientific Work of the Late Spencer Pickering, FRS*, edited by T. M. Lowry and Sir John Russell (London: Harrison).
Pope, Sir William (1930). *Science and Modern Industry* (London: British Science Guild).
Poynting, J. H. (1896). "Osmotic Pressure." *Philosophical* 42: 289–300.
Price, Michael H. (1994). *Mathematics for the Multitude?* (Leicester: Mathematical Association).
R. T. G., (1930). "Obituary of Sir William McCormick (1859–1930)." *Nature* 125 (12 April): 569–71.
Ramsay, William (1891). "The Teaching of Chemistry: Three Lectures to the College of Preceptors." *Educational Times* 44: 228–30, 270–73.
Ramsay, William (1914a). "Reply to German Professors." *Times*, 21 October: 10.
Ramsay, William (1914b). "Science and the State." *Nature* 94 (29 October): 221.
Ramsay, William, and James Walker (1893). "The Conditions Determinative of Chemical Change: Some Comments on Prof. Armstrong's Remarks," *Nature* 48 (20 July): 267–68.
Rayner-Canham, Marelene, and Geoffrey Rayner-Canham (2017). *A Chemical Passion: The Forgotten Story of Chemistry in British Independent Girls' Schools, 1820s-1930s* (London: UCL-IOE Press).
Rayner-Canham, Marelene, and Geoffrey Rayner-Canham (2020). *Pioneering British Women Chemists* (London: World Scientific).
Report of the Reorganization Commission for Milk (1933). London: Ministry of Agriculture and Fisheries.
Rice, Richard E. (2004), "Henry Armstrong on the Offensive: Association as an Alternative to Dissociation." *Ambix* 51: 1–21.
Rice, Richard E. (2011). "Hydrating Ions in St. Petersburg and Moscow, Ignoring Them in Leipzig and Baltimore." *Bulletin for the History of Chemistry* 27: 17–25.
Richmond, H. D. (1908). "Women Chemists." *Chemical News* 98: 58–59.
Roberts, Gerrylynn K. (1976). "The Establishment of the Royal College of Chemistry." *Historical Studies in the Physical Sciences* 7: 437–85.
Robertson, Sir Charles Grant (1928). "Biology for the Empire." *Times*, 24 December: 11.

Robins, E. C. (1885). *Papers on Technical Education, Applied Science Buildings, Fittings and Sanitation* (London: J. Davy).
Robins, E. C. (1887). *Technical School and College Buildings* (London: Whittaker).
Robinson, Robert (1935). "Kenneth Frankland Armstrong." *JCS* 1935: 1892.
Robinson, Robert (1976). *Memoirs of a Minor Prophet: Seventy Years of Organic Chemistry* (London: Elsevier).
Rocke, Alan J. (1993). *The Quiet Revolution: Hermann Kolbe and the Science of Organic Chemistry* (Berkeley: University of California Press).
Rocke, Alan J. (2010). *Image and Reality: Kekulé, Kopp, and the Scientific Imagination* (Chicago: University of Chicago Press).
Rodd, Ernest H. (1940). "Henry Edward Armstrong." *JCS* 1940: 1418–39.
Rodd, Ernest H. (1968). *Henry Edward Armstrong and Charles E. Browne* (Horsham: privately printed).
Roscoe, Henry (1908). "Women and the Fellowship of Scientific Societies." *Times*, 3 July: 15.
Ross, Sydney (1962). "Scientist, the Story of a Word." *Annals of Science* 18: 65–85.
Ross, William (1858). *The Teacher's Manual of Method, or the General Principles of Teaching and Schoolkeeping* (London: Longman).
Rowe, Arthur William (1900). "The Zones of the White Chalk of the English Coast: Kent and Sussex." *Proceedings of the Geologists' Association* 16: 289–367.
Rowlinson, John S. (2012). *Sir James Dewar, 1842–1923: A Ruthless Chemist* (Farnham: Ashgate).
Rowold, Katharina (2010). *The Educated Woman: Minds, Bodies, and Women's Higher Education in Britain, Germany, and Spain.* New York: Routledge.
Rugg, Harold (1927). *The School Curriculum, 1825–1880* (Bloomington, IL: Public School Publishing).
Russell, Colin A. (1971). *The History of Valency* (Leicester: Leicester University Press).
Russell, Colin A. (1986). *Lancastrian Chemist: The Early Years of Sir Edward Frankland* (Milton Keynes: Open University Press).
Russell, Colin A. (1996). *Edward Frankland: Controversy and Conspiracy in Victorian England* (Cambridge: Cambridge University Press).
Russell, Thomas H. (1903). *The Planning and Fitting Up of Chemical and Physical Laboratories* (London: Batsford).
Russell, William James (1897). "On the Action Exerted by Certain Metals and Other Substances on a Photographic Plate." *PRS* 61: 424–33.
Schiemenz, Günther P. (1993). "A Heretical Look at the Benzolfest." *BJHS* 26: 195–205.
Schuster, Arthur (1932). *Biographical Fragments* (London: Macmillan).
Scott, Alexander (1915). "A Consultative Council in Chemistry." *Nature* 95 (8 July): 523–24.
Secord, James E. (2002). "Quick and Magical Shapes of Science." *Science* 297: 1648–49.
Selleck, R. J. W. (1968). *The New Education: The English Background 1870–1914* (London: Isaac Pitman).
Selleck, R. J. W. (1972). *English Primary Education and the Progressives, 1914–1939* (London: Kegan Paul).

Senning, Alexander (2007). *Elsevier's Dictionary of Chemoetymology* (Amsterdam: Elsevier).
Servos, John W. (1990). *Physical Chemistry from Ostwald to Pauling: The Making of an American Science* (Princeton, NJ: Princeton University Press).
Sharpe, P. R. (1971). "'Whiskey Money' and the Development of Technical and Secondary Education." *Journal of Education Administration and History* 4: 31–33.
Sheppard, F. H. W. (1975). *Survey of London*, vol. 38: *The Museums Area of South Kensington and Westminster* (London: Athlone Press).
Shipley, Arthur (1921). "Oil from Ships: Effects of Discharge in the Sea." *Times*, 21 January: 6.
Shulman, L. S., and E. R. Keisler (1966). *Learning by Discovery: A Critical Appraisal* (Chicago: Rand McNally).
Smith, David M. (1997). *Nutrition in Britain: Science, Scientists and Politics in the Twentieth Century* (London: Routledge).
Smith, H. Bompas (1914). "Education at the British Association." *Educational Times* 67: 466–94.
Smithells, Arthur (1893). "Flame Controversy." *Nature* 49 (28 December): 198.
Smithells, Arthur (1904). "Prof. Armstrong's Educational Campaign." *Nature* 69 (28 January): 289–90.
Smithells, Arthur (1906). "School Training for Home Duties of Women." *Brit. Ass. Reports* 1906: 781–84.
Soddy, Frederick (1908–1912). *The Interpretation of Radium* (London: John Murray, 1908); 2nd ed. 1909; 3rd ed. 1912.
Soddy, Frederick (1917–1919). "On the Complexity of the Elements." *PRI* 22: 121.
Soddy, Frederick (1935). "Foreword." In *The Frustration of Science*, edited by Sir Daniel Hall (London: George Allen): 7.
Solvay (1922). "The Solvay Institute of Chemistry." *Nature* 109 (3 June): 718–19.
Spencer, Herbert (1878). *Education: Intellectual, Moral, and Physical* (London: Williams and Norgate).
Stanford, R. V. (1920). "The British Association at Cardiff." *Nature* 106 (2 September): 12–13.
Stanley, Matthew (2020). "Lodge and Mathematics." In *A Pioneer of Connection: Recovering the Life and Work of Oliver Lodge*, edited by James Mussell and Graeme Gooday (Pittsburgh, PA: University of Pittsburgh Press): 87–103.
Statham, William Edward (1842). *First Steps in Chemistry* (London: Statham).
Stewart, A. S. (1910). "The Meaning of 'Ionisation.'" *Nature* 83 (3 March): 6.
Stöckhardt, Julius Adolph (1850). *Principles of Chemistry* (London: Henry G. Bohn).
Streatfeild, W. (1912). "The City & Guilds of London Technical College, Finsbury." *Chemical World* 1: 373–77.
Strutt, R. J. (1903). "Radioactivity of Ordinary Materials." *Nature* 67 (19 February): 369–70.
Strutt, R. J. (1968). *Life of John William Strutt, Third Baron Rayleigh*, augmented ed. (Madison: University of Wisconsin Press).
Sutton, Leslie, and Mansel Davies (1996). *A History of the Faraday Society* (London: Royal Society of Chemistry).
Terakawa, T., and W. H. Brock (1978). "The Introduction of Heurism into Japan." *History of Education* 7: 35–44.

Thompson, D. (1958). "General Science: Its Origin and Growth." *School Science Review* 40: 109–22.
Thomsen, Julius (1882–1886). *Thermochemische Untersuchungen*, 4 vols. (Leipzig: Barth).
Thomson, J. J. (1918). *Natural Science in Education* (London: HMSO): section 42.
Thomson, J. J. (1923). Review of R. J. Strutt, fourth Baron Rayleigh, *John William Strutt: Third Baron Rayleigh* (London: E. Arnold, 1923), in *Nature* 115 (6 December): 814.
Thorpe, Thomas Edward (1901). "Presidential Address." *JCS* 79: 877.
Tilden, Philip Armstrong (1954). *True Remembrances: The Memoirs of an Architect* (London: Country Life).
Tilden, William A. (1878). "On the Theory of Solution and Crystallisation." *Proceedings of the Bristol Naturalists' Society* 2: 249–63.
Tilden, William A. (1905). "Presidential Address." *JCS* 87: 546–50.
Tilden, William A., et al. (1908a). "Women and the Fellowship of the Chemical Society." *Nature* 78 (9 July): 226–28.
Tilden, William A. (1908b). Women and the Chemical Society." *Nature* 79 (24 December): 221.
Travers, M. W. (1925). "The Discovery of Argon." *Nature* 115 (24 January): 121–22.
Travers, M. W. (1928). *The Discovery of the Rare Gases* (London: Edward Arnold).
Travis, Anthony S. (1993). *The Rainbow Makers: The Origins of the Synthetic Dyestuffs Industry in Western Europe* (Bethlehem, PA: Lehigh University Press).
Travis, Anthony S., Willem J. Hornix, Robert Bud, and Peter Reed. (1992). "The British Chemical Industry and the Indigo Trade." *BJHS* 25: 113–25.
Trench, Maria (1888). *Richard Chenevix Trench: Letters and Memorials*, 2 vols. (London: K. Paul).
Trench, Richard Chenevix (1853). *On the Study of Words* (London: John W. Parker).
Trench, Richard Chenevix (1892). *On the Study of Words*, 22nd ed. (London: Kegan Paul).
Trenn, Thaddeus J. (1977). *The Self-Splitting Atom* (London: Taylor and Francis).
Usherwood, T. S., and C. J. A. Trimble (1913). *A First Book of Practical Mathematics* (London: Macmillan).
Van Praagh, Gordon (1949). *Chemistry by Discovery* (London: John Murray).
Van Praagh, Gordon (1973). *H. E. Armstrong and Science Education* (London: John Murray).
Van Praagh, Gordon (1992). *The Teaching of Science at Christ's Hospital since 1900* (Crawley, Sussex: privately printed).
Van Praagh, Gordon (2001). *Encounters with Stuff: Adventures of a Chemist* (Durham: Pentland Books).
Van Tiggelen, Brigitte, and Danielle Fauque (2012). "The Formation of the International Association of Chemical Societies." *Chemistry International* 34: 8–11.
Varcoe, Ian (2000). "Practical Proposals by Scientists for Reforming the Machinery of Scientific Advice." *BJHS* 33: 109–14.
Vassall, Arthur (1920). "Science for All." *School Science Review* 2: 241.
Walden, Paul (1910). "Is Water an Electrolyte?" *Transactions of the Faraday Society* 6: 71–78.
Walker, E. G. (1933). "Finsbury Technical College." *Central* 30 (July 1933): 35–48.
Walker, James (1910). "Jubilee of the Theory of Electrolytic Dissociation." *Nature* 82 (3 February): 401–02.

Wanklyn, J. Alfred, and E. T. Chapman (1868). *Water Analysis* (London: Trübner).

Wanklyn, J. Alfred, E. T. Chapman, and Miles H. Smith. (1868). "Note on Frankland and Armstrong's Memoir on the Analysis of Potable Waters." *JCS* 21: 152–60.

Waring, Mary (1979). *Social Pressures and Curriculum Innovation: A Study of the Nuffield Foundation Science Teaching Project* (London: Methuen).

Watchurst, Edgar G. (1974). "The Journal of the Chemical Society, 1862–1900: Enquiries into Some Aspects of Nineteenth-Century Chemical Publishing" (MSc thesis, University of Bristol).

Watson, E. R. (1918). *Colour in Relation to Chemical Constitution* (London: Longmans).

Watson, Katherine D. (2002). "Temporary Hotel Accommodation? The Early History of the Davy-Faraday Research Laboratory." In *The Common Purposes of Life*, edited by Frank A. J. L. James (Aldershot: Ashgate): 191–224.

Westaway, Frederick W. (1929). *Science Teaching: What It Was, What It Is, and What It Might Be* (London: Blackie).

Wetzel, Walter (1991). *Die Naturwissenschaften und chemische Industrie in Deutschland* (Stuttgart: Steiner).

Whetham, W. C. D. (1896). "The Theory of Dissociation into Ions." *Nature* 55 (17 December): 151–52.

Whitworth, Adrian (1985). *A Centenary History: A History of the City and Guilds College, 1885 to 1985* (London: Imperial College).

Williams, Ernest E. (1896a). *Made in Germany* (London: Heinemann).

Williams, Ernest E. (1896b). *The German Menace* (London: Henry).

Williams, Ernest E. (1917). *Liberty: Anti-Prohibitionist Essays* (London: Everleigh Nash).

Wilson, Arnold (1939). "Henry Edward Armstrong Memorial Fund." *Scotsman*, 21 March: 11.

Wilson, David (1983). *Rutherford: Simple Genius* (London: Hodder and Stoughton).

Wilson, Jennifer (2012). "Celebrating Michael Faraday's Discovery of Benzene." *Ambix* 59: 241–65.

Witt, O. N. (1876). "Kenntniss des Baues und der Bildung farbender Kohlenstoffverbindungen." *Berichte der Deutschen Chemischen Gesellschaft* 9: 522–27.

Wohl, Anthony S. (1983). *Endangered Lives: Public Health in Victorian Britain* (Cambridge, MA: Harvard University Press).

Wormell, R. (1900). "Unstable Questions of Method in the Teaching of Elementary Science." *Educational Times*, 1 June: 240–43.

Worthington, A. M. (1881). *An Elementary Course of Practical Physics* (London: Rivingtons).

Worthington, A. M. (1886). *A First Course of Physical Laboratory Practice* (London: Longmans Green).

Wright, Rebecca, H. Shin, and F. Trentmann (2013). *From World Power Conference to World Energy Council*. London: World Energy Council. https://www.worldenergy.org/assets/downloads/A-Brief-History-of-the-World-Energy-Council.pdf.

Young, F. G. (1954). "Sir Jack Drummond." *Obituary Notices of Fellows of the Royal Society* 9: 99–129.

Index

Abel, Frederick, 53–54, 94, 99
Acworth, Joseph John, 48
Adams, W. G., 31
Adkin, Joseph Fletcher, 29
Adkin, Robert, 4, 28
Adlam, G. H. J., 228
agriculture and horticulture, x, 12, 57–59, 212–16, 239, 244, 256–57, 297–300, 303. *See also* Lawes Agricultural Trust
Albert, Prince, ix, 52, 245, 302
alcoholic beverages, 6, 35, 57, 83, 218–21, 291
Amundsen, Roald, 198
analytical chemistry, water analysis, 9–11, 57
Appleyard, Roll, 284
Archimedes, 132
Armstrong, Annie (Mrs. D. W. L. MacGregor) (daughter of HEA), 29, 66, 81, 142–43, 200, 307
Armstrong, Edith Emilie (Mrs. Stephen Miall) (daughter of HEA), 66, 142–43, 200, 280, 307
Armstrong, Edward Frankland ("Frank") (son of HEA), x, 4, 20, 110, 123, 126, 143, 264, 274, 306, 308; biography, 28, 53, 64–65, 139, 142, 200, 294; enzymes, 82; osmosis, 171, 179–80
Armstrong, Ethel Mary Turpin (daughter-in-law of HEA), 64
Armstrong, Frances Louisa Lavers ("Louisa") (wife of HEA), 5, 19, 26, 28–29, 51, 64, 110, 142, 188, 200, 269–70, 281, 296, 300, 302
Armstrong, Harold Lavers (son of HEA), 65, 142, 154, 200, 294, 307
Armstong, Harriet (sister of HEA), 3, 28
Armstrong, Henry Clifford ("Clifford") (son of HEA) 28, 65–66, 142–43, 200, 294, 307
Armstrong, Henry E. (medical officer, no relation to HEA), 216
Armstrong, Henry Edward (HEA): and BAAS (*see* British Association for the Advancement of Science); and Central Technical College, 50–93 passim (*see also* Central Technical College); centric benzene formula, 69–71, 77–78, 92, 275, 309 (*see also* benzene, naphthalene, and other aromatic compounds); and Chemical Society (of London), ix, xi, 22, 27, 46, 50, 82, 94–127 passim, 182, 192–94, 201–8, 223, 237–44, 254, 263–64, 271–72, 279, 302; color chemistry, 75–78; crystallography, 78–80; death, 306; dyes, 212–16; education, 5–18; electrons and electronic theories (*see* electronic theories in chemistry); environmentalism, 246–49, 255, 297; enzymes, 81–83; *Essays on the Art and Principles of Chemistry*, 89, 170, 251; eugenics, 124 (*see also* eugenics); and Finsbury College, 30–49 passim; food chemistry (*see* food, nutrition, and vitamins); fuels, 209–12; heurism (*see* heuristic method); hydrone, 164–70,

Armstrong, Henry Edward (HEA) (*cont.*) 174–76, 251, 259, 267; *Introduction to the Study of Organic Chemistry*, 27, 46; ions (*see* ions, ionists, and ionization); legal consulting, 21–22, 132; and London Institution, 23–24, 48–49; marriage, 28–29; osmosis, 179–81; periodic systems and classification, 89–92; physical chemistry, 153–82 passim, 260, 262, 296; radioactivity, 83–89; residual affinity and reverse electrolysis, xi, 70, 83–89, 146, 153–182 passim, 259, 265–68; and St. Bartholomew's Hospital, 22–23; scientific method, 19–22; *The Teaching of Scientific Method and Other Papers on Education*, 235; terpenes, 80–81; United States and Americans, relations with, 57–59, 96–97, 122, 148–50, 176–79, 193, 219, 245, 264, 292; wartime work, 68, 195, 200–223; youth, 3–5
Armstrong, Kenneth Frankland (grandson of HEA), 64, 294
Armstrong, Louis (brother of HEA), 3, 29
Armstrong, Mary (mother of HEA), 3, 15, 29
Armstrong, Mary (sister of HEA), 3
Armstrong, Nora (daughter of HEA), 66, 81, 142, 200, 307
Armstrong, Richard (father of HEA), 3, 5, 15–19, 26, 29
Armstrong, Richard Robins ("Robin") (son of HEA), 28, 49, 65–66, 142, 200, 307
Arnold, Matthew, 145
Arrhenius, Svante, x, 152–65, 170, 173, 181, 251, 254, 262
Asquith, Herbert, 207
Atkinson, W. J., 122
Austen, Jane, 280, 290
Ayrton, Edith, 280
Ayrton, Hertha Marks, 41, 280–82
Ayrton, Mathilda Chaplin, 280
Ayrton, William Edward, 32, 34–43, 38, 46–47, 53, 55, 99, 123–24, 150, 210, 280–82

BAAS. *See* British Association for the Advancement of Science
Baeyer, Adolf von, 12, 16, 71, 77, 106, 203–4, 212
Baker, Herbert Brereton, 84, 155, 166, 192, 198, 201, 265, 269
Balfour, Arthur, 195
Balfour, Gerald, 123
Bamberger, Eugen, 204
Banting, Frederick, 258–59
Bardwell, Dwight, 266
Barger, Florence, 125
Barlow, William, 79–80, 93, 199, 259, 268
Barry, John Wolfe, 62
Basset, Alfred, 202
Bates, Henry, 4
Bateson, William, 295
Bazarov, Aleksandr, 13
B-Club, 24, 110
Beale, William P., 79, 112
Beilby, George Thomas, 209, 211
Beilstein, Friedrich, 106–7
Bell, Andrew, 133
Benn, Ernest, 268, 279
benzene, naphthalene, and other aromatic compounds, 17, 48–49, 51–52, 67–78, 80, 92–93
Bernal, John Desmond, 274
Berthelot, Marcellin, 106, 250, 279–80
Berthollet, Claude-Louis, 172
Best, Charles, 259
Bindewald, Hans, 76
biochemistry and medical chemistry, 8–9, 16, 81–83, 256–61
Bjerrum, Niels, 169
Black, Joseph, 222
Boisbaudran, Paul Émile Lecoq de, 90
Bone, William A., 155, 209, 250
Bosch, Carl, 298
Boyle, Robert, 178
Brabrook, Edward W., 63
Bradley, John, 144, 244
Bragg, William and Lawrence, 93, 169, 250–51, 259, 264, 268, 278
Brande, William Thomas, 264
Bredt, Julius, 81
Briggs, J. F., 80

British Association for the Advancement of Science (BAAS), 81, 95, 237, 271; Australian meeting in 1914, 195–200, 202; debates over heurism at, 225–28, 230–33; debates over solution chemistry at, 158–63, 172; friends from, 26, 49; HEA's speeches, papers, and discussions at, 71, 75, 77, 86, 89, 122, 124–27, 135–37, 140–41, 145, 147, 151, 154–55, 167; leadership in, 52, 59–61, 66, 191, 207–10, 243–44, 301; Section B, 24, 61, 122, 135–36, 209
Brodie, Benjamin Collins, Jr., 26–27, 59, 96
Bromwell, Frederick, 33
Brönsted, Johannes, 169
Brophy, A. F., 40
Brough, John Cargill, 24–25
Brown, Alexander Crum. *See* Crum Brown, Alexander
Brown, Frederick, 49
Brown, Horace Taberer, 6–7, 121, 221, 279, 306
Brown, James Campbell, 31–32
Brown, John, 65
Brown, Reginald, 214
Browne, Charles E., 134, 143–44, 309
Brühle, Julius Wilhelm, 77
Bryant, C. L., 224, 228
Bryant, E. G., 119
Bunsen, Robert, 10, 12, 16, 23, 26, 79
Burke, Edmund, 134
Burn, Joshua Harold, 300

Cain, John, 172, 174, 194
Caldwell, Robert John, 173–74
Campbell, Norman, 169, 240
Cannizzaro, Stanislao, 242
Carlton, Margaret, 265
Carlyle, Thomas, 4, 21, 60
Caro, Heinrich, 6, 286
Carroll, Lewis, 166, 267, 271, 285
Central Technical College, xi, 30–31, 34, 36–37, 39, 46–47, 50–93 passim, 114, 132–40, 152, 171, 182, 360
Chadwick, Edwin, 33
Chamberlain, Neville, 257
Chapman, A. Chaston, 238

Chapman, Ernest Theophron, 6, 11
Chemical Society (of London), ix, xi, 22, 27, 46, 50, 82, 94–127 passim, 182, 192–94, 201–8, 223, 237–44, 254, 263–64, 271–72, 279, 302
Chesterton, Gilbert Keith, 178, 218
Churchill, Winston, 288–89
City and Guilds of London Institute, 30–32, 34, 51, 52, 192–93
City Guilds of London (livery companies), 30, 33–41, 46–49, 52–55, 61–64, 123, 139, 192–93, 301, 308
Clausius, Rudolf, 165, 172
Cleve, Per, 106
Clifford, W. K., 265
Clifton, Edward, 35–36, 38
Clodd, Edward, 307
Colgate, Richard, 308
Combe, George, 130, 148
Comte, Auguste, 220, 244
Conant, James Bryant, 294
Cooke, Josiah Parsons, 59
Cope, William, 269
Coventry, Bernard, 213
Crafts, James, 14
Crichton-Brown, James, 218
Crompton, Holland A., 41, 159
Crookes, William, 9, 83–85, 94, 108–9, 114, 136, 202, 205–6, 215, 283
Cross, Charles, Frederick, 291
Crossley, Arthur W., 118, 120, 206
Crowther, James, 273–74
Crum Brown, Alexander, 72, 94
Cundall, Tudor, 147
Curie, Marie, 83, 115–17, 120
Curtius, Theodor, 204

Dalton, John, 59
Darmstaedter, Ludwig, 13
Darwin, Charles, 20, 63, 178, 196, 243–44, 264
Davies, Robertson, xii
Davis, William Alfred, 193, 214–16, 239
Davy, Humphry, 276
Debus, Heinrich, 26
Debye, Peter, 169, 181–82, 254
Department of Science and Art (DSA), 36, 43, 129–30, 133

Department of Scientific and Industrial Research (DSIR), 207, 211, 222, 273, 278
Devonshire Commission, 130–31
Dewar, James, 26, 104, 107–9, 237, 240, 253, 275–79, 283, 294
Dewey, John, 149–50
Divers, Edward, 121
Dixon, Harold, 60, 154, 170
Donnelly, John, 131
Dougal, Margaret, 97
Dreaper, William Porter, 192, 194
Driffield, Vero, 253–54
Drummond, Jack, 258, 298–99
Dugdale, Thomas C., 269
Dumas, Jean-Baptiste, 90
Dunn, John Thomas, 137
Dunstan, Wyndham, 82, 99–100, 137
Dyer, Bernard, 206
dyes and dyeing, 6, 22, 36, 191, 205, 212–16, 239, 242, 264, 284–87

Eddy, Mary Baker, 268
Eggar, W. D., 224, 228
Einstein, Albert, 254, 267
electronic theories in chemistry, x, 52, 71–72, 93, 97, 153–82 passim, 259, 262, 240–41, 259
environmentalism, 246–49, 255, 297
enzymes, 52, 64, 81–83, 166, 173, 180, 200, 216, 219–20, 259, 265
eugenics, 124, 126–27, 218, 244, 282, 294, 298, 300, 309. *See also* Galton, Francis
Evans, Clare de Brereton, 79
Evans, John Castell, 39, 44–47, 133
Eyre, John Vargas, xii, 52, 216, 285, 299

faculty psychology, 129, 147 (defined), 148, 150–51, 225, 228, 230, 235
Faraday, Michael, 38, 73, 129, 154, 176, 259, 262–64, 270, 276–77, 284, 310
Farrar, Frederic, 129–30
Fawcett, W. M., 31
Fenton, Henry, 28
Finsbury College, 27, 29, 30–49 passim, 52–53, 55, 62, 69, 81, 94, 132–33, 136–37, 149, 211, 245, 260

Fischer, Emil, 53, 64–65, 82, 120, 180, 202–4, 274
Fischer, Otto, 48
Fisher, 1st Baron (John Arbuthnot), 212, 301
Fittig, Rudolf, 14
FitzGerald, George, 166, 172
Flurscheim, Bernhard J., 73
food, nutrition, and vitamins, ix, 23, 35, 59, 147, 216–20, 247, 256–59, 291, 298–300, 303
Forster, Martin, 51, 57, 63, 81, 110, 118, 120, 194, 203–4, 206, 221
Foster, George Carey, 23, 31
Foster, Michael, 65
France, Anatole, 260
Frankland, Edward, 16, 17, 21, 23, 44, 51, 92, 135, 150, 193, 297, 302; biographical details, 9–10; collaboration with HEA on water analysis, 9–11, 131; obituary by HEA, 99–101; promotion of HEA's career by, 22, 24, 26, 35; as teacher of HEA at Royal College of Chemistry, 6–12, 274; X-Club, 129–30
Frankland, Percy, 61, 206
Franklin, Edward Curtis, 305
Freeman, Arnold, 220–21
Fresenius, Carl Remigius, 45
Freund, Ida, 116
Friedel, Charles, 14, 106
Friend, Gerald E., 299
Friswell, R. J., 77
fuels, 35, 57, 209–12, 239–40, 247, 250, 298
Funk, Casimir, 256

Gall, Franz, 148
Galton, Francis, 124, 127, 196, 294
gender. *See* women and gender relations
geology, 8, 225, 246, 301–2
Gibson, C. S., 272
Gibson, J., 72
Gilbert, Joseph Henry, 58, 94
Gladstone, John Hall, 94, 106, 130–31, 136–37, 140, 146
Glanville, Joseph, 179
Godwin, George, 33

Gordon, George, 108
Gordon, Hugh, 140–41
Gore, George, 305
Gorst, Harold Edward, 288
Gosse, Edmund, 296
Gould, Barbara Ayrton, 281
Graebe, Carl, 13, 106, 203–4, 286
Graham, N. C., 74
Green, A. H., 77
Gregory, Richard, 118–22, 178, 192, 226–33, 240, 249, 252–53, 263, 269, 281–83, 304
Griess, Peter, 6, 14
Groth, Paul Hendrich von, 203–4
Grove, William Robert, 24–25
Groves, Charles, 46, 49, 96, 98
Guthrie, Frederick, 290

Haber, Fritz, 205, 215, 298
Hale, G. D., 304
Hall, Daniel, 273
Hamer, F. E., 269
Hankel, Wilhelm, 13
Harcourt, Augustus George Vernon, xi, 112–13, 118, 137
Harden, Arthur, 102–3, 251
Harkins, William D., 266
Harrow, George, 49
Hartley, Harold, 191, 233, 302, 308
Hartley, Walter, 76, 97, 114
Hartog, Philip, 102–3
Hayward, Frank, 151, 220–21
Heath, Grace, 140
Heaton, Charles William, 112–13
Heller, William M., 135, 139–41, 227, 233
Helmholtz, Hermann von, 76, 154, 162
Henderson, George G., 207, 242
Hendrick, Ellwood, 293
Henrici, Olaus, 53
Henson, Herbert Hensley, 218
Herbart, Johann Friedrich, 148–51, 197, 220
Herschel, Alexander, 79
heuristic method, x, xii, 22, 34, 214, 221, 223; HEA's ideas about and methods on, 135–39, 260–61, 301; assessment of HEA's work on, 150–52, 236, 309–10; at Central Technical College,

55–56; family experiments, 142–44; at Finsbury College, 41–45, 137; opposition to, 146–48, 224–36; roots and early development of, 132–35; successes of, 144–46; in USA, 148–50
Hill, M. D., 224–25
Hill, Octavia, 246
Hirst, Thomas Archer, 129
Hitler, Adolf, 303
Hodgkins, Thomas George, 275–78
Hodgkinson, William, 118
Hoff, Jacobus Henricus (Henry) van 't, x, 12, 65, 171; HEA's opposition to, 158–62, 165–66, 170, 173–74, 179–81, 305; on osmosis, 173, 158; on stereochemistry, 79; on theory of solutions, 156–58
Hofmann, August Wilhelm von: at Benzolfest, 78; in Berlin, 16–17, 39, 53, 99; reputation of, 4, 12, 99, 111, 302; water analysis by, 10; at Royal College of Chemistry, 4, 6–8, 10, 12, 51, 62, 79, 95
Hooker, Joseph, 129
Hopkins, Frederick Gowland, 126, 256, 259, 272
Howorth, Henry, 105
Hübner, Hans, 49, 69
Hückel, Erich, 169, 181–82, 254
Hüfner, Gustav, 13
Humphrey, Herbert A., 111, 211
Hurter, Ferdinand, 253–54
Huxley, Thomas Henry, 163; committee work, 130, 144; influence, 30, 64, 131, 137, 150, 196, 262–63, 274, 289–90; as teacher of HEA, xi, 4, 8, 130, 289; in X-Club, 129
hydrone, 164–70, 174–76, 251, 259, 267

Imperial College, xi, 31, 61, 140, 192, 198, 200–201, 108–9, 221–22, 250, 289, 301
Inge, William, 260
Ingold, Christopher, x, 72, 135, 140, 179, 240–41, 262, 310
Ingold, Hilda Usherwood, 140
ions, ionists, and ionization, x, 88, 93, 105, 111, 146, 153–182 passim, 251, 254, 262, 268

Jackson, H. W., 63
jargon, x, 167, 260, 306
Jeans, James, 267, 279
Jevons, William Stanley, 128–29, 210
Johnson, Samuel, xii
Jones, Francis, 137, 147
Jones, Harry Clary, 182
Jowett, Hooper, 118

Kahlenberg, Louis, 163, 178
Kamerlingh Onnes, Heike, 276
Kant, Immanuel, 133
Keeble, Frederick, 4, 52, 299, 308
Kekulé, August, x, 4, 12, 16–18, 48, 67–69, 78, 92, 296, 302
Kekulé, Charles (Carl), 4, 17
Kelvin, 1st Baron (William Thomson), 31, 86
Kendall, James, 156, 161–62, 181, 307
Kimmins, Charles William, 225–26
King's College London, 20, 26, 31, 56, 94, 208, 222, 274
Kingsley, Charles, 8, 64, 129
Kipping, Frederick, 51, 57, 81, 94
Knapp, Karl, 13–14
Knop, Wilhelm, 13
Kolbe, Hermann, 11–12; as teacher of HEA, x–xi, 11–17, 22, 26, 48, 131; his reputation as harsh critic and influence on HEA, x, 51, 57, 67, 92, 147, 170, 193, 268, 305–6; scientific work, 9, 21, 68, 245, 302
Körner, Wilhelm (a.k.a. Guglielmo Koerner), 48, 68–69

Ladenburg, Albert, 14
La Fontaine, Henri, 272
Lagueur, B. *See* Miall, Stephen
Lang, Stephen S., 218
Lankester, Ray, 105, 203, 228, 243
Lapworth, Arthur, 8, 51, 57, 81, 210
Lapworth, Charles, 8
Larmor, Joseph, 103–4, 202–3, 281
Latter, Oliver H., 224–27
Laurie, Arthur Pillars, 261
Lavers, Thomas (HEA's father-in-law), 5, 19
Lavoisier, Antoine Laurent, 59, 174

Lawes, John Bennet, 58
Lawes Agricultural Trust, 57–59, 66, 82, 122, 180, 191, 214–16, 251, 258, 292, 305
Leather, Walter, 215
LeBel, Joseph Achille, 79
Leonard, J. H., 226
Levinstein, Herbert, 242
Levinstein, Ivan, 242
Lewes, Vivian Byam, 209
Lewis, Gilbert Newton, 259, 266–67
Lieben, Adolf, 106
Liebermann, Carl, 75
Liebig, Heinrich, 77
Liebig, Justus von, 7, 11–12, 24, 38, 77, 215, 244, 256, 302
Ling, Arthur Robert, 37
Lister, Joseph, x, 105, 221
Liveing, George, 26, 28, 253
Liversidge, Arthur, 256
Livingstone, Richard, 301
Lockyer, Norman, 9, 49, 59–60, 119, 130, 252–53
Lockyer, Thomasina Mary, 252
Lockyer, Winifred, 252
Lodge, Oliver, 27–28, 47, 87–89, 158–60, 163, 202, 226, 254, 266–69, 309
Loeb, Jacques, 82, 253
London Institution (LI), 11, 24–30, 33, 35, 48–49, 52, 57, 69, 81, 131, 133, 195, 274, 300
Lonsdale, Kathleen, 78
Lowry, T. Martin, 51, 57, 79, 81, 85–86, 89, 155–56, 169, 175, 179, 251–52, 259
Lubbock, John, 129, 264
Ludwig, Carl, 12, 14
Lyons, Henry, 274

MacGregor, Dugald William Lionel (HEA's son-in-law), 66
Macmillan, Frederick, 304
Magnus, Laurie, 146
Magnus, Philip, 39–40, 46–47, 54, 74, 132, 146, 222, 226
Mather, Thomas, 282
Mathieson, C. M., 257
Matthiessen, Augustus, 23, 49
Maxwell, James Clerk, 31

McCormick, William, 207
McKenzie, F. A., 122
McLean, Ida. *See* Smedley, Ida
McLeod, Herbert, 7, 25
Meiklejohn, John Miller Dow, 133–35
Meldola, Raphael, 41, 46–48, 53, 59, 117, 120, 137, 206
Mellor, Joseph William, 155, 222–23, 249
Mendeleev, Dmitrii, 45–46, 59, 87–90, 159
Messel, Rudolf, 277
Meyer, Lothar, 69
Miall, Stephen (HEA's son-in-law), xii, 66, 142, 178–79, 191, 238, 286, 306–7
Midgley, Thomas, 247
Miers, Henry A., 79, 112
Mill, John Stuart, 195
Miller, Alexander Kenneth, 47, 49
Miller, William Allen, 46
Mitchell, Ada, 125–26
Mond, Ludwig, 201, 277, 280, 293
Mond, Robert, 279
Moody, Gerald, 56, 123, 197, 221
Moore, Bernard, 222–23
Morgan, G. T., 41
Morse, Harmon Northrup, 174
Morton, James, 264
Moseley, Henry, 198
Mosely, Alfred, 122–24, 127, 149–50
Mottram, Vincent H., 274
Moureu, Charles, 279–80
Muir, Matthew Moncrieff Pattison, 59, 107, 137
Müller, Hugo, 17, 24, 94, 99, 202, 205

Nature (journal), 118, 121, 168, 225, 279, 284–85; Norman Campbell and, 240; Richard Gregory and, 233, 263, 269, 281, 283; HEA letters, reviews, and essays in, x, 49, 59–61, 88, 106, 140, 160–61, 178, 191, 197, 237–38, 240, 242, 249–52, 265–67, 279–98, 303–4, 309; Norman Lockyer and, 119, 59–60, 130–31; obituary of HEA in, 307; obituary of Hertha Ayrton by HEA in, 280–82
Nelson, Muriel, 125
Nernst, Walther, 162, 203–4

Newlands, John, 90
Naumann, K. F., 13
Newton, Sir Isaac, 268
Nicol, William, 159
Norrish, Ronald, 155–56
Noyes, Arthur, 96
Noyes, William A., 97

Odling, William, 17, 24, 99, 101, 282
Oldham, F. M., 228
Onsager, Lars, 169
Orr, John Boyd, 300
Ostwald, Wilhelm, x, 93, 158, 161–63, 167, 170, 175–76, 181–82, 203–4, 251, 272, 309
O'Sullivan, Cornelius, 7
Otler, Paul, 272
Overskou, Thomas, xii
Owen, Richard, 32

Page, T. E., 226
Palladino, Eusapia, 88
Parker, Francis W., 148–49
Parker, Richard I., 115
Parker, Thomas, 210
Parkhurst, Helen, 150
Pasteur, Louis, and pasteurization, x, 79, 221, 244, 257–58, 298
Paternò, Emanuele, 106
Pearson, Karl, 124, 137, 288–89
Pedler, Alexander, 110
Pepper, John, 5
periodic systems of the elements, 83–89, 136
Perkin, Arthur George, 214
Perkin, William Henry, Jr., 94, 202, 206, 208, 282–83
Perkin, William Henry, Sr., 53, 94, 98–99, 110, 112, 114, 135, 214, 286
Perry, John, 32, 38, 40–42, 149, 197, 281
Pestalozzi, Johann Heinrich, 132, 134, 151
Pettenkofer, Max, 90
Pfeffer, Wilhelm, 156
pharmaceuticals, 23, 238, 300
Philip, James Charles, 192
Piaget, Jean, 132
Pickering, Spencer, 98, 154, 159, 251–52
Playfair, Lyon, ix, 9, 99

Pope, William J., 41, 50–51, 57, 79–81, 93, 199, 238, 242–43, 253, 257, 264–65, 268, 280, 308
Poynting, J. H., 160
Praagh, Gordon Van. *See* Van Praagh, Gordon
Preece, William, 210
Price, T. S., 272
Prout, William, 87, 252, 310
psychology, faculty. *See* faculty psychology

Ramsay, Andrew, 8
Ramsay, William, x, 86–87, 98, 101–9, 113–14, 120, 127, 147, 151, 155, 162–63, 202, 206
Rāy, Prafulla Chandra, 196–97
Rayleigh, 3rd Baron (John William Strutt), 87, 107–9
Rayleigh, 4th Baron (Robert John Strutt), 83–84, 108–9
Read, John, 41
Reay, Donald Mackay, 11th Lord, 123
Remsen, Ira, 106, 245
Renan, Ernest, 279
Rennie, Edward, 195
Richmond, Henry Droop, 119
Robins, Edward Cookworthy, 31–34, 37–40, 49, 53–54
Robinson, Robert, x, 57, 241, 262, 284, 294, 310
Rodd, Ernest H., 52, 80, 308
Romanes, George, 105
Roosevelt, Theodore, 123
Roscoe, Henry: assists HEA, 25–28, 32, 53, 216; Chemical Society leadership, 94, 96, 112–13, 118–19; science and pedagogy, 135, 141, 144, 146, 207, 225, 253
Ross, Ronald, 88
Rothamsted. *See* Lawes Agricultural Trust
Rowe, Arthur, 8, 246, 297
Royal College of Chemistry, x, 4–10, 12, 25, 35, 44, 49, 51, 62, 79, 131, 154, 221, 263, 274, 289
Royal College of Science, xi, 54, 61–62, 74–76, 100, 114, 131, 136, 192, 201, 221, 250, 289–90

Royal Institution: Arrhenius at, 163; Bragg at, 250; Dewar at, 26, 107; Faraday at, 10, 129, 263–64; HEA invited by Frankland to, 274; HEA applies for professorship at, 26; HEA lectures at, 89, 285; HEA gives Hodgkins Trust Lectures, 274–78; HEA leadership of, 52, 191, 201, 271, 274–78; Helmholtz at, 154; Rayleigh at, 107; Tyndall at, 8, 129, 274
Royal School of Mines, xi, 7–8, 61, 131, 289
Royal Society (of London): gender issues in, 41, 118, 280–81; German scientists excluded from during WWI, 202–3; HEA critiques of, ix, 102–5, 111, 178, 206–8, 223, 237, 258, 272–74, 294; HEA elected FRS, 103; HEA leadership in, 51, 90, 163, 191, 201, 205, 271, 304; HEA obituary in, 308; HEA papers in RS publications, 13, 70, 82, 84, 92, 95, 154, 159, 164, 171–73, 179, 181–82, 259, 265, 279; HEA students elected FRS, 51, 57, 81–82, 251
Royal Society of Arts, ix, 30, 191, 209–10, 215, 219, 237, 239, 245, 271, 280, 298, 300
Rue, Warren de la, 24
Rumford, Benjamin Thompson, Count, 276
Ruskin, John, 4, 21, 165, 246, 255, 260, 265
Russell, John, 251
Russell, William James, 23, 84, 94, 137
Rutherford, Ernest, 83, 86, 89, 92, 198, 272–73

Sadler, Michael, 122–23, 226
St. Bartholomew's Hospital, 8, 10, 23–24, 28–29, 49, 94, 131
Sakurai, Joji, 306
Saleeby, Caleb, 218
Salisbury-Jones, F., W., 210
Sanderson, Frederick William, 140, 228, 231, 294
Schorlemmer, Carl, 26
Schuster, Arthur, 108, 202–3, 207

Index 357

scientific method, 19–22, 130, 132, 152, 230, 263, 265, 276–77, 291, 302
Scott, Alexander, 83, 97–98, 203, 206–7
Scott, Robert, 197–98
Scott, Sir Walter, 6
Senter, George, 167–68
Shaddock, Hugh A., 266
Shaw, George Bernard, 178–79, 194
Shenstone, William, 137
Sidgwick, Nevil, 305–6
Simpson, George C., 266
Smedley, Ida, 116, 126–27
Smith, Henry Bompas, 197
Smithells, Arthur, 137, 145, 226, 235
Soddy, Frederick, 83, 86, 89, 92, 243, 273–74, 277, 294
Solomon, Maurice, 50–51
Solvay, Ernest, 265
Sonnenschein, Adolph, 133
Sörensen, Søren, 177
Soyer, Alexis, 282
Spencer, Frederick, 134
Spencer, Herbert, 63, 129, 263
Spurzheim, Johann, 148
Stanford, R. V., 243
Stapledon, Reginald G., 297, 303
Statham, W. E., 5
Steiner, Rudolf, 220
Stenhouse, John, 17
Stewart, Alfred, 168–69
Stöckhardt, Adolf, 5
Stokes, Henry N., 96
Strachey, Lytton, 277, 294
structural theory, 9, 17, 48, 53, 67–69, 80, 92, 136
Stuart, Charles Maddock, 139, 142, 214, 226

Tagore, Rabindranath, 296
Thomas, John, 264
Thompson, Sylvanus P., 46, 53
Thomsen, Julius, 69, 250
Thomson, John Millar, 94
Thomson, J[oseph] J[ohn], 85, 92, 108, 176, 228–33
Thomson, William. See Kelvin, 1st Baron
Thorne, Will, 218
Thorpe, Jocelyn Field, 192, 201

Thorpe, Thomas Edward, 26, 32, 49, 54, 62, 69, 97–98, 100, 118, 120, 192
Tilden, Philip Armstrong, 54
Tilden, William, 26, 47, 54, 60, 62, 75, 81, 100, 105, 114–21, 127, 136, 159, 173, 225
Tite, William, 35
Tollens, Bernhard, 14
Townend, Donald, 155, 250
Traube, Wilhelm, 178
Travers, Morris W., 108–9
Trench, Richard Chenevix, 5, 20–22, 132
Tripp, Edward Howard, 57–58
True, Alfred Charles, 58
Tyndall, John, xi, 8, 10, 22, 129–30, 274

University College London, 31, 35, 113–14, 179, 222, 258, 299
University of London, 42–43, 55–56, 61, 65, 127, 192, 208, 222, 288
Unwin, William C., 53–54
Usherwood, Hilda. See Ingold, Hilda Usherwood
Usherwood, Thomas S., 139–40

Valentin, William, 7
Van Praagh, Gordon, 144, 234, 309
Vassall, Arthur, 225, 228–29, 231, 233
Vries, Hugo de, 156

Walden, Paul, 175
Walker, James, 155, 161–62, 168, 175, 181
Wallace, Alfred Russell, 4
Wallach, Otto, 81, 203–4
Walter, L. Edna, 140
Wanklyn, Alfred, 11, 24–25
Warington, Robert, Jr., 58, 101
Warington, Robert, Sr., 101
Waterhouse, Alfred, 31–32, 53
Waterlow, Sydney, 33
Watney, John, 33
Watson, Edwin, 77
Watts, Henry, 26, 95–96, 98, 194, 207
Watts, William Whitehead, 301
Webb, Aston, 31
Wegener, Alfred, 301
Wells, H. G., 240, 249–50, 290, 301–2
Werner, Alfred, 73
Westaway, Frederic W., 234

Whetham, William, 160
Whewell, William, 129
Whiteley, Martha, 116, 127, 140, 200
Whitman, Walt, 283
Wichelhaus, Hermann, 7
Wiley, Harvey Washington, 58–59, 292
Williams, Ernest, 60, 218–19
Williamson, Alexander William, 12, 17, 23, 35, 95, 165, 174, 302
Willstätter, Richard, 77, 203–4, 283–84
Wilson, C. T. R., 85, 267
Wilson, James, 129–30
Witt, Otto, 75
Wöhler, Friedrich, 24
women and gender relations, xi–xii, 41, 64, 78, 97, 110–127 passim, 196–97, 218, 221, 224, 262, 266, 280–82, 290–92, 309
Wood, Trueman, 34
Wordsworth, William, 260
Worley, Frederick, 171
Wormell, Richard, 33–35, 39
Worthington, Arthur, 138, 141, 228
Wright, Charles Alder, 23
Wright, Lewis, 209
Wynne, William, 51, 74–75, 96–97, 100, 307

Yellowlees, Lesley, 127

Zangwill, Israel, 280
Zincke, Theodor, 17, 57, 76